The Extra-Virgin Olive Oil Handbook

The Extra-Virgin Olive Oil Handbook

Edited by

Claudio Peri
University of Milan, Milan, Italy

WILEY Blackwell

This edition first published 2014 © 2014 by John Wiley & Sons, Ltd

Registered office: John Wiley & Sons, Ltd, The Atrium, Southern Gate, Chichester, West Sussex, PO19 8SQ, UK

Editorial offices: 9600 Garsington Road, Oxford, OX4 2DQ, UK
The Atrium, Southern Gate, Chichester, West Sussex, PO19 8SQ, UK
111 River Street, Hoboken, NJ 07030-5774, USA

For details of our global editorial offices, for customer services and for information about how to apply for permission to reuse the copyright material in this book please see our website at www.wiley.com/wiley-blackwell.

Library of Congress Cataloging-in-Publication Data

The extra-virgin olive oil handbook / [edited by] Claudio Peri.
pages cm
Includes bibliographical references and index.
ISBN 978-1-118-46045-0 (cloth)
1. Olive oil–Handbooks, manuals, etc. I. Peri, C. (Claudio), editor of compilation.
TP683.E98 2014
664′.362–dc23

2013039742

A catalogue record for this book is available from the British Library.

Wiley also publishes its books in a variety of electronic formats. Some content that appears in print may not be available in electronic books.

Cover image: Noam Armonn www.shutterstock.com
Cover design by www.hisandhersdesign.co.uk

Typeset in 10/12 Times by Laserwords Private Limited, Chennai, India.
Printed and bound in Singapore by Markono Print Media Pte Ltd

1 2014

Contents

List of Contributors

Cristina Alamprese, Department of Food, Environmental and Nutritional Sciences, University of Milan, Milan, Italy

Lamberto Baccioni, Agrivision, Florence, Italy

Enrico Bertolotti, BTS Business & Technic Systems srl, Milan, Italy

Mario Bertuccioli, Department of Agricultural, Food and Forestry System Management, University of Florence, Florence, Italy

Luana Ilarioni, Department of Agricultural, Food and Environmental Sciences, University of Perugia, Perugia, Italy

Alessandro Leone, Department of Science of Agriculture, Food and Environment, University of Foggia, Foggia, Italy

Sara Limbo, Department of Food, Environmental and Nutritional Sciences, University of Milan, Milan, Italy

Manuela Mariotti, Department of Food, Environmental and Nutritional Sciences, University of Milan, Milan, Italy

Ardian Marjani, Ardian Marjani & C Sas, Milan, Italy

Erminio Monteleone, Department of Agricultural, Food and Forestry System Management, University of Florence, Florence, Italy

Luigi Nasini, Department of Agricultural, Food and Environmental Sciences, University of Perugia, Perugia, Italy

Claudio Peri, University of Milan, Milan, Italy

Luciano Piergiovanni, Department of Food, Environmental and Nutritional Sciences, University of Milan, Milan, Italy

Primo Proietti, Department of Agricultural, Food and Environmental Sciences, University of Perugia, Perugia, Italy

Stefano Rossi, Life Cycle Engineering, S.r.l., Torino, Italy

Antonia Tamborrino, Department of Agro Environmental and Territorial Sciences, University of Bari, Bari, Italy

Alan Tardi, University of Gastronomic Sciences, Pollenzo, Italy

Bruno Zanoni, Department of Agricultural, Food and Forestry System Management, University of Florence, Florence, Italy

Acknowledgements

I would like to express my deep gratitude to Sr. Mary Frances Traynor, teacher of English at the University of Perugia (Italy), for her care in correcting, improving and sometimes reshaping the English text. Her knowledge and experience in the food chemistry and technology field made her contribution much more valuable than simple language editing. She has been a precious collaborator in detecting mistakes and inconsistencies.

I am also indebted to Valérie Ganio Vecchiolino, a student at the University of Gastronomic Sciences in Pollenzo (Italy), who drew plant and designs with great care, precision and patience.

Introduction

This handbook deals with the basic science and technical aspects of extra-virgin olive oil, from harvesting the olives to processing, storing and using the oil at the consumer's table. It is divided into three parts: the product, the process and the process control system. One chapter gives some fundamental information about the best culinary uses of olive oils. Some important physical and physical-chemical parameters are summarized in the appendix and a detailed subject index indicates where major topics can be found in the handbook.

The main purpose of the handbook is to guide those involved in the extra-virgin olive oil chain in making the most appropriate decisions about product quality and operating conditions in the production and distribution processes. The approach of the handbook is mainly educational, providing guidelines for good extra-virgin olive oil practice. Basic information about various phenomena is presented in an easy-to-understand form, while systematic methods for choosing the most appropriate operating conditions are suggested. The instructive approach is evident in many parts of the text: (i) in the presentation of the principles, methods and examples of quality and safety management; (ii) in the presentation of methods to calculate mass and cost balances because they are considered as important as evaluating the chemical and sensory characteristics of the oil; (iii) in the choice of time temperature relationships in olive storage, in olive paste malaxation and in olive oil storage. The semi-log plots in Chapters 9, 12 and 15 are a contribution to critical points management based on sound scientific principles; (iv) in the presentation of the health-promoting properties of extra-virgin olive oils and in the choice of the analytical parameters for their evaluation; (v) in the discussion and presentation of sensory profiles as essential tools of product style and differentiation.

The second purpose of the handbook relates to quality as the guiding factor of managerial and operating strategies. A producer of olive oil has different options. The first is to put aside quality and focus on yield increase and cost reduction. Overripe olives are harvested by letting them drop to the ground and then collected mechanically; in this case 'lampante' oil is produced and sold to refineries to produce refined olive oil. This approach has proven profitable in some cases and is sometimes unavoidable due to olive spoilage, pest attack or lack of appropriate harvesting equipment. The handbook describes the olive oil refining process and explains why

The Extra-Virgin Olive Oil Handbook, First Edition. Edited by Claudio Peri.

refined olive oil should be considered as a good, reliable and useful product among the vegetable oils.

Another option is to produce an olive oil that meets extra-virgin standards. This requires considerable care to ensure that the olives are healthy and undamaged and that proper operating conditions are used in the milling and handling operations. With this choice, the quality of the olive oil reaches a high level.

The handbook points out that the standards of extra-virgin olive oil can be further improved beyond the present legal requirements and levels of excellence can be achieved. Attention is focused on neglected but critical issues such as: (i) residence time distribution in olive paste malaxation; (ii) hygienic design of plants and equipment and, most of all, (iii) in Chapter 15 on olive oil storage and handling, where the principles of the quality-proximity matrix are presented and thoroughly discussed.

It can be said that producing a 'lampante' olive oil does not require special care and skill. Producing an extra-virgin olive oil is a much more challenging and demanding task. Finally, only excellent operators can make excellent extra-virgin olive oils available to the final consumer. The goal of this handbook is to guide the operators in the olive oil chain towards excellence, all the way from the olive grower to the restaurant chef. With a chapter on the culinary uses of extra-virgin olive oil, we would like to activate a new alliance between excellent producers, retailers and chefs for the production and use of excellent extra-virgin olive oil. We hope to contribute to spreading a new consumer culture about this exceptionally good, healthy and natural product, which is so old in its millenarian tradition, so young in present-day processing technologies and so well tailored to the health, taste and dietary needs of new and traditional consumers around the world.

Part I
The product

1

The extra-virgin olive oil chain

Claudio Peri
University of Milan, Milan, Italy

Abstract

This chapter presents the classification and commercial denomination of six olive oil categories recognized in international law (two virgin and four refined). Extra-virgin olive oil is the highest quality olive oil. The extra-virgin olive oil chain, is presented as a sequence of five processes: (i) olive tree cultivation, (ii) olive harvesting and processing, (iii) oil storage, bottling and distribution, (iv) selling bottled oil and (v) oil use in culinary preparations. Processes (ii) and (iii), the subject matter of this handbook, are further presented as a sequence of unit operations. The main steps and conditions determining oil quality and yield are outlined.

1.1 The legal classification and denomination of olive oils

When talking or reading about olive oil, the first point to be clarified is the category of olive oil that is being discussed. Ignoring the category that the oil belongs to can be a source of confusion and misunderstanding and can lead to mistakes in buying, tasting or using it. Figure 1.1 is a flow-chart of the classification and denomination of the various categories of virgin and refined olive oils as globally agreed (see Council Regulation (EC) No. 1234/2007 of 22 October 2007, (Single CMO Regulation), consolidated version 2013-01-26, Annex XVI).

The six categories highlighted by the grey background are suitable for human consumption.

The flow-chart starts with the olive milling process, whose products are the 'virgin' olive oils. Two of them, namely extra-virgin and virgin, are allowed for consumption. The third category, lampante, becomes edible only after a physical-chemical refining process and it is called 'refined olive oil'.

On the other hand, the pomace, which is the solid residue from the milling process, still contains a small amount of olive oil that is impossible to extract by mechanical means. It can be extracted with solvents; the raw oil from this extraction

The Extra-Virgin Olive Oil Handbook, First Edition. Edited by Claudio Peri.

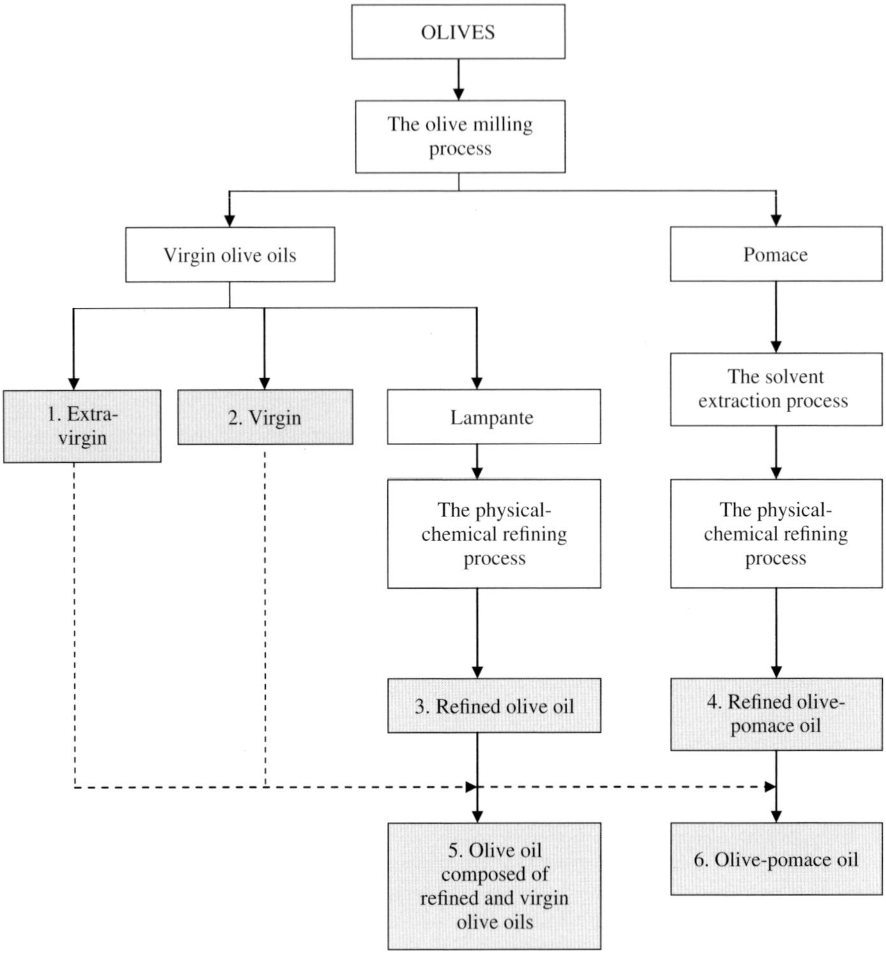

Figure 1.1 The flow-chart of virgin and refined olive oils.

is refined with a process very similar to that applied to lampante oil. The refined oil derived from pomace is called 'refined olive-pomace oil'.

Both the 'refined olive oil' and the 'refined olive-pomace oil' can be mixed with extra-virgin or virgin olive oil in various undefined proportions in order to improve their flavour. These are called, respectively, 'olive oil composed of refined and virgin olive oil' and 'olive-pomace oil'.

Chapter 17 gives a short presentation of the refining process. It is important that olive oil producers, retailers and consumers know the difference in technological and compositional terms between a virgin oil and a refined oil.

Regarding quality, extra-virgin olive oil is higher in quality than virgin olive oil and refined olive oil is higher in quality than refined olive-pomace oil. Refined olive oil is very mild and almost neutral in taste: it is very good for cooking, frying and for preserving canned vegetables or meat or fish. Extra-virgin olive oil is flavourful and tasty. A picture of the culinary uses of olive oil and especially excellent extra-virgin olive oils is given in Chapter 24.

1.2 The subject of this handbook

Focusing on extra-virgin olive oil opens a wide panorama because the oil varies depending on cultivar, climate and soil, and the conditions of the production-extraction-storage-and-distribution process. Extra-virgin olive oils can be of common or good or excellent quality. The purpose of this handbook is to discuss the technological and management conditions that allow an operator of the extra-virgin olive oil chain to improve the quality of the product, which is finally served at the consumer's table.

1.3 The extra-virgin olive oil chain

The extra-virgin olive oil chain can be divided into a series of five processes: (i) olive tree cultivation, (ii) olive harvesting and milling, (iii) oil storage, bottling and distribution, (iv) oil selling and (v) oil use in culinary preparations (Table 1.1).

These five processes have different structural and operational requirements, different marketing policies and different economies of scale. They are therefore usually managed and owned by different companies.

Table 1.1 The processes of the extra-virgin olive oil chain.

Process	Input	Output	Ownership and responsibility
(i) Olive tree cultivation	Olive trees (plus soil, atmospheric conditions, machinery, services, information, work, ...)	Olives	Agricultural companies
(ii) Olive harvesting and milling	Olives (plus the mill, services, information, work ...)	Batches of oil	Milling companies
(iii) Oil storage, packaging and distribution	Batches of oil (plus packaging plant and materials, services, information, work ...)	Oil in bottles or other suitable containers	Packaging and distribution companies
(iv) Sale of packaged oil	Packaged oil (plus sales facilities, services, information, work ...)	Packaged oil sold to the final user	Retail companies
(v) Culinary use of oil	Packaged oil (plus cooking facilities, services, information, work, ...)	Oil in culinary preparations	Restaurants, foodservice and families

Processes (ii) and (iii) represent the core content of this handbook. There is some discussion about process (i) in Chapters 5 (olive tree cultivars) and 7 (olive harvesting), whereas Chapter 24 gives some general indications about the use of extra-virgin olive oil in culinary preparations.

1.3.1 Compact versus complex chain organization

The most compact organization of an extra-virgin olive oil chain entails a direct connection between only two parts (or modules): the first is represented by the producer and the second by the final consumer. In this case, which is very common in olive oil producing regions, a producer who is responsible for the chain from the field to the package, sells his oil directly to the final consumer, either a family or a restaurant. This organization is typical of traditional markets in a narrow area close to production, but sometimes it is also implemented in a global market and across continents. It is common to find commercial agreements between a restaurant in Los Angeles or Tokyo and a producer in Andalusia or Tuscany.

On the other hand, very complex chain organizations are implemented in large-scale and global businesses with multiple inputs and outputs connecting the five processes listed in Table 1.1.

Traceability of product origin and identity is easy in the case of the compact chain, whereas it may be very difficult or impossible in complex chain organizations.

1.3.2 The extra-virgin olive oil processes

As chains can be considered sequences of processes, processes can similarly be considered as sequences of unit operations. Processes are interconnected in series in a chain, so unit operations are interconnected in series in a process with the output of a unit operation being the input of the following one (Peri *et al.* 2004). Table 1.2 presents the unit operations of processes (ii) and (iii).

1.4 Yield and quality

The primary objective of an extra-virgin olive oil company is to maximize oil yield and quality. Obtaining the largest quantity of oil with a high level of quality is the ultimate measure of process effectiveness and efficiency.

Contrary to the situation with other agricultural products, yield and quality are not competitive in extra-virgin olive oil production, but independent or concurrent parameters. Conditions determining quality losses also determine yield losses. Figure 1.2 represents the critical steps and conditions determining extra-virgin olive oil yield and quality.

The extra-virgin olive oil chain is divided into two parts. In the first part, corresponding to olive tree cultivation, the basic condition for success is olive integrity. If, due to climatic conditions or pest attack, olives are seriously damaged, the unavoidable consequence is an irreversible loss of yield and quality.

Table 1.2 The unit operations of extra-virgin olive oil processes.

Preliminary activities	Unit operation	Ancillary activities
Monitoring of olive maturity. Supply and maintenance of harvesting nets, crates and equipment	**Harvesting**	
	Storage and transportation of olives	
Mill plant maintenance, cleaning, and start trial	**Olive reception at the mill plant**	
Standards agreed upon between the olive grower and the milling company	**Visual inspection, control of origin and olive integrity**	Decisions in case of nonconformity to standards
	Milling batches, identification and weighing	Record of milling batches
Deleafing at the olive grove site. Supply of potable water	**Olive cleaning and washing**	Disposal of solid residues and dirty water
	Olive milling or pitting	In case of pitting: discharge and use of olive stones
	Olive-paste malaxation	Monitoring and control of the time-temperature relationship
	Solid-liquid and liquid-liquid separation	Pomace to treatment and use. Wastewater to disposal
Supply of filter aids or filter pads	**Oil filtration**	Disposal of exhausted filtering material
Agreed upon standards of oil quality and yield	**Oil weighing, chemical and sensory evaluation**	Decisions in case of nonconformity
Maintenance of storage facilities	**Storage batches formation and identification**	Standard documentation of storage batches
	Oil storage	Waste disposal
Customers' orders and requirements	**Oil blending, packaging batches formation**	Chemical and sensory evaluation of packaging batches. Record of packaging batches
Maintenance and supply of packaging material	**Packaging**	Waste disposal
	Shipment of consignment to customers	

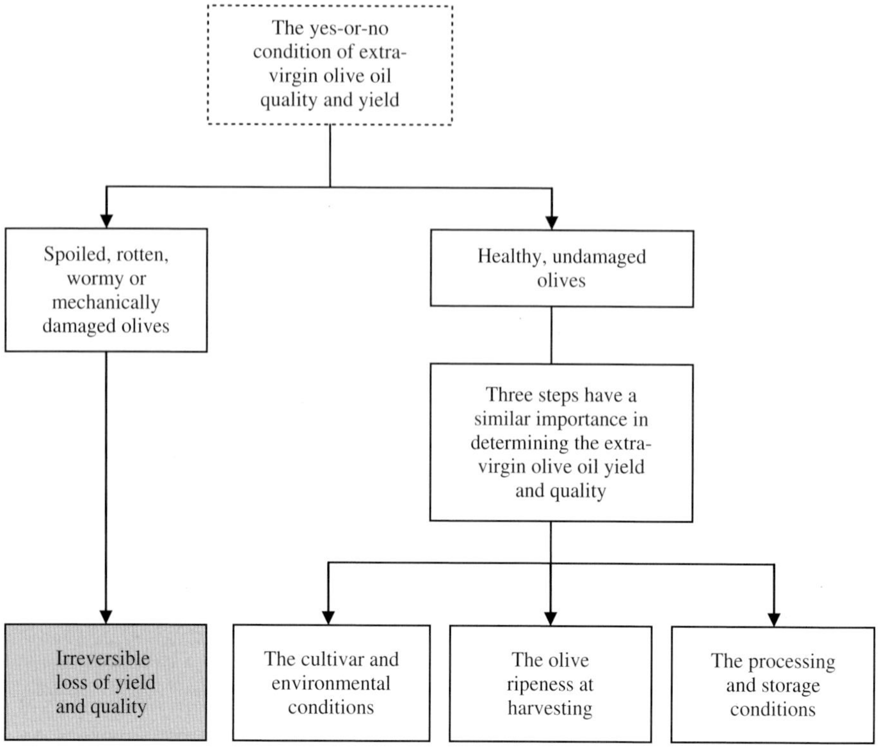

Figure 1.2 Critical steps and conditions for yield and quality of extra-virgin olive oil.

If, on the other hand, olives are undamaged and healthy, the final result is determined by three factors of similar importance: (i) the cultivar and the environment (climate and soil); (ii) the ripeness of olives at harvesting; (iii) the oil processing and storage conditions.

Processing of the olives and storage of the oil are the last and hence the decisive steps affecting oil quality. The product of the best, healthy and undamaged olives can be a very common oil or even a bad oil, as a consequence of errors and carelessness in the processing and distribution steps.

A point that should be kept in mind is that the loss of quality of virgin olive oil is irreversible. Feedback control is not possible and amplification instead of slowdown of the negative effects takes place: spoiled olives produce spoiled oil and spoiled oil tends to spoil further at a much faster rate than good oil. Process control should be based on prevention. The only corrective action available in the case of a spoiled oil is downgrading and refining it.

Reference

Peri, C., Lavelli, V. and Marjani, A. (2006) *Sistemi di gestione per la qualità nei processi e nelle filiere agro-alimentari*, Hoepli, Milan.

2

Virgin olive oil: definition and standards

Manuela Mariotti
Department of Food, Environmental and Nutritional Sciences, University of Milan, Milan, Italy

Abstract

Basic information is given about virgin olive oil standards according to European legislation. Standards are divided into two groups: (i) quality standards aimed at classifying extra-virgin, virgin and inedible 'lampante' olive oil and (ii) authenticity standards aimed at identifying oil adulteration by mixing virgin olive oil with refined olive oil or oil of other kinds. The importance and meaning of free acidity, peroxide value, UV absorption values and sensory defects are discussed.

2.1 The legal definition of virgin olive oil

Definitions and standards for virgin olive oil are primarily based on European legislation, especially Commission Regulation (EC) No. 1019/2002 of 13 June 2002, on marketing standards for olive oil, and Commission Regulation (EC) No. 702/2007 of 21 June 2007, on the characteristics of olive oil and on the relevant methods of analysis. Other standardization organizations, such as the International Olive Council (IOC; www.internationaloliveoil.org/) and The Codex Alimentarius Commission (www.codexalimentarius.org/), take part in defining olive oil standards, but European legislation is the first and main reference worldwide. The European Community recognizes several categories of olive oil, each with its particular qualities and market value.

Virgin olive oil is defined by EC Regulation No 1019/2002 (Art. 3) as follows: 'Virgin olive oil is the olive oil obtained directly from olives and solely by mechanical means.' Conformity with this definition is the basic authenticity requirement for virgin olive oil.

The Extra-Virgin Olive Oil Handbook, First Edition. Edited by Claudio Peri.

Three categories of virgin olive oil are further defined, based on quality criteria, as: 'extra-virgin olive oil', 'virgin olive oil' and 'lampante olive oil'. Lampante (literally 'lamp oil' according to its use in ancient times) is a virgin olive oil obtained from bad fruit or careless processing and it is of such a low quality that it cannot be used for human consumption and must be refined in order to become edible.

Regulation (EU) No 29/2012, dated 13 January 2012, codifies the substantial amendments that have taken place since regulation No 1019 on olive oil marketing standards was introduced in 2002. One of those, Council Regulation (EC) No 1234/2007 of 22 October 2007, establishes a common organization of agricultural markets and specific provisions for certain agricultural products, olive oil included. According to this regulation, 'virgin olive oils' are defined as:

"oils obtained from the fruit of the olive tree solely by mechanical or other physical means under conditions that do not lead to alterations in the oil, which have not undergone any treatment other than washing, decantation, centrifugation or filtration, to the exclusion of oils obtained using solvents or adjuvants having a chemical or biochemical action, or by re-esterification processes and any mixture with oils of other kinds."

Commission Regulation (EC) No 702/2007 of 21 June 2007 defines the analytical and sensory standards of all the categories of olive oils, virgin or refined. Only the standards related to the three categories of virgin olive oil are presented here. These standards have been established by law as indicators of oil quality and authenticity.

Quality standards are analytical parameters that allow virgin olive oils to be classified according to a scale of quality. In general, these parameters indicate oil spoilage. Therefore, it is assumed that the lower their values, the higher the quality of the oil.

Authenticity standards are analytical parameters that allow an oil to be declared as 'virgin', in compliance with the definition reported above. In general, these parameters indicate the presence of refined olive oil (violation of the condition of 'solely by mechanical means') or other vegetable oils (violation of the condition of being obtained 'directly from olives').

2.2 Quality standards of virgin olive oil

Quality standards of virgin olive oil can be divided into two groups: chemical and sensory standards.

2.2.1 Chemical quality standards

The chemical standards that must be evaluated for classifying the quality levels of virgin olive oils are reported in Table 2.1. Quality standards are useful to verify hydrolytic and oxidative degradation that takes place in the olives and the oil

Table 2.1 Chemical standards of virgin olive oil.

Chemical standard	Extra-virgin	Virgin	Lampante
Free acidity (%)	≤ 0.8	≤ 2.0	> 2.0
Peroxide index ($mEqO_2/kg$)	≤ 20	≤ 20	–
K_{232}	≤ 2.50	≤ 2.60	–
K_{270}	≤ 0.22	≤ 0.25	–
ΔK	≤ 0.01	≤ 0.01	–

Spectrophotometric values in the UV at 232 and 270 nm depend on conjugated double bonds derived from oxidation (they are therefore quality indicators) or from refining (therefore, they are also authenticity indicators).

during processing and storage. Olive oil producers should concentrate on the quality standards presented in Table 2.1, making a coherent decision about the processing procedure and conditions according to the level of quality they want to achieve.

Free acidity

Hydrolysis of triglycerides due to lipolytic enzymes (lipases) causes free fatty acids and monoglycerides or diglycerides to be released from the triglycerides. The products of the lipolytic reaction are tasteless and odourless and therefore no sensory defects can be perceived. Hence, it is not correct to refer to 'acidity' as a flavour sensation of an olive oil. Sometimes, the sensation of pungency is mistakenly interpreted as 'acidity'.

The lipolytic reaction is due to the endogenous lipases that are naturally present in the olive. When the integrity of the olive is lost due to mechanical action, lipases that are present in the pulp or in the seed cells come into contact with the oil, originally contained in specialized vacuoles. At this point, lipolysis starts and free fatty acids are produced. The reaction accelerates with increase in temperature and is a function of the time of contact between the lipases and the oil. Lipases are hydrophilic and they are active only in the presence of an aqueous phase. When water is separated, by decanting and centrifugation, lipolysis slows down or is totally stopped if the water and cell residues are completely separated from the oil. This is the reason why filtering the oil, removing the suspended materials and partially reducing the amount of water, is important.

In any case, the lipolytic reaction due to the endogenous lipases in the olives is relatively slow. Oil obtained from healthy fruit, regardless of the cultivar and processed just after harvesting, have very low values of free acidity. Free acidity rapidly increases in the presence of moulds and micro-organisms, which produce large quantities of very active lipases (exogenous lipases). In broken, dirty, unhealthy olives, lipase activity causes a rapid increase in free acidity beyond the limits for extra-virgin or virgin olive oils, with an obvious loss in quality and value. A further and very rapid acceleration of this reaction takes place due to olive fly attacks. The intestines of the olive flies and their excrement, in fact, contain very high concentrations of lipases that cause a very rapid increase in free acidity. It is most unfortunate that this happens when the fruit is still on the tree. Thus at harvesting time, the damage has been irreversibly done. Other factors affecting the integrity of olives are attacks by parasites, mechanical crushing and bruising, extended contact with soil, delayed harvesting (over-ripeness), prolonged heaping and storage before processing.

Free acidity is expressed as the percentage of free fatty acids on the basis of oleic acid, which is the main fatty acid of olive oil. Each producer should be able to determine free acidity at the milling site, not only to verify the quality of the oil, but also to avoid mixing good and bad oil.

The legal limit of 0.8% for extra-virgin olive oil is not very demanding. A good oil should have a free acidity value less than 0.5% and an excellent oil less than 0.3%.

Peroxide value and spectrophotometric absorption in the UV

The two main spoiling reactions of olive oil are lipolysis and lipid oxidation. Lipolysis can be easily estimated as free acidity, but oxidation is more difficult and complex to evaluate.

Assessment of the degree of olive oil oxidation is based on determinations of both the primary and secondary products of oxidation. The primary stage of oxidation is the formation of hydroperoxides from polyunsaturated fatty acids, through a radical mechanism (see Chapter 6).

Peroxides are primary oxidation products and they are used as indicators of oil quality and stability. Their value increases, reaches a maximum and then decreases because of their further degradation into secondary products of oxidation such as aldehydes, ketones and conjugated dienes. These substances, that are formed at an advanced stage of oxidation, are responsible for the rancid flavour of the oil.

Thus, the peroxide value is a measure of the degree of oxidation of the oil at an early stage of oxidative spoilage, long before a rancid smell or taste becomes perceivable. An increase in the peroxide value should be considered as a warning signal that oxidation is taking place.

Spectrophotometric values: Specific absorbances (conventionally indicated as K) are measured in the UV region, at the wavelengths corresponding to the maximum absorption (about 232 and 270 nm) of secondary products formed in the autoxidation process. An increase in absorption at K_{232} and K_{270}, may also be due to the presence of conjugated dienes and trienes, which are formed in oils that have been heated during the refining process. Conjugated dienes contain two double bonds that alternate with single bonds. A conjugated triene contains three alternating double bonds.

Therefore, a high spectrometric value can be considered an indicator of oxidation or of adulteration of the oil.

Peroxide and spectrophotometric values are not easy to evaluate (see framed note below). However, they are so important that a good producer seeking high quality must have one or both of the analyses carried out as an essential tool of process control.

Lipid oxidation is greatly accelerated by lipolysis because free fatty acids are more easily oxidizable than fatty acids linked in a triglyceride molecule. This means that there is a synergy between lipid hydrolysis and oxidation in accelerating oil spoilage. Oil oxidation is also greatly accelerated by the presence of oxidative enzymes (lipoxidases, lipoxygenases) that are naturally present in the olive pulp and seed cells, and much more in case of moulds and fly spoilage. The removal of water may prevent or eliminate enzymatic oxidizing activity. However, unlike lipolysis, oxidation can also take place in the absence of water by a purely chemical, autocatalytic mechanism.

The legal limit of the peroxide value is 20 meqO_2/kg for extra-virgin oil, which is a very poor standard, as a good oil must have a peroxide value of less than 12, and an excellent oil less than 8. The same can be said for K_{232}, the most reliable spectrophotometric indicator of oil oxidation. The limit of 2.50 is not very selective; it should be lower than 2.10 for a good extra-virgin olive oil and lower than 1.90 for an excellent extra-virgin olive oil.

Analytical methods

Official methods of analysis of olive oils are thoroughly described in European legal prescriptions (Commission Regulation (EEC) No 702/2007 amending Commission Regulation (EEC) No 2568/91 on characteristics of olive oil and olive-residue oil and on relevant methods of analysis, 21 June 2007).

According to the official procedure for free acidity, the (filtered) sample is dissolved in a mixture of solvents and the free fatty acids present are titrated with an ethanolic solution of potassium hydroxide using phenolphthalein as indicator. Results are expressed as a percentage by weight of oleic acid. This method is relatively slow (7–8 samples/h) and solvent consuming (100–150 mL/sample).

The peroxide value is the quantity of those substances in the sample, expressed in terms of milliequivalents of active oxygen per kilogram (meq/kg), which oxidize iodide. The method is based on treatment of the sample in solution in acetic acid and chloroform, with a solution of potassium iodide. The freed iodine is titrated with a standardized sodium thiosulphate solution in the presence of a starch solution as indicator.

This method is slow and relatively complex, so well-trained technicians are needed to obtain reliable results.

For the analysis of specrophotometric values, the oil is dissolved in spectrophotometric pure iso-octane and extinction of the solution is determined at the specified wavelengths (232 and 270 nm) with reference to pure solvent. Specific extinctions are calculated from the spectrophotometric readings.

Absorptions are expressed as specific extinctions of a 1% solution of the oil in the specified solvent, in a thickness of 1 cm.

This determination is also relatively complex and requires a spectrophotometer for measuring extinction in the ultraviolet between 220 and 360 nm, with the possibility of reading individual nanometric units.

Simple, portable, easy-to-use, rapid analytical apparatuses are available for free acidity, peroxide value and, in some cases, total phenolic compounds as well an oil stability index. These are very useful tools for online control of the product. Attention and care are needed for maintenance and calibration and periodical testing versus the official methods.

2.2.2 Sensory quality standards

The sensory analysis as a method for the legal recognition and classification of virgin olive oil was proposed by the International Olive Council in June 1987, and recognized by the European Commission in July 1991 (Commission Regulation (EEC) No 702/2007 amending Commission Regulation (EEC) No 2568/91 on characteristics of olive oil and olive-residue oil and on relevant methods of analysis, 21 June 2007). The method was further modified and replaced by subsequent amendments and finally updated by Commission Regulation (EC) No 640/2008 of 4 July 2008 amending Regulation (EEC) No 2568/91 on sensory characteristics of olive oil and the relevant methods of analysis. The sensory standards for classifying the various levels of quality of virgin olive oil are reported in Table 2.2 (see Chapter 4).

Table 2.2 Sensory standards of virgin olive oil.

Sensory standard	Extra-virgin	Virgin	Lampante
Median defect (Md)	0	≤ 3.5	> 3.5
Median fruity (Mf)	> 0	> 0	–

Nonconformity with the requirement of a fruitiness note greater than zero is rare, as it is almost impossible to have a virgin olive oil without any odour and taste. Therefore, conformity to this requirement is very easy to achieve. Nonconformity to the prescription of zero defects, on the other hand, is frequent and may be the consequence of minor mistakes in processing. Therefore, conformity to this requirement is difficult to achieve. In addition, due to some difficulties in standardizing and obtaining reproducible results from different panels of tasters, sensory standards are difficult to be universally defined. The inclusion of sensory requirements among the legal standards for virgin olive oil was a 'daring' decision because sensory analysis can detect very low concentrations of good or bad oil components and because it is difficult to obtain reproducible results of positive and negative sensory attributes.

Probably, when establishing the sensory standards, the legislators did not take enough care in considering the difference between this requirement and the others. In fact, while the chemical standards measure concentrations of negative components at the level of *per cent* or *per thousand*, sensory defects are easily detected by human odour and taste receptors at concentrations that may be in the range of ppm (parts per million) or even ppb (parts per billion). It has been found that when mixing 1 mL of rancid oil with 10 L of a perfect oil (1:10 000 dilution), a defect is easily perceived by trained sensory assessors, while no significant change occurs in the values of the chemical indices. In addition, a variation in free acidity or peroxide value or K_{232} requires a massive evolution of spoilage reactions; on the contrary, the presence of a smell or flavour of staleness or rancidity may be due to a very minor contamination of a good oil. Thus, it is not surprising that some oils that have very good chemical standards may fail to be classified as extra-virgin olive oil because of the presence of sensory defects. These defects often arise from the presence of dead spots in the milling plant (for example, improper plant design) or to uneven distribution of residence times at critical temperature conditions (see the

Table 2.3 Authenticity standards of virgin olive oil: fraudulent mixing with refined olive oil.

Standard	Extra-virgin	Virgin	Lampante	Comment
Erythrodiol and uvaol (%)	≤ 4.5	≤ 4.5	≤ 4.5	The concentration of these components is higher in refined olive oil derived from pomace.
Waxes (ppb)	≤ 250	≤ 250	≤ 300	
Total *trans*-oleic isomers (%)	≤ 0.05	≤ 0.05	≤ 0.10	These components are formed in the refining process. They indicate mixing with refined olive oil and also seed oils.
Total *trans*-linoleic +*trans*-linolenic isomers (%)	≤ 0.05	≤ 0.05	≤ 0.10	
Stigmastediene (mg/kg)[a]	≤ 0.10	≤ 0.10	≤ 0.50	
2-glyceril monopalmitate[b]	≤ 0.9 or 1.0 depending on palmitic acid concentration	≤ 0.9 or 1.0 depending on palmitic acid concentration	≤ 0.9 or 1.0 depending on palmitic acid concentration	This is an unnatural arrangement of fatty acids in a triglyceride; it derives from esterified olive oil.

Notes: [a]Decreasing the limit value for stigmastediene in virgin olive oil makes it possible to achieve better separation of virgin and refined olive oil. [b]Experts have concluded that quantification of the percentage of 2-glyceryl monopalmitate is more precise for the detection of esterified oils (Commission Regulation (EC) No 702/2007.)

hygienic design in Annex 21.1 in Chapter 21). Small quantities of olive paste or oil undergoing intense spoilage, in fact, may act as sensory contaminants of large quantities of good oil.

The identification of olive oil defects through sensory analysis has the disadvantage of being a lengthy and expensive methodology, whose final result depends on many factors, especially the training and experience of the panelists. In recent years, instrumental methods have been proposed, based on the analysis of volatile compounds by dynamic headspace high-resolution gas chromatography. This is a more precise and reliable technique, but has two critical limitations: in the first place it requires the use of complex and expensive analytical instruments by highly skilled analysts; in the second place, establishing a reliable relationship between the instrumental response and the perceived sensations of a panelist or a consumer is still a matter of research and long-term experience. An interesting alternative for rapid, in-line evaluation of both negative and positive odour notes is the use of sensors to detect the volatile compounds present in the headspace of an oil container (the electronic nose). An interesting review on the different techniques applied to olive oil aroma analysis, with their advantages and disadvantages, has been published by Escuderos *et al.* (2007).

Table 2.4 Authenticity standards of virgin olive oil: fraudulent mixing with other vegetable oils.

Standard	Extra-virgin	Virgin	Lampante	Comment
ECN42[a]	≤ 0.2	≤ 0.2	≤ 0.3	A higher value of this standard indicates mixing with seed oils rich in linoleic acid
Fatty acids (%) (myristic, linolenic, arachidic, eicosenoic, behenic, lignoceric and others)	A maximum value depending on the fatty acid (range: 0.05–1.0)	A maximum value depending on the fatty acid (range: 0.05–1.0)	A maximum value depending on the fatty acid (range: 0.05–1.0)	The fatty acid composition of oils is considered as a fingerprint of an oil-bearing species. Therefore the presence and concentration of some fatty acids indicates mixing with other vegetable oils
β-sitosterol	≥ 93.0	≥ 93.0	≥ 93.0	93.0% β-sitosterol is the minimum for olive oil; mixing with other oils results in a lower concentration of this sterol

Note: [a]ECN – equivalent chain number. The ECN is the actual carbon number minus twice the number of double bonds per molecule. Olive oil, different from most seed oils, has mainly triglycerides with ECNs 44, 46, 48, and 50; triglycerides with ECN40 and ECN42 are absent or found in trace amounts, respectively. Therefore, evaluation of ECN42, which varies according to the content of glycerol trilinoleate, is an effective tool to detect more unsaturated oils.

2.3 Authenticity standards of virgin olive oil

Authenticity standards are used to detect frauds that may derive from: (i) the use of prohibited additives or technological adjuvants; (ii) the use of prohibited technological operations as, for example, deacidifying and deodorizing by vacuum evaporation and steam stripping, or (iii) mixing with refined olive oil or other vegetable oils.

The main chemical standards that are evaluated for detecting extra-virgin olive oil adulteration are reported in Tables 2.3 and 2.4.

Reference

Escuderos, M.E., Uceda, M., Sánchez, S. and Jiménez, A. (2007) Olive oil sensory analysis techniques evolution. *European Journal of Lipid Science and Technology* **109**, 536–546.

3

The composition and nutritional properties of extra-virgin olive oil

Manuela Mariotti[1] and Claudio Peri[2]

[1] *Department of Food, Environmental and Nutritional Sciences, University of Milan, Milan, Italy*
[2] *University of Milan, Milan, Italy*

Abstract

Chapter 3 gives basic information about the composition and nutritional properties of extra-virgin olive oil. As triglycerides make up 97 to 99% of extra-virgin olive oil, the main chemical-physical characteristics of the oil depend on the composition of the triglyceride moiety. However, the minor components give an invaluable contribution to sensory and health-promoting properties. It is mainly the presence of these components that differentiates extra-virgin olive oil from all other edible oils

3.1 Triglycerides and fatty acids

Extra-virgin olive oil essentially includes two groups of chemical compounds:

- triglycerides: 97–99% wt
- minor components: 1–3% wt

Triglycerides mainly contain a monounsaturated fatty acid (oleic acid), a fair amount of polyunsaturated fatty acids (linoleic and α-linolenic) and a slight amount of saturated fatty acids (palmitic and stearic).

The minor components are a complex mixture of polar, nonpolar and amphiphilic substances: hydrocarbons, tocopherols, phenolic compounds, sterols, chlorophyll, carotenoids, terpenic acids, monoglycerides and diglycerides, free fatty acids, esters

The Extra-Virgin Olive Oil Handbook, First Edition. Edited by Claudio Peri.

and other volatiles. They contribute in a particular way to the sensory and health-promoting properties of extra-virgin olive oil.

Extra-virgin olive oil is separated from an aqueous medium so it still contains a very small, but essential amount of water. The water saturation threshold of extra-virgin olive oils is 300–400 mg per kg of oil, but they often have higher amounts, ranging from 300 to 1200 mg per kg of oil. Water is present in micro-droplets, less than one-tenth of a μm in diameter, impossible to separate by centrifugation. These microdroplets are associated with and stabilized by water-compatible, polar or amphiphilic substances of the minor components group.

Triglycerides belong to lipids, organic compounds that do not mix with water. They derive from the combination of three fatty acid molecules with one molecule of glycerol. Glycerol is a short 3-carbon chain alcohol that serves as the frame to which the three fatty acids can attach themselves with an 'ester' link. Fatty acids consist of chains 4 to 30 carbons long, with an acidic group at one end: it is this group that binds to glycerol to make a glyceride. Natural fatty acids usually have an even number of carbon atoms, as their synthesis *in vivo* is based on the assembly of a variable number of acetyl-CoA, a 2-carbon molecule (O'Keefe 2008).

Figure 3.1 shows the structural formula of a fatty acid molecule (stearic acid). It consists of a long chain of 18 carbon atoms (C) with their four valence bonds. Two bonds create the basic connection of the chain, whereas the other two are saturated by hydrogen atoms (H). At one end of the molecule there is a special group in which the carbon atom is linked to an oxygen atom (O) and to a hydroxyl group (OH). The resulting COOH group is called the 'acidic group'. At the other end, the carbon atom is linked to three hydrogen atoms forming a 'methyl group' (CH_3).

The position of carbon atoms is usually identified by a number in the sequence that starts from the acidic group and ends at the methyl group. The acidic group is also called *α* (alpha), the first letter of the Greek alphabet. The methyl group is called ω (omega), the last letter of the Greek alphabet.

Figure 3.2 represents the same molecule of Figure 3.1 but with a different graphical convention, which is called the 'skeletal' formula. Carbon atoms are represented as black dots and hydrogen atoms, directly connected to the carbon atoms, are not indicated: each carbon atom is understood to be associated with enough hydrogen atoms to give the carbon atom four bonds. An even simpler representation is the

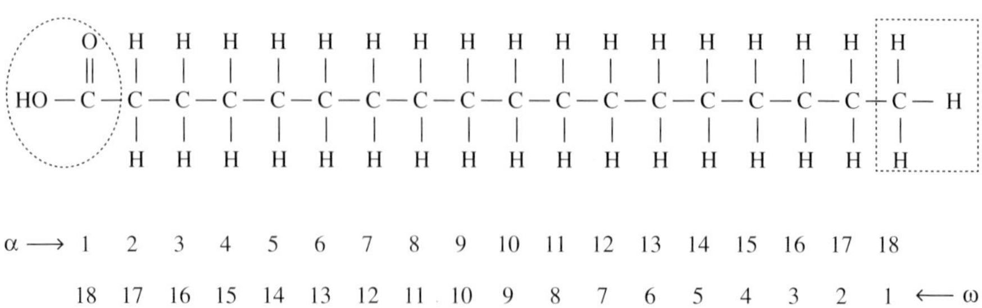

Figure 3.1 The structural formula of stearic acid.

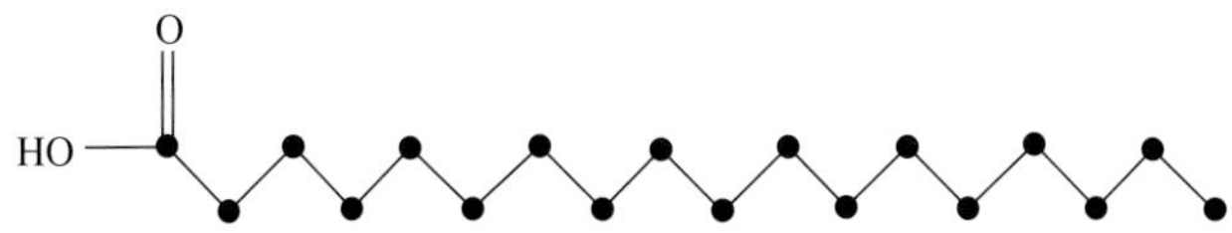

Figure 3.2 The skeletal formula of stearic acid.

skeletal formula without the black dots. In this type of representation, it is implied that the carbon atoms are located at the corners and ends of the line segments.

Figure 3.3 shows how a molecule of glycerol combines with three molecules of the fatty acid (in this case stearic acid), giving a triglyceride molecule (in this case 'tristearin') and three molecules of water.

In some fatty acids there are double bonds that derive from the elimination of two hydrogen atoms from two adjacent carbon atoms. Figure 3.4 shows a fatty acid derived from the stearic acid of Figure 3.2, by removing two hydrogen atoms from carbons 9 and 10, in the middle of the stearic molecule.

When double bonds are present, fatty acids are defined as 'unsaturated'. The unsaturated fatty acid represented in Figure 3.4 is the most important fatty acid of olive oil and it is called 'oleic acid'. It represents from 65 to 85% of all fatty acids in olive oil.

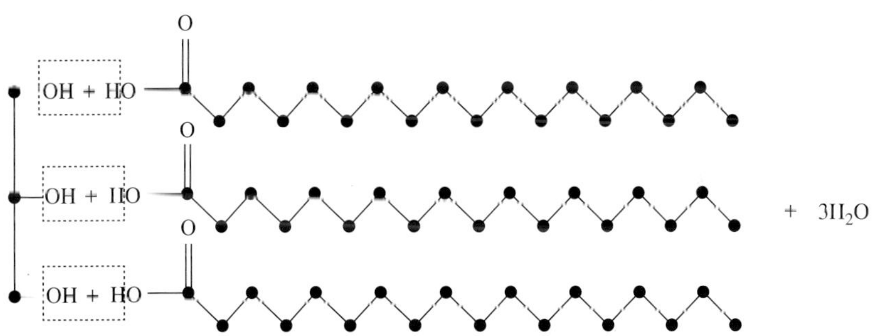

Figure 3.3 The triglyceride of stearic acid (tristearin).

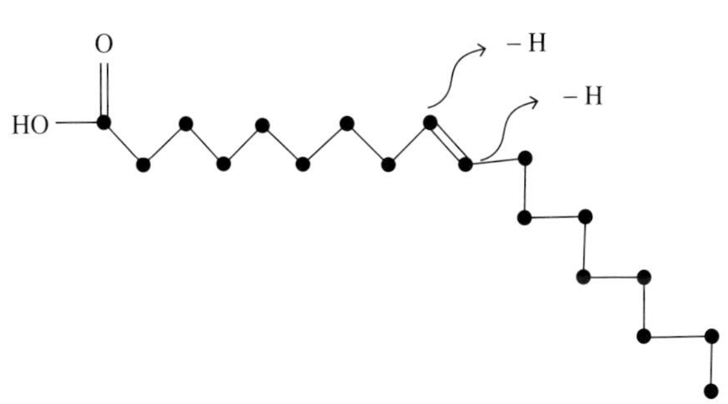

Figure 3.4 The skeletal formula of oleic acid.

A very particular structural change that takes place in the presence of double bonds is the bending of the fatty acid molecule, as shown in Figure 3.4.

The same fatty acid molecule can have several double bonds, as illustrated in Figure 3.5, in which four fatty acids are shown, all with 18 carbon atoms but with a different number of double bonds:

- stearic acid is a saturated fatty acid
- oleic acid is a monounsaturated fatty acid (acronym: MUFA)
- linoleic and α-linolenic acids are polyunsaturated fatty acids (acronym: PUFA) with two and three double bonds, respectively.

Double bonds are the most reactive position in a fatty acid molecule, especially if multiple double bonds are 'conjugated', which means 'separated by a single CH_2 group', which is the case with both linoleic and α-linolenic acid. Double bonds can react with oxygen, thus spurring the oxidative spoilage of oil, or they can react with hydrogen, thus re-establishing a saturated condition.

If one double bond of a natural PUFA is saturated by chemical reaction with two hydrogen atoms, the resulting unsaturated fatty acid has a linear structure. Therefore, if a double bond of α-linolenic acid is transformed into linoleic acid by saturation of a double bond, the resulting linoleic acid has a linear structure. Similarly, if a natural linoleic acid is transformed into oleic acid by saturation of a double bond, the resulting molecule of oleic acid has a linear structure (Figure 3.6). In order to distinguish these forms, all natural forms of unsaturated (and bended) fatty acids are identified with the prefix *cis*-, while all *trans*-formed, artificial, unsaturated fatty acids are identified with the prefix *trans*-.

However having an equal number of carbon atoms and the same number of double bonds, trans-isomers have physical, chemical and biological characteristics that are more similar to saturated fatty acids than to unsaturated fatty acids. Thus, trans-oleic acid is more similar to stearic acid than to cis-oleic acid.

Trans oils increase the risk of coronary heart disease by raising the level of LDL cholesterol and lowering the level of 'good' HDL cholesterol. The presence of trans-isomers in an extra-virgin olive oil is a clear sign of fraud and can be easily detected through analytical methods that are common practice nowadays.

Due to the bending of the molecule, the '*cis*' structure makes it more difficult for these fatty acids to solidify into compact crystals, so at a given temperature unsaturated fatty acids are softer than saturated fatty acids. In other words, unsaturated fatty acids have lower melting points in comparison to saturated fatty acids. Table 3.1 shows the four C18 fatty acids that are present in olive oil. Despite the fact that they have very similar molecular formulas and molar masses, they have different degrees of unsaturation and very different melting points. It is especially worth noting that stearic acid, a saturated fatty acid, has a melting point that is much higher than the human body temperature (37 °C or 98.5 °F) and therefore is solid in the body, whereas all the others are unsaturated fatty acids and have a melting point lower than the body temperature and therefore are liquid in the body.

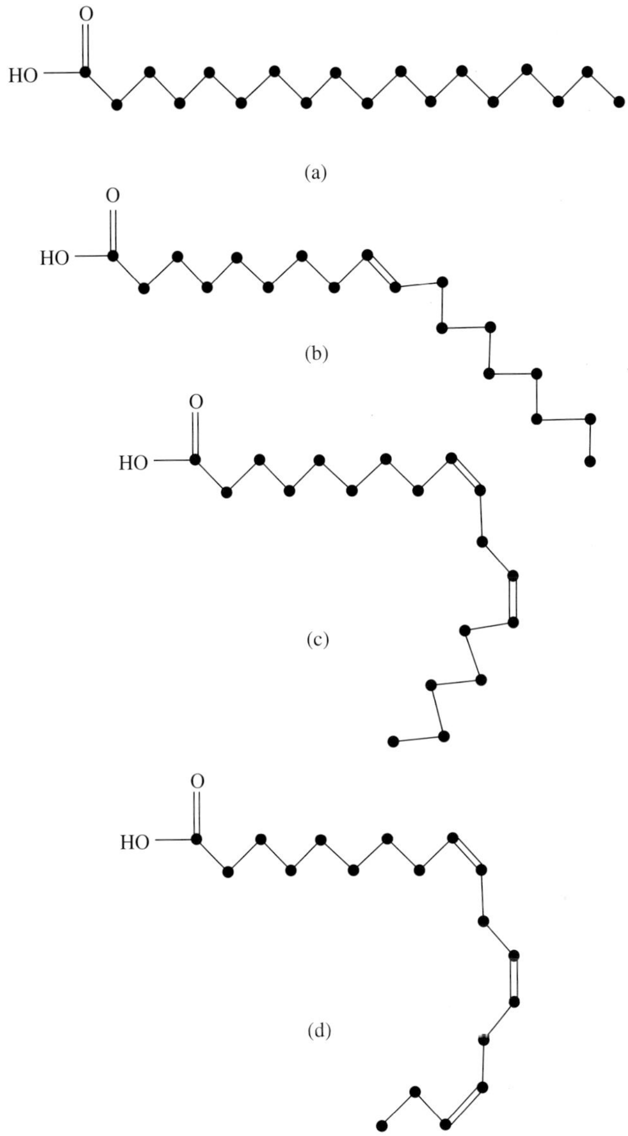

Figure 3.5 The molecules of stearic acid (a), *cis*-oleic acid (b), *cis*-linoleic acid (c) and *cis*-α-linolenic acid (d).

Lipid oxidation is influenced by many factors: the presence and concentration of oxygen, temperature, light and metal catalysts, but, most of all by the degree of unsaturation and the presence of conjugated double bonds.

The velocity of oxidative reactions is more than proportional to the number of double bonds as is evident by comparing the data in Table 3.2.

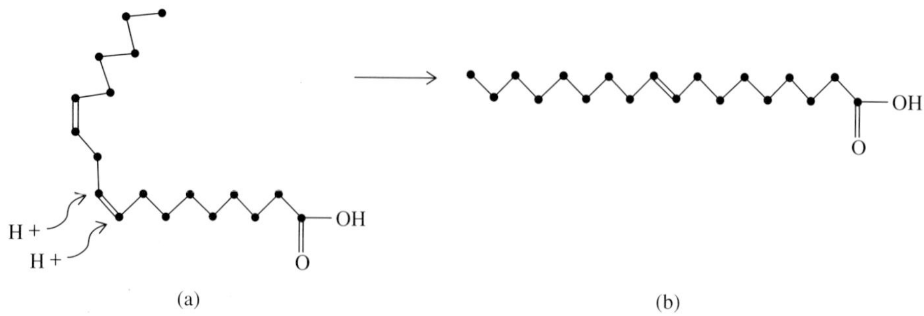

Figure 3.6 Transformation from *cis*-linoleic (a) to *trans*-oleic acid (b) by hydrogenation.

Table 3.1 Melting points of C18 fatty acids.

Fatty acid	Molecular formula	Molar mass	Melting point (°C)	Melting point (°F)
Stearic acid	$C_{18}H_{36}O_2$	284	69 °C	156.2
Oleic acid	$C_{18}H_{34}O_2$	282	13 °C	55.4
Linoleic acid	$C_{18}H_{32}O_2$	280	−5 °C	23
α-linolenic acid	$C_{18}H_{30}O_2$	278	−11 °C	12.2

Table 3.2 Relative velocity of oxidative degradation.

Fatty acid	The relative velocity of oxidative reactions
Stearic acid	0
Oleic acid	1
Linoleic acid	64
α-linolenic acid	100

3.2 The nutritional role of olive oil triglycerides and fatty acids

Oils and fats are the nutrients with the highest caloric value (9 kcal/g). Excess fat in the diet results in accumulation of fat in the adipose tissue. This aspect of their nutritional contribution, however, is only partial. In fact, they have essential structural roles in the skin, retina, nervous system (the brain is the body's organ with the highest concentration of lipids), and biological membranes. They are precursors of hormones and the vehicle for the absorption of liposoluble vitamins (Kritchevsky 2008).

It is interesting to observe the percentage distribution of fatty acids in olive oil in Table 3.3. Polyunsaturated fatty acids with 18 carbon atoms (linoleic and α-linolenic acid) play crucial roles in cell structure and function. They cannot be synthesized

Table 3.3 Distribution (%) of the fatty acids in the triglycerides in olive oils.

Fatty acids	Percentage of total fatty acids in olive oil
Monounsaturated fatty acids, oleic acid	65–83
Saturated fatty acids	8–14
Polyunsaturated (ω-6), linoleic acid	6–15
Polyunsaturated (ω-3), α-linolenic acid	0.2–1.5

by the body and therefore must be part of our diet; hence, they are referred to as 'essential fatty acids' (EFA). Nutritional studies have shown that they have preventive effects on cardiovascular diseases and nutritionists have identified them as ω-6 (linoleic acid) and ω-3 (α-linolenic acid) depending on the position of the first double bond in their molecule, counting from the ω end (Figure 3.5). However, these two important fatty acids are metabolically and functionally distinct, and often have opposing physiological functions in cell membranes. In some cases it has been found that their high reactivity and susceptibility to oxidation may represent a health risk. Therefore, a suitable balance of these essential fatty acids is important for good health and normal development. The ratio of monounsaturated to polyunsaturated fatty acids, and in particular ω-6 to ω-3 fatty acids in olive oil, is close to the optimal ratio recommended by nutritionists.

Most of all, the profile of olive oil fatty acids is characterized by the abundant presence of oleic acid, whose characteristics and functions are summarized in the box below.

Oleic acid

The data in Tables 3.1 and 3.2 demonstrate the unique characteristics of oleic acid compared to the other C18 fatty acids. Like the other unsaturated fatty acids, it has a melting point that is lower than the human body temperature, an essential requisite for preventing accumulation on artery walls (atherosclerosis) and for guaranteeing cell membrane fluidity. At the same time, it is much more resistant to oxidation than the other unsaturated fatty acids. This is essential for preventing oxidative damage to critical cell structures.

These characteristics make oleic acid an almost ideal food component and particularly useful in a number of biological functions, for example: (i) lowering blood pressure; (ii) ensuring the free flow of blood by reducing the clogging and hardening of arteries; (iii) lowering the levels of low-density lipoprotein (LDL) or bad cholesterol, while increasing the levels of high-density lipoprotein (HDL) or good cholesterol; (iv) strengthening cell-membrane integrity and helping to repair cells and damaged tissues; (v) fighting cancer, especially breast cancer; (vi) relieving symptoms of asthma and (vii) an ingredient in cosmetics, serving as a moisturizer, giving soft, supple skin.

3.3 Minor components and antioxidants in extra-virgin olive oil

Free radicals are highly reactive oxygen species. They are formed in the body during normal metabolism and, at a higher rate, upon exposure to environmental factors such as cigarette smoke and pollutants, or as a consequence of disease and traumatic events.

If free radicals are not intercepted and neutralized, they can cause serious damage to essential molecules such as DNA, protein, polyunsaturated fatty acids, especially those in the phospholipids of cell membranes, and lipoprotein. As a consequence, free radicals are closely associated with a range of disorders including cancer, arthritis, atherosclerosis, Alzheimer's disease, diabetes, and aging. The body reacts to oxidative threat with internal defence mechanisms and with molecules derived from food (tocopherols, carotenoids, phenolic compounds).

Since the discovery of the health benefits of the Mediterranean diet (Willet *et al.* 1995), interest in health protection from the daily consumption of extra-virgin olive oil has increased enormously (Visioli *et al.* 2002; Covas *et al.* 2006; Cicerale *et al.* 2009; Viola and Viola 2009; Pelucchi *et al.* 2010). This has also stimulated studies on the relationship between the health-promoting properties and quality of extra-virgin olive oils (Lavelli 2002; Servili and Montedoro 2002).

3.3.1 Hydrocarbons

Hydrocarbons are organic compounds that contain only carbon and hydrogen atoms. The major hydrocarbon in olive oil is squalene (skeletal formula in Figure 3.7); its name derives from the fact that it is extracted from the liver oil of sharks (in Latin *squalus*).

Squalene is a triterpene hydrocarbon that exerts antioxidant activity by reacting with oxygen radicals and oxygen-reactive species, thus protecting the skin against UV rays (something like a biological filter). Squalene has also been cited for its immune-stimulating properties and for its antineoplastic effects on colon, breast and prostate cancers.

Olive oil is a major source of squalene in the diet. Extra-virgin olive oil contains 200–700 mg of squalene per 100 g of oil, while refined olive oil contains about 25% less. Other useful hydrocarbons are present in extra-virgin olive oil, as for example β-carotene (pro-vitamin A), even if in small quantities.

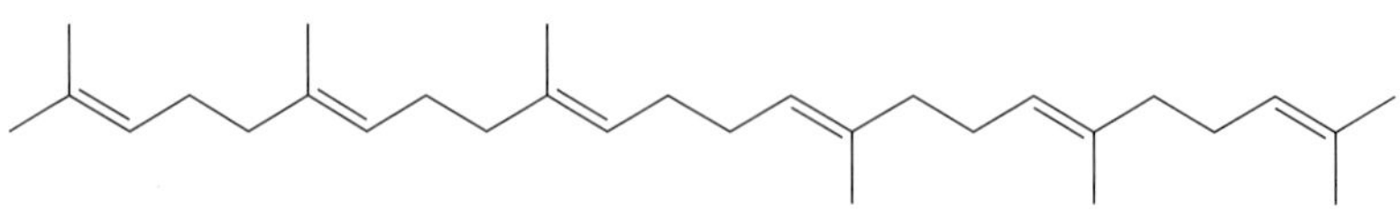

Figure 3.7 Squalene.

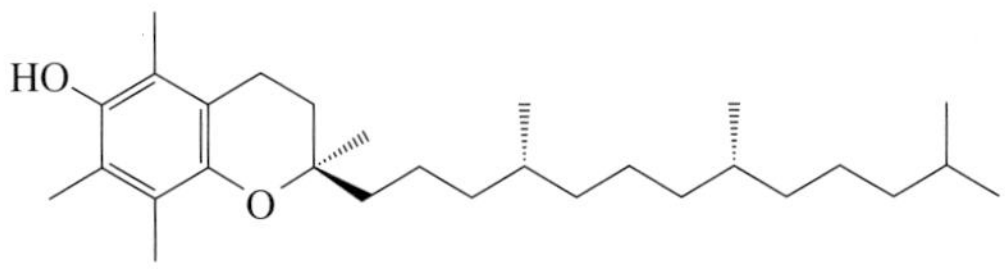

Figure 3.8 α-tocopherol.

3.3.2 Tocopherols

Tocopherols are fat-soluble alcohols that function as vitamin E, especially α-tocopherol (spacial formula in Figure 3.8), a very important antioxidant. Alpha-tocopherol is uniquely able to intercept free radicals and prevent chain reactions of lipid destruction at the cell membrane level. It also protects low-density lipoprotein (LDL) from oxidation. Oxidized LDL is implicated in the development of cardiovascular diseases. Extra-virgin olive oil contains 150 to 250 mg/kg of α-tocopherol with an optimal vitamin E-to-polyunsaturated fatty acid ratio of 1.5–2.0.

3.3.3 Phytosterols

Sterols are unsaturated alcohols present in the fatty tissues of plants (phytosterols) and animals. Although these compounds represent a minor part of lipids in vegetable oils, their quantification can be useful to establish the origin of an oil and to reveal intentional adulterations. The amount in extra-virgin olive oil varies from 100 to 250 mg/100 g of oil, of which 90–95% is β-sitosterol (spacial formula in Figure 3.9). Cholesterol is absent.

In addition to their cholesterol-lowering actions, mounting evidence suggests that phytosterols act against cancer of the lung, stomach, ovary and estrogen-dependent human breast cancer. *In vitro* studies using cell culture models have shown that β-sitosterol may have anticarcinogenic effects with regard to cancer of the prostate, colon, breast and stomach.

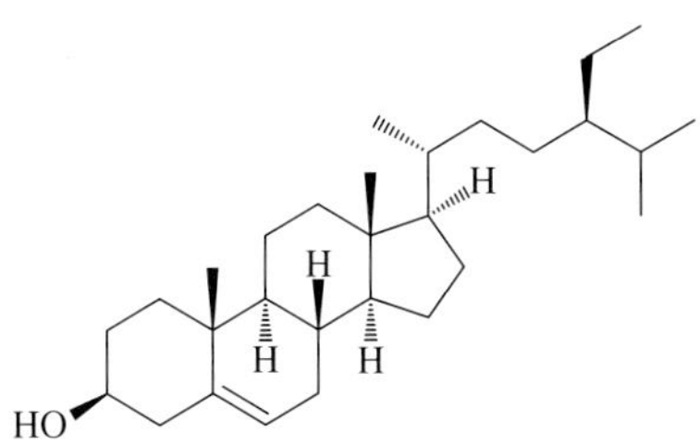

Figure 3.9 β-sitosterol.

The total sterol content and determination of the amount of individual sterols (cholesterol, brassicasterol, campesterol, stigmasterol, Δ-7-stigmastenol, and β-sitosterol) gives an indication of authenticity (see Table 2.4 in Chapter 2).

3.3.4 Phenolic compounds

These compounds are increasingly attracting the attention of researchers. The most important are the 5-hydroxytyrosol and its elenoic acid ester, oleuropein (Figure 3.10), the latter being an exclusive constituent of olive leaves and olive oil.

The sugar part of the oleuropein molecule is removed by enzymatic digestion during the malaxing operation, thus producing oleuropein aglycones. These compounds are partially soluble in oil and therefore they pass from the olive paste to the oil. This is a critical step for olive oil quality because these compounds have a high antioxidant potential and a bitter taste, which is a typical and positive sensory attribute of extra-virgin olive oils.

Oleuropein and its metabolite hydroxytyrosol have powerful antioxidant activity *in vitro* and *in vivo* associated with anti-inflammatory action. Oleuropein, in particular, has several pharmacological properties. Other simpler phenolic compounds present in lower amounts, such as caffeic acid, vanillic acid and ferulic acid have a protective effect on α-tocopherol and lignans, a class of phenols with protective effects against colon and breast cancer.

In summary, from the recent and extensive scientific literature, the following biological activities have been attributed to the phenolic compounds of extra-virgin olive oil: (i) direct antioxidant activity; (ii) protection of α-tocopherol antioxidant activity; (iii) binding of metal ions that favours radical formation; (iv) inhibition of platelet aggregation; (v) reduction of plasma cholesterol levels; (vi) inhibition of LDL oxidation; (vii) increased immune activity; (viii) anti-inflammatory activity; (ix) decreased cancer growth; (x) anti-allergic activity; (xi) skin protection.

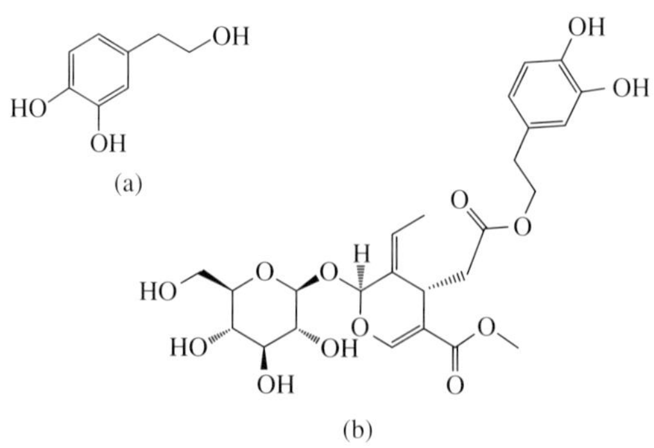

Figure 3.10 5-hydroxytyrosol (a) and oleuropein (b).

Evaluation of the phenolic fraction and its composition provides important information in terms of oil quality, stability and nutritional value. The most reliable method is based on high-performance liquid chromatography (HPLC) (Servili *et al.* 1999). Interesting information can be obtained from chemical or enzymatic methods for assessing antioxidant potential (Lavelli 2002). These are, however, complex and time-consuming methods, also requiring sophisticated analytical equipment and therefore they cannot be used as process control tools. In fact, evaluation of the phenolic content of extra-virgin olive oil can be used to characterize milling batches or to establish blending proportions, or to correlate an olive maturity index to the quality profile of the oil. A very commonly used method is based on the Folin–Ciocalteau colorimetric assay of total phenolics, which is simple but necessitates a well-equipped laboratory. Recently, the use of portable, easy-to-use, rapid analytical apparatuses has spread with success for online control duties among olive oil producers and millers. However, the problem of setting up a rapid and precise method of phenolic compounds analysis of olive oil that is in good agreement with the more reliable HPLC method, is still open (Garcia *et al.* 2013).

3.4 The colour and odour components of extra-virgin olive oil

The *colour* of extra-virgin olive oil ranges from green to yellow due to the prevalence of chlorophyll or carotenoids, respectively. The green colour of early-harvested olive oil, which is particularly intense in oil from some cultivars (e.g. Correggiolo), is very appealing to many consumers. Chlorophyll is a molecule very sensitive to light, and careful storage to protect the oil from oxygen and light helps maintain the green colour for a longer time. A rapid loss of the green colour is a sensitive indicator of poor storage conditions.

The *odour* of extra-virgin olive oil is a much more complex subject. Over a hundred volatile compounds have been identified by gas chromatography and mass spectrometry in extra-virgin olive oil: aldehydes, alcohols, esters, hydrocarbons, ketones, furans and others. Only a few of them have a real impact on the perceived odour, like, for example, those derived from hexanal (green, grassy), *trans*-2-hexenal (green, bitter), and 1-hexanol and 3-methylbutan-1-ol, which are the major volatile compounds of olive oil. Other, mostly unknown molecules, which are present in very low concentrations (in the range of parts per billion), are responsible for the odour and flavour notes that characterize extra-virgin olive oils. They arise from complex relationships between the genetics of the olive tree (cultivar), the origin in terms of soil and climate, and the conditions of the milling process. Notes of tomato leaf or ripe tomato, green or ripe olive, artichoke, fresh almond, apple, citrus, freshly mown grass and many others give an essential contribution to the sensory profile of the extra-virgin olive oil. They may be considered as the most important aspect of quality in the culinary arts and for consumer appreciation. In the end, these components, which are unknown, are very difficult to identify and to explain in terms of fruit genetics and metabolism, but they have the greatest impact in differentiating excellent from common oils and hence in the commercial success of an oil or a brand.

3.5 Conclusion

It can be concluded that extra-virgin olive oil is a complex, unique food with significant nutritional and health-promoting properties:

1. Unlike any other edible oil, it contains phenolic compounds with a high antioxidant potential. Most of these compounds are an exclusive metabolic product of the olive tree and the olive fruit.
2. Concerning the nonpolar antioxidants, extra-virgin olive oil is a rich mixture containing squalene, α-tocopherol, β-sitosterols and other minor compounds such as β-carotene. They play different and synergistic antioxidant roles.
3. The polar and nonpolar antioxidants in extra-virgin olive oil assure reciprocal protection from oxidation. Thus, polyphenols are known to protect α-tocopherol and simple phenolics to protect the more complex phenolic molecules from oxidation. This is what is defined as *in vitro* antioxidant activity, whereas the *in vivo* antioxidant activity consists in fighting and inactivating free radicals in the body. The second property is closely connected to the first.
4. The fatty acid mix of olive oil triglycerides also has interesting properties, as for example, an optimal ratio of monounsaturated-to-polyunsaturated, of ω-6-to-ω-3 fatty acids.
5. The abundant amount of oleic acid assures a high resistance of olive oil to oxidation and further strengthens the health-promoting properties of olive oil.

The complexity of the overall picture should discourage an oversimplification of the functions and benefits of olive oil. Focusing on phenolics without considering the nonpolar antioxidants would be a mistake. Emphasizing the presence of oleic acid without giving appropriate weight to the presence of polyunsaturated essential fatty acids would also be a mistake. In general, trying to extract and isolate some health-promoting components to make pills or concentrated solutions is a dramatic underestimation of the importance of equilibriums and synergistic effects that make the natural product an unmatched source of pleasure and health.

A good extra-virgin olive oil is a tasty, healthy ingredient in the diet. An average consumption of 20 g per day of a very good extra-virgin olive oil can, and, in fact, should be part of the diet from infancy to old age, as suggested by the Mediterranean diet pyramid.

Balance is everything

'Balance is everything' was one of the mantras that were used by John Wooden, the best basketball coach ever (Hill and Wooden 2001). He used to summarize the concept as follows: 'I must have offensive balance, defensive balance, squad balance, emotional balance, mental balance, balance, balance, balance.' John Wooden did not just talk about the need for balance in basketball; everything in his life suggested that he lived this value as well.

Balance is certainly a critical feature of good diet and nutrition, as an unbalanced diet is the most common mistake and risk in nutrition.

Extra-virgin olive oil is a very special case and almost a prototype of the requirements for balance:

- In a good olive oil there is an almost perfect balance of MUFA and PUFA, all of them at concentrations perfectly suited for the human diet. There is also an almost perfect balance of PUFA and α-tocopherol, maximizing vitamin E effectiveness.
- In an excellent extra-virgin olive oil there is a balance of hydrophilic and lipophilic antioxidants, which is the most convincing cellular basis of their synergy and effectiveness.
- Referring to the previous point, excess water should be avoided, as it is a condition for oil spoilage, but, at the same time, water is needed to guarantee the presence of precious minor polar compounds, with a useful biological activity and an essential role in sensory quality. Balance is the condition.
- Oxygen and oxidative damage must be avoided, but some oxidation is necessary for the formation of particular flavour compounds. Processing under inert gas has proved to maximize the antioxidant concentration, but it impairs flavour formation. Thus, an optimal balance of health and flavour is a matter of balancing oxygen availability/oxygen exclusion in the olive oil process.
- Olive harvesting should be carried out not too early and not too late. The balance between oil yield and oil quality requires an optimal choice of the harvesting time.

Dealing with complexity and excellence always requires balance and olive oil is a most convincing case in point.

References

Cicerale, S., Conlan X.A., Sinclair A.J. and Keast, R.S.J. (2009) Chemistry and health of olive oil phenolics. *Critical Reviews in Food Science and Nutrition* **49** (3), 218–236.

Covas, M.I., Ruiz-Gutièrrez, V., De La Torre, R. *et al.* (2006) Minor components of olive oil: evidence to date of health benefits in humans. *Nutrition Reviews* **64** (10), 20–30.

Garcia, B., Coelho, J., Costa, M. *et al.* (2013) A simple method for the determination of bioactive antioxidants in virgin olive oils. *Journal of the Science of Food and Agriculture* **93**, 1727–1732.

Hill, A., Wooden, J. (2001) *Be Quick – But Don't Hurry*, Simon & Schuster, New York.

Kritchevsky, D. (2008) Fats and oils in human health, in *Food Lipids: Chemistry, Nutrition and Biotechnology* (eds C.C. Akoh and D.B. Min), CRC Press, Boca Raton, FL.

Lavelli, V. (2002) Comparison of the antioxidant activities of extra-virgin olive oils. *Journal of Agricultural and Food Chemistry* **50** (26), 7704–7708.

López-Miranda, J., Pérez-Jiménez, F., Ros, E. *et al.* (2010), Olive oil and health: summary of the II international conference on olive oil and health consensus report, Jaén and Córdoba (Spain) 2008. Nutrition, *Metabolism and Cardiovascular Disease* **20**(4), 284–294.

O'Keefe, S.F. (2008) Nomenclature and classification of lipids. In *Food Lipids: Chemistry, Nutrition and Biotechnology* (eds C.C. Akoh and D.B. Min), CRC Press, Boca Raton, FL.

Pelucchi, C., Bosetti, C., Negri, E. *et al.* (2010) Olive oil and cancer risk: an update of epidemiological findings through 2010. *Current Pharmaceutical Design* **17**, 805–812.

Servili, M., Baldioli, M., Selvaggini, R. *et al.* (1999) High-performance liquid chromatography evaluation of phenols in olive fruit, virgin olive oil, vegetation water, and pomace and 1D- and 2D-nuclear magnetic resonance characterization. *Journal of the American Oil Chemists' Society* **76** (7), 873–882.

Servili, M. and Montedoro, G. (2002) Contribution of phenolic compounds to virgin olive oil quality. *European Journal of Lipid Science and Technology* **104**, 602–613.

Viola P. and Viola, M. (2009) Virgin olive oil as a fundamental nutritional component and skin protector. *Clinics in Dermatology* **27**, 159–165.

Visioli, F., Poli, A. and Galli, C. (2002) Antioxidant and other biological activities of phenols from olive and olive oil. *Medicinal Research Reviews* **22**, 65–75.

Willet, W.C., Sacks, F., Trichopoulou, A. *et al.* (1995), Mediterranean diet pyramid: a cultural model for healthy eating. *American Journal of Clinical Nutrition* **61**, 1402S–06S.

4

The sensory quality of extra-virgin olive oil

Mario Bertuccioli and Erminio Monteleone
Department of Agricultural, Food and Forestry System Management, University of Florence, Florence, Italy

Abstract

Description of the sensory quality of extra virgin olive oils implies the analysis of different sensory and affective responses. The qualities of fruitiness, bitterness and pungency and the absence of defects are necessary to qualify oils as extra virgin. They should be analysed using the methods and standards defined by the International Olive Council and adopted in EU legislation. Descriptive analysis conducted according to the procedures reported in this chapter aims to describe similarities and differences among oils in order to identify different sensory styles. Sensory profiles should ensure a complete description of sensory attributes and a vocabulary to be used in communication between oil producers and retailers, culinary experts and consumers. The sensory functionality of extra-virgin olive oils with varied sensory styles should be investigated using the Temporal Dominance of Sensations (TDS) method and a measure of consumer liking for oil-food pairings.

4.1 Introduction

Most of the attention on the sensory characteristics of olive oil is currently focused on how to evaluate whether a given oil is free of defects and how categories of virgin olive oil are classified. The International Olive Council standards for sensory evaluation of oils are the recognized standard used to classify oils in categories such as extra virgin or virgin or 'lampante'. These standards are based on evaluation of both 'negative' and 'positive' attributes (International Olive Council 2005 and 2011).

Negative attributes – sensory defects – cannot be present in extra-virgin olive oil and, therefore, screening for them should be considered a sensory prerequisite of

The Extra-Virgin Olive Oil Handbook, First Edition. Edited by Claudio Peri.

extra-virgin olive oil. Positive sensory attributes are bitterness, pungency and fruity notes and their intensity is determined to define extra-virgin olive oil sensory characteristics (Chapter 2).

In many international 'competitions', the evaluation procedures used by experts to rank the sensory quality of extra-virgin olive oils are characterized by a partial flavour description and an overall quality judgment (from poor to excellent), deriving from the evaluation of the perceived complexity, balance and finish. These competitions contribute to creating a sort of 'model' that is then used to score the sensory quality of oils. There is more than one reason to be critical of this approach.

It appears obvious that sensory evaluations aimed at classifying oils accordingly to 'models' of ideal sensory quality contradict the importance of discovering the sensory diversity among oils. The frequent result of this approach is the possible marginalization of some interesting, but extreme, sensory characteristics. Conversely, producers need to emphasize variety and sensory style. The term 'style' does not imply a quality model, but rather a sensory profile that describes an oil (or a group of oils) as different from other oils.

A further and important reason to be critical of this approach is the consideration that oils are never consumed alone. They are used as ingredients in preparing dishes and are paired with other foods; thus, they should not be judged for their sensory attributes alone, but for their sensory functionality in food combinations. Sensations marking the sensory characteristics of a given oil may not be experienced in conditions of normal use (Monteleone 2010; Dinnella *et al.* 2012). Extra-virgin olive oil quality should therefore be considered in relation to the capacity of the oil to modify the sensory properties of a dish and to enhance the acceptability of the food to which it is combined.

The contents of this chapter are divided into three parts:

- The first part concerns the evaluation of sensory defects and of the three basic positive sensory notes of bitterness, pungency and fruitiness, according to the legal standards and method for grading virgin olive oils.
- The second part concerns the description of the sensory style of an extra-virgin olive oil defined as a sensory profile. This is the pivotal step in sensory analysis. It assures a complete description of the sensory attributes and the vocabulary to be used in communication between oil producers and retailers, culinary experts, and consumers.
- The third part presents the most recent evolution in sensory studies on extra-virgin olive oil. It aims at establishing the sensory performance of the oil when used as a condiment or as an ingredient in foods and culinary preparations

4.2 The official evaluation of defects and positive sensory attributes

This part concerns the basic sensory evaluation according to European Law and the International Olive Council standards (Commission Regulation (EC) No 640/2008 of 4 July 2008 amending Regulation (EEC) No 2568/91 on sensory characteristics of olive oil and the relevant methods of analysis; International Olive Oil Council 2011). The official scorecard in Figure 4.1 shows how judges are asked to evaluate negative and positive attributes, also giving a semi-quantitative appreciation of the perceived intensity on a 10 cm long scale from zero (no perception) to 10 (extremely strong perception).

According to the European regulation, oils should be graded according to the median of defects and the median of the 'fruity' perception (Chapter 2).

The median of the defects is defined as the median of the defect perceived with the greatest intensity. The median of the defects and the median for 'fruity' are expressed

INTENSITY OF PERCEPTION OF DEFECTS:

Fusty/muddy sediment

Musty-humid-earthy

Winey-vinegary-acid-sour

Frostbitten olives (wet wood)

Rancid

Others (specify)

INTENSITY OF PERCEPTION OF POSITIVE ATTRIBUTES:

Fruity — greenly, ripely

Bitter

Pungent

Name of taster:

Sample code:

Date:

Comments:

Figure 4.1 The scorecard for the official evaluation of the sensory characteristics of virgin olive oil (Commission Regulation (EC) No 640/2008 of 4 July 2008 amending Regulation (EEC) No. 2568/91 on sensory characteristics of olive oil and the relevant methods of analysis).

to one decimal place, and the value of the robust variation coefficient that defines them must be no greater than 20%. The following grading applies:

- *extra-virgin olive oil*: the median of the defects is 0 and the median for fruity is above 0
- *virgin olive oil*: the median of the defects is above 0, but not more than 3.5, and the median for fruity is above 0
- *lampante olive oil*: the median of defects is above 3.5.

Considering that in extra-virgin olive oil a median of zero of the 'fruity' perception is extremely rare (extra-virgin olive oils always have at least a hint of fruity flavour), the real sensory pre-condition for grading an oil as extra virgin is the absence of defects.

Table 4.1 presents a glossary of terms identifying sensory defects, and Table 4.2 presents the terms identifying the three basic positive attributes according to the European Regulation (Commission Regulation (EC) No 640/2008 of 4 July 2008 amending Regulation (EEC) No 2568/91 on sensory characteristics of olive oil and the relevant methods of analysis).

The sensory procedure for the evaluation of defects and the basic positive sensory attributes is out of the scope of this handbook and is described in detail in the official documents and in many references cited in this chapter. Good knowledge and practical experience of this procedure must be available in all extra-virgin olive oil factories, as sensory tasting (either amateur or professional) should be considered as a routine on-line evaluation tool in the extra-virgin process.

In order to give some hints about the method and goal of achieving the highest possible standardization and reproducibility, a few excerpts from the official documents are reported in the box below.

Excerpts from the official EU and International Olive Oil Council texts

About the Panel

The panel consists of a panel head and from eight to twelve tasters. The panel head must be a soundly trained expert in the various types of oils. He or she is responsible for the panel and its organization and operation, coding and presentation of the samples to the tasters and collection and processing of the data. He or she selects the tasters, sees to their training and checks that their performance remains at a suitable standard. The tasters must be selected and trained on account of their skill in distinguishing between similar samples. The International Olive Council's manual on selection, training and monitoring of qualified virgin olive oil tasters must be followed ...

About the use of the profile sheet by tasters

The profile sheet to be used by the tasters is shown in Figure 4.1. Tasters must smell and then taste the oil submitted for examination, marking the intensity of their perception of each negative and positive attribute on the 10-cm scale representing his/her judgment of intensity ...

... If a taster perceives the fruitiness to be of a green or ripe character, he or she must tick the corresponding box on the profile sheet.

If a taster perceives any negative attributes not listed on the profile sheet, he or she must note it under ‘Other’, using the term or terms that describe them best from among those defined in Table 4.1 ...

About processing of data by the panel head

The panel head collects the profile sheets and scrutinizes the intensities assigned to the various attributes. In the event of an anomaly, he or she will ask the tasters to re-examine their sheets and, if necessary, repeat the test . . .

. . .

The panel head may certify that the oil meets the conditions for the term ‘green’ or ‘ripe’ only if at least 50% of the panel perceived that the fruitiness had this character and noted it down . . .

. . .

If the median of a positive attribute other than ‘fruity’ is above 5.0, the panel head must note this in the analysis certificate.

The method for the calculation of the median and confidence intervals is described in detail with examples in the official documents.

In order to provide some further information to consumers about positive attributes, the regulation allows the use of the following terms for fruitiness or bitterness or pungency on the package label:

- ‘*intense*’ may be used when the median of the attribute concerned is greater than 6
- ‘*medium*’ may be used when the median of the attribute concerned is between 3 and 6
- ‘*light*’ may be used when the median of the attribute concerned is less than 3.

The regulation also allows use of the terms:

- ‘*well balanced*’ when the median of the bitter and/or pungent attributes is two points higher than the median of fruitiness
- ‘*mild*’ when the median of the pungent and bitter attributes is 2 or less.

Table 4.1 Glossary of terms for sensory defects of virgin olive oil (Commission Regulation (EC) No 640/2008 of 4 July 2008 amending Regulation (EEC) No. 2568/91 on sensory characteristics of olive oil and the relevant methods of analysis).

Fusty/muddy sediment	Characteristic flavour of oil obtained from olives piled or stored in such conditions as to have undergone an advanced stage of anaerobic fermentation, or of oil which has been left in contact with the sediment that settles in the bottom of tanks and vats and which has also undergone a process of anaerobic fermentation.
Musty-humid	Characteristic flavour of oils obtained from fruit in which large numbers of fungi and yeasts have developed as a result of storage in humid conditions for several days.
Oily winey-vinegary acid-sour	Characteristic flavour of certain oils reminiscent of wine or vinegar. This flavour is mainly due to aerobic fermentation of the olives or of the olive paste and leads to the formation of acetic acid, ethyl acetate and ethanol.
Metallic	Flavour reminiscent of metal, characteristic of oil that has been in prolonged contact with metallic surfaces during milling, malaxing, pressing or storage.
Rancid	Flavour of oils that have undergone an intense process of oxidation.
Frostbitten olives (wet wood)	Characteristic flavour of oils extracted from olives that have been injured by frost while on the tree.
Heated or burnt	Characteristic flavour of oils caused by excessive and/or prolonged heating during processing, particularly by thermo-mixing of the paste in unsuitable conditions. It is typical of paste sticking to the walls of the malaxer wall for several hours.
Hay–wood	Characteristic flavour of certain oils from dry olives.
Rough	Thick, pasty mouthfeel sensation produced by certain old oils.
Greasy	Flavour of oil reminiscent of diesel, grease or mineral oil.
Vegetable water	Flavour acquired by the oil as a result of prolonged contact with vegetation water, which has undergone fermentation.
Brine	Flavour of oil extracted from olives that have been preserved in brine.
Earthy	Flavour of oil from olives that have been collected with earth or mud on them and not washed.
Grubby	Flavour of oil from olives that have been heavily attacked by the grubs of olive fly (*Bactrocera oleae*).
Cucumber	Characteristic flavour of oil kept too long in hermetically sealed containers, notably in tins, attributed to the formation of 2,6 nonadienal.

These information are useful to consumers for recognizing and choosing their preferred oil style but they have a low discriminating potential.

In fact, the compounds determining the three basic sensations, especially bitterness and pungency, have a similar origin and fate in the production process. In other words, an intensely pungent oil can often be also intensely bitter and intensely fruity, whereas an oil with low bitterness is rarely intensely pungent or fruity. Furthermore, these attributes say nothing about the odour and flavour sensations, which are the most characterizing sensory attributes in the consumer's perception and preference. Developing a more complete description of the sensory profile would allow much more meaningful communication to consumers.

Table 4.2 Positive sensory attributes of extra-virgin olive oil (Commission Regulation (EC) No 640/2008 of 4 July 2008 amending Regulation (EEC) No. 2568/91 on sensory characteristics of olive oil and the relevant methods of analysis).

Fruity	Range of smells (depending on cultivar, degree of maturity at harvest, and processing conditions) characteristic of oil from healthy fresh fruit, green or ripe, perceived directly as odour when smelling the oil and/or retronasally as flavour when tasting the oil in the mouth.
	Fruitiness is qualified as *green* if the range of smells is reminiscent of green fruit and is characteristic of oil from green fruit.
	Fruitiness is qualified as *ripe* if the range of smells is reminiscent of ripe fruit and is characteristic of oil from green and ripe fruit.
Bitter	Characteristic primary taste of oil from green olives or olives turning colour. Bitterness is detected by the circumvallate papillae on the 'V' region of the tongue.
Pungent	Tingling sensation characteristic of oils produced at the beginning of the season, mainly from olives that are still green. It can be perceived throughout the mouth cavity, particularly in the throat.

4.3 The sensory profile

The evaluation of the sensory profile is the pivotal step for defining the sensory style of an extra-virgin olive oil. The most widely used method is Descriptive Sensory Analysis (DSA) (Lawless and Heymann 1998). Some details providing useful insight into the DSA procedure are given in Annex 4.1.

4.3.1 The problem of terminology

The problem of terminology standardization is of critical importance. Generally, each panel develops the sensory language itself. This is a consensus-building process aimed at defining the attributes that the panel will use to express perceptions. Tasters familiarize themselves with the product space and generate attributes that describe the differences among products. Before assessing products, tasters participate in a series of language sessions managed by a panel leader.

Table 4.3 reports the terms used to describe the sensory characteristics of extra-virgin olive oils according to three key publications: Mojet and de Jong (1994); Delgado and Guinard (2011b) and Monteleone *et al.* (2012). These papers were selected for two main reasons: (i) descriptive studies were carried out in different countries and in qualified research sensory laboratories; (ii) descriptive terms were generated to profile oils with varied origins (country and variety). There are several descriptors that are recurrent in the three lists.

In order to achieve the best reproducibility of sensory assessment, precise definitions should be given for every sensory descriptor and reference standards should be used to select, train, tune and compare the performance of sensory judges. Table 4.4 is an example of standard definitions and references of the sensory descriptors proposed by Monteleone *et al.* (2012).

Table 4.3 Lists of sensory descriptors of extra-virgin olive oils as reported in three key papers from 1994 to 2012. Descriptors of defects or off-flavours are not reported.

Descriptor	Reference		
	Mojet and de Jong (1994)	Delagdo and Guinard (2011b)	Monteleone *et al.* (2012)
grassy	+	+	+
green fruit (green olives, green banana; green apple)	+	+	+
ripe fruit (olives, banana, apple)	+	+	+
tropical fruit	-	+	-
hay	+	-	-
tea	-	+	-
tomato leaf	+	+	+
tomato fruit	+	-	+
herbs	-	+	-
citrus	-	+	+
floral	+	+	-
spicy	-	+	-
nutty	+	+	-
butter	-	-	-
mint	-	-	-
perfumy	+	-	-
almond	+	+	+
briny	+	-	-
artichoke	+	-	+
bitter	+	+	+
thick	+	+	+
rough	+	-	-
pungent	+	+	+
peppery	+	+	+
astringent	+	+	+

More complex representations of sensory descriptors were proposed in the form of a 'sensory wheel' for the first time by Mojet and de Jong (1994) and, more recently, by Richard Gawel. Gawel's wheel lists 72 different terms that can be used to describe the complex range of aromas and tastes found in virgin olive oils (Gawel 2007). These forms, which are interesting for research, are far too complex to be suitably used in process control and product optimization.

4.3.2 The evaluation of the sensory profile

The approach to extra-virgin olive oil sensory profiling that was published by Bertuccioli in 1994 remains the most convincing and reliable method for application in process optimization as well as in standard evaluation of the extra-virgin olive oil sensory characteristics (Bertuccioli 1994). The essential results of this research are discussed below as a model for sensory profile studies of extra-virgin olive oil.

Table 4.4 Glossary and references used in descriptive analysis of extra-virgin olive oils varying in origin (country and variety) according to Monteleone *et al.* (2012).

Descriptor	Definition	Reference[a]
Green olive	Odour associated with freshly milled green olives	100 g of olive paste from fresh green olives in 100 ml of seed oil. The standard should be presented to the panelists within 60 min from preparation.
Ripe olive	Odour associated with black (ripe) olives malaxed for 30 min.	olive pastes from black (ripe) olives malaxed for 30 min
Grassy	Odour associated with fresh-cut grass	14 µl of 1-cis-3-hexenol in 100 ml of seed oil. The standard should be presented to the panelists within 4 h from preparation.
Tomato fruit	Aroma of ripe tomato	100 g of fresh and ripe 'pachino' tomatoes in 100 ml of seed oil. The standard should be presented to the panelists within 60 min from preparation.
Tomato leaf	Aroma of tomato leaves	fresh tomato leaves
Apple	Aroma of golden apple	20 g of skin and 20 g of pulp of a ripe Golden Delicious apple in 100 ml of seed oil. The standard should be presented to the panelists within 60 min from preparation.
Citrus	aroma of lemon/orange	2 g of lemon skin and 2 g of orange skin in 100 ml of seed oil. The standard should be presented to the panelists within 60 min from preparation.
Artichoke	Aroma of artichoke	12 g of artichoke heads in 100 ml of seed oil. The standard should be presented to the panelists after 60 min from preparation.
Astringency	Dryness in the mouth	water solution of aluminium potassium sulphate (0.3 g/L)
Bitterness	Bitter taste	water-oil emulsions of quinine dihydrochloride solutions (intensities from weak to strong 50, 100, and 200 ppm). The model oil should be prepared by using an odourless and tasteless food-grade seed oil and a food-grade water-oil emulsifier
Pungency	Leaving a burning sensation in the back of the throat.	verbal description
Viscosity	Thin–thick	verbal description

Note: [a] Odourless and tasteless food-grade seed oil is used to prepare references in oil. Fresh ingredients were weighed in a coffee filter bag, sealed, stapled, and dipped into the oil. The standards were prepared at least 3 to 4 days in advance to let the oil absorb the aromas of the ingredients.

Figure 4.2 is the scorecard proposed by Bertuccioli (1994). A limited number of sensory descriptors, which are generally recognized as significant, are listed with an interesting and almost complete overlapping of perceptions as aromas by nose and flavours by mouth. Four oil samples are suggested as the full utilization of the scorecard.

In Table 4.5, columns report the intensity evaluations of the eight sensory attributes selected by the panel as appropriate for evaluating the sensory profiles of four extra-virgin olive oils, generically identified with letters A, B, C and D. The last column reports the least significant difference (LSD) values for each attribute.

The data are analysed by analysis of variance (ANOVA) to determine if a significant difference exists among the samples and if the judges are reproducible. Using tests such as Fischer's least significant difference (LSD) or Duncan's multiple range test, olive oils that differ significantly can be identified. Further details about performing ANOVA and interpreting the results are available in the bibliography (O'Mahony 1986).

	SAMPLE CODE			
	______	______	______	______
	INTENSITY SCORE (1 = low intensity 9 = high intensity)			
COLOUR				
1 – yellow	______	______	______	______
2 – green	______	______	______	______
AROMA				
3 – tomato	______	______	______	______
4 – green olive	______	______	______	______
5 – ripe black olive	______	______	______	______
6 – cut grass	______	______	______	______
7 – artichoke	______	______	______	______
8 – apple	______	______	______	______
9 – yeast	______	______	______	______
TASTE				
10 – bitter	______	______	______	______
MOUTHFEL				
11 – pungent	______	______	______	______
12 – astringent	______	______	______	______
FLAVOUR BY MOUTH				
13 – tomato	______	______	______	______
14 – green olive	______	______	______	______
15 – ripe black olive	______	______	______	______
16 – cut grass	______	______	______	______
17 – artichoke	______	______	______	______
18 – apple	______	______	______	______

Figure 4.2 An example of a scorecard for a parallel Descriptive Analysis evaluation of four Extra-virgin Olive Oils (*Source*: Bertuccioli 1994. Reproduced with permission from Grasas Y Aceites).

Table 4.5 Overall means of eight attributes rated by a descriptive analysis panel for four EVOO samples (*Source*: Bertuccioli 1994. Reproduced with permission from Grasas Y Aceites).

Descriptor	Extra-virgin olive oil sample				LSD
	A	*B*	*C*	*D*	
Yellow	7.4	7.3	6.6	7.9	0.61*
Ripe black olive	3.9	3.2	4.7	3.0	1.03NS
Green olive	4.2	5.1	3.1	4.6	1.14*
Cut grass	2.7	5.3	2.2	3.1	0.58***
Apple	1.2	1.1	1.0	5.4	0.23***
Tomato	5.4	4.4	1.2	1.4	0.38***
Bitter	6.1	6.7	6.3	6.8	0.33NS
Pungent	6.1	6.7	5.9	6.2	0.5NS

Note: LSD = least significant difference; * significant at $p < 0.05$; *** significant at $p < 0.001$; NS = not significant.

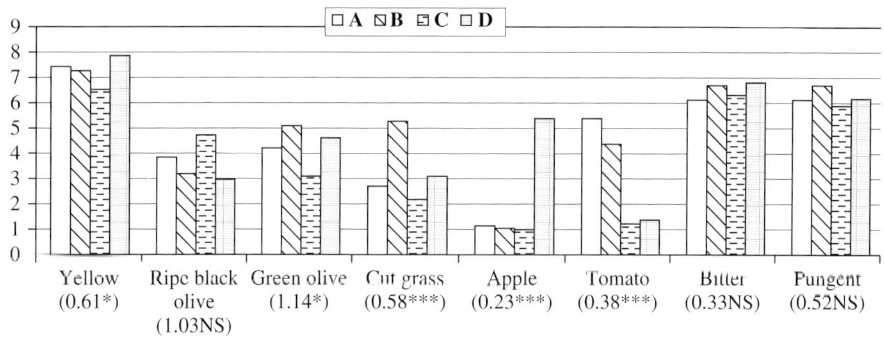

Figure 4.3 Bar graph of intensity scores from data in Table 4.5 (*Source*: Bertuccioli 1994. Reproduced with permission from Grasas Y Aceites). Comparison of sensory profiles of extra-virgin olive oils and least significant difference LSD (in brackets). *Note*: * significant at $p < 0.05$; *** significant at $p < 0.001$; NS = not significant.

Figure 4.3 shows a traditional bar graph corresponding to the intensity scores of the sensory attributes in Table 4.5.

The same data are reported in polar coordinates (spider plot) in Figure 4.4. Distances from the centre are proportional to the intensity rating of the sensory attributes. By connecting the mean ratings of intensities scored by the judges, the sensory profiles of the oils are obtained (Bertuccioli 1994).

By comparing the spider plots of the four oils, it can be observed that the sensory notes of bitterness and pungency have a very limited discriminating power among the four oils. In fact, the intensities of these two (important) sensory notes are very similar. Likewise, the fruity notes (both ripe and green olive) do not change much and are also similar in the four oils. What makes a striking difference among the four oils are the odour and flavour notes: one of the oils, (A), is characterized by the tomato flavour; another, (B), by the cut grass flavour; another, (D), by a strong apple flavour, while oil C, which is very similar to the others for bitterness, pungency and

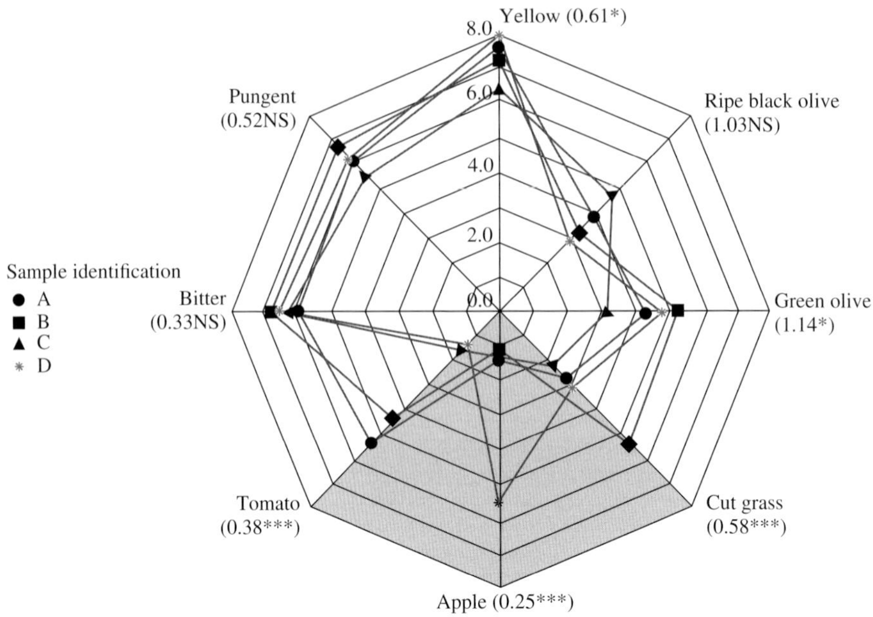

Figure 4.4 Spider plot of data in Table 4.5 (*Source*: Bertuccioli 1994. Reproduced with permission from Grasas Y Aceites). Comparison of sensory profiles of extra-virgin olive oils and least significant difference LSD (in brackets). *Note*: * significant at $p < 0.05$; *** significant at $p < 0.001$; NS = not significant. The background in grey points out the most characterizing part of the sensory profile.

fruitiness, has no characterizing notes of flavour or odour and hence it has a lesser style and personality. These differences are what consumers most understand and like. It may be concluded that only DSA can differentiate the perceived quality and style of an excellent extra-virgin olive oil.

Today, the DSA approach is also suggested by official organizations. In 2005, the International Olive Council issued a document on methods to be used for the organoleptic assessment of extra-virgin olive oils for Protected Designation of Origin (PDO), granting a typicality status (International Olive Oil Council 2005). This document declares that the PDO authority shall select the characteristic descriptors of the designation of origin (10 at the most) from those defined and reported in Table 4.6, and shall incorporate them into the profile sheet of the method.

4.3.3 Other methods of data analysis and presentation

When a large number of oils are examined by descriptive analysis, comparison through simple tables or graphs is not useful. Principal component analysis (PCA) (Martens and Martens 2001) is used to interpret large data sets.

Principal component analysis is a multivariate statistical technique that reduces the multiple dimensionality of the data by showing relationships among the

Table 4.6 Sensory descriptors for the designation of origin of extra-virgin olive oils (International Olive Oil Council 2005).

Direct or retronasal aromatic olfactory sensation	
Almond	Olfactory sensation reminiscent of fresh almonds
Apple	Olfactory sensation reminiscent of fresh apples
Artichoke	Olfactory sensation of artichokes
Camomile	Olfactory sensation reminiscent of camomile flowers
Citrus fruit	Olfactory sensation reminiscent of citrus fruit (lemon, orange, bergamot, mandarin and grapefruit)
Eucalyptus	Olfactory sensation typical of *Eucalyptus* leaves
Exotic fruit	Olfactory sensation reminiscent of the characteristic odours of exotic fruit (pineapple, banana, passion fruit, mango, papaya, etc.)
Fig leaf	Olfactory sensation typical of fig leaves
Flowers	Complex olfactory sensation generally reminiscent of the odour of flowers, also known as floral
Grass	Olfactory sensation typical of freshly mown grass
Green pepper	Olfactory sensation of green peppercorns
Green	Complex olfactory sensation reminiscent of the typical odour of fruit before it ripens
Green fruit	Olfactory sensation typical of oils obtained from olives that have been harvested before or during colour change
Herbs	Olfactory sensation reminiscent of herbs
Olive leaf	Olfactory sensation reminiscent of the odour of fresh olive leaves
Pear	Olfactory sensation typical of fresh pears
Pine kernel	Olfactory sensation reminiscent of fresh pine kernels
Ripe fruit	Olfactory sensation typical of oils obtained from olives that have been harvested when fully ripe
Soft fruit	Olfactory sensation typical of soft fruit: blackberries, raspberries, bilberries, blackcurrants and redcurrants
Sweet pepper	Olfactory sensation reminiscent of fresh sweet red or green peppers
Tomato	Olfactory sensation typical of tomato leaves
Vanilla	Olfactory sensation of natural dried vanilla powder or pods, different from the sensation of vanillin
Walnut	Olfactory sensation typical of shelled walnuts
Gustatory sensations	
Bitter	Characteristic taste of oil obtained from green olives or olives turning colour; it defines the primary taste associated with aqueous solutions of substances like quinine and caffeine
Sweet	Complex gustatory-kinaesthetic sensation characteristic of oil obtained from olives that have reached full maturity
Qualitative retronasal sensation	
Retronasal persistence	Length of time that retronasal sensations persist after the sip of olive oil is no longer in the mouth
Tactile or kinaesthetic sensations	
Fluidity	Kinaesthetic characteristics of the rheological properties of the oil, the set of which are capable of stimulating the mechanical receptors located in the mouth during the test
Pungent	Biting tactile sensation characteristic of oils produced at the start of the crop year, primarily from olives that are still unripe

attributes and the oils in a two-dimensional space. When all the samples are placed in such a space, their positions reflect their similarities and differences. Samples that are close together are similar in character and those that are far apart are different. By exploring the position of the axes with respect to the samples, it is possible to see which attributes indicate that the samples are similar or different.

An example of such a map based on the assessment of 32 extra-virgin oils of different origin, using the terminology extracted from the sensory wheel, is presented in Figure 4.5 (Bertuccioli 1994; Lyon and Watson 1994; Delgado and Guinard 2011(a) and (b)).

For the first two principal components, which account for 70% (34% + 36%) of the total variation in the data, the loadings for the attributes are shown as vectors, together with the scores for means of each sample.

Bitter and green sensations are correlated with each other, as suggested by the small angle between their vectors. They are negatively correlated with the first principal component (PC1) to which they are closely aligned. Ripe black olive (odour and flavour) is positively correlated with both PC1 and PC2. Yellow is positively correlated with PC1. Cut grass is negatively correlated with both PC1 and PC2. Tomato and artichoke are negatively correlated with PC2.

From the location of the extra-virgin olive oils shown in Figure 4.5, inferences can be made about the olive variety of these oils. The first principal component (PC1) separates the oils of the Frantoio and Picual varieties; the second principal component (PC2) separates the oils of the Coroneiki, Frantoio, Moraiolo and Picual varieties.

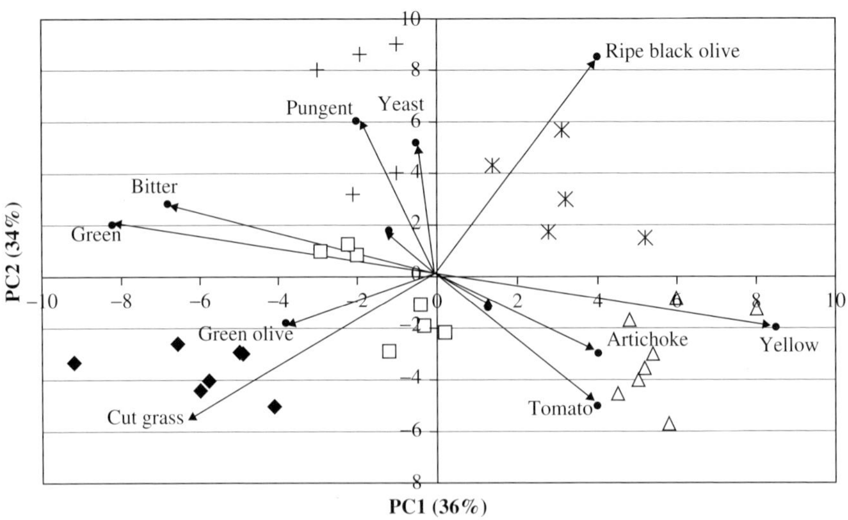

Figure 4.5 Principal component analysis of Extra-virgin Olive Oils. Projection of sensory data on principal components I e II. Attribute loadings (vectors) and mean factor scores for oils from Frantoio (♦), Moraiolo (□), Picual (Δ), Arbequina (*) and Coroneiki (+) (*Source*: Bertuccioli 1994. Reproduced with permission from Grasas Y Aceites).

4.4 Sensory performance of extra-virgin olive oil-food pairing

This subject has been the focus of five annual meetings of the International Conference on Excellence in Olive Oil, 'Beyond Extra Virgin' from 2007 to 2011 (Annex 18.1). For the first time, application of the time dominance of sensation (TDS) method to the extra-virgin olive oil-food pairing was presented by Monteleone and thoroughly discussed in a multidisciplinary approach involving olive oil experts, sensory scientists and experts in culinary arts (Drescher 2010; Monteleone 2010). The TDS method seems to be more appropriate than descriptive analysis to study sensory interactions occurring when tasting food and for developing new approaches to study oil-food pairing (Dinnella *et al.* 2012).

4.4.1 The consumer approach to extra-virgin olive oil sensory characteristics

Considering that oils are never used alone, consumer tasting of extra-virgin olive oil as such can spur incongruous or negative reactions. It seems clear that consumer liking for oils is negatively correlated to the bitterness, pungency and astringency perceived. Furthermore, the more the consumers are unfamiliar with extra-virgin olive oil, the stronger is this negative relationship (Caporale *et al.* 2006). This is not surprising at all. As reported by Tuorila and Recchia (2013), bitterness, pungency and astringency are inherently unpleasant for humans and the appreciation of these core qualities of good extra-virgin olive oil requires learning. In fact, despite initial rejection, later in life, bitterness is accepted as a sensation characterizing many food products, for example coffee, beer, many wines, grapefruit and dairy products. Similar remarks can be made concerning the perception of pungency and astringency. These perceptions are initially disliked, but later in age they become part of the desired qualities of spicy foods. Tuorila and Recchia (2013) propose a list of conditions that contribute strongly to the liking of the flavour of high quality extra-virgin olive oils:

- Repeated exposure is a necessary condition for learning to like the flavour of olive oil. Social and educational reinforcement, especially the opinions and comments of gourmets and experts.
- Momentary situational and social reinforcement lead to conditioning and thereby to the shift in liking.
- Pairing with appropriate dishes.
- Health considerations (for example, discovering that the molecules that are responsible for bitterness and pungency are the same that have the most significant health-promoting, antioxidant potential) may lead to a shift in liking.

Considering the points listed above, it appears obvious that the success of extra-virgin olive oils in relation to their sensory characteristics requires that both producers and researchers focus their attention on the culinary uses of oils with varied sensory styles.

People struggle with the balance between neophilia (interest in what is new) and the need for diverse sensory experiences versus caution and neophobia, lest we consume something dangerous to our health (Lawless 2000). As a consequence, we have a sensory struggle between the harmony of the foods to be consumed and the desire for sensory contrast. Oils with varied sensory styles as key elements in dish preparations and oil-food pairings can play a tremendous role in governing the need for balance and contrast.

4.4.2 The temporal dominance of sensations (TDS) method

Combining an oil with a specific sensory style with food is not simply adding aromas to aromas or tastes to tastes. In fact, combining different odours and tastes results in complex interactions leading to perceptual phenomena described as mixture suppression (individual sensory stimulus is perceived as less intense in a blend than when experienced alone), adaptation (lowering of sensory system responsiveness after exposure to a constant stimulus) and release from suppression (following adaptation to one stimulus in a mixture, other stimuli are less suppressed and perceived at increased intensity) (Lawless 2000; Keast and Breslin 2002).

The multidimensionality of the perceptual space over time is well represented by the TDS method (Pineau *et al.* 2009). It consists in presenting the panelist with the complete list of attributes on a computer screen. Thereafter, the panelist is asked to assess which of the attributes is perceived as dominant (the most striking perception at a given time). During the testing of a product, the panelist is free to select an attribute several times. Conversely, another attribute may not be selected at all. In the course of the evaluation, when the panelist considers that the dominant attribute has changed, he or she has to select the new dominant attribute, and so on, until the perception ends. For each run, this method enables the collection of a sequence of sensory attributes quoted at different times during the tasting.

The product perception pattern is represented by curves reporting the frequency with which the sensations reported in a list of several attributes are considered as dominant during food consumption by a trained panel. This descriptive method allows the investigation of qualitative changes perceived during eating and explicitly considers sensory interactions taking place during food consumption (Labbe *et al.* 2009; Lenfant *et al.* 2009; Meillon *et al.* 2010). The size of a TDS panel is similar to that of a descriptive panel.

The application of the method requires dedicated software for data acquisition, now available on the market. This software also allows for the analysis of the data and the creation of TDS curves for each product.

As an example of the application of TDS, some data by Dinnella *et al.* (2012) are reported and briefly discussed. Two oils differing in their sensory profile were considered (Figure 4.6). The two oils were added (at 10% w/w) to a tomato sauce

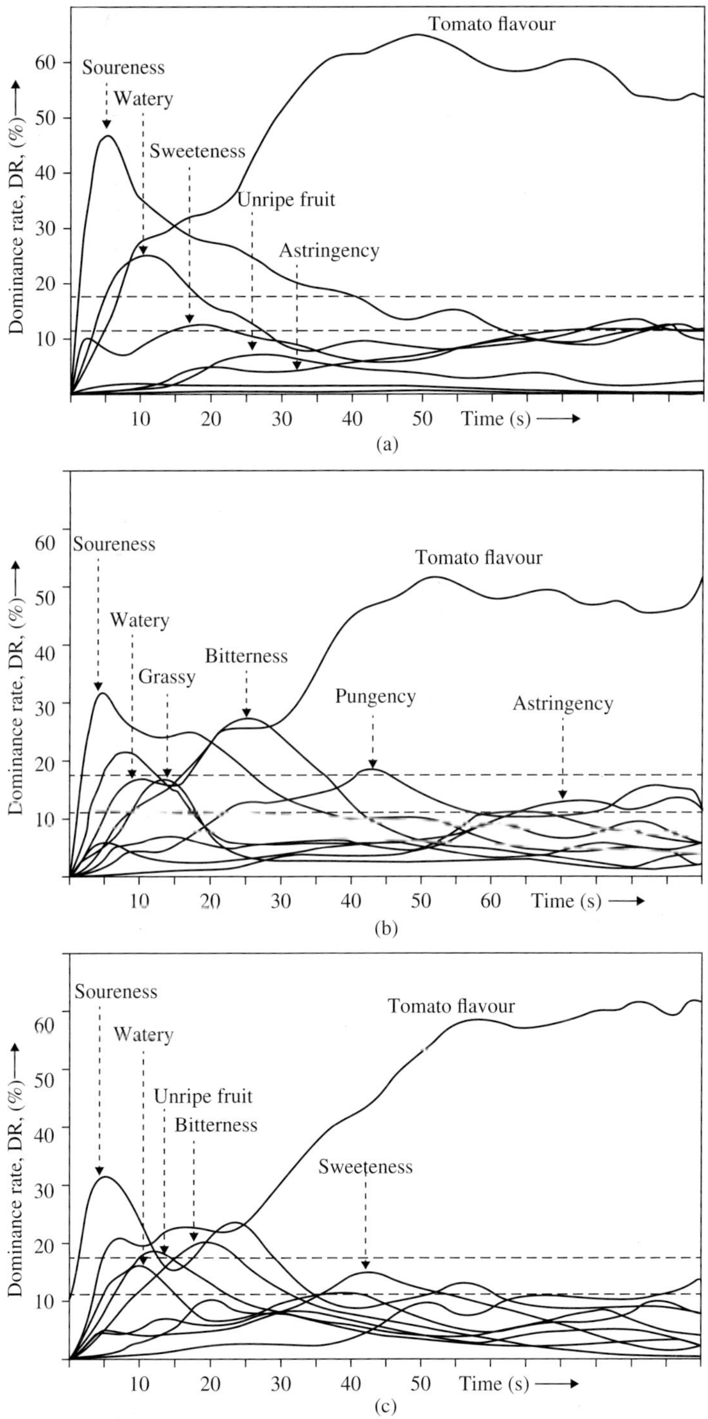

Figure 4.6 TDS response of tasting (*Source*: Dinnella *et al.* 2012. Reproduced with kind permission from Elsevier).

and the resulting combinations analysed by the TDS method. Consumer responses on liking and perceived freshness of oils, tomato sauce and their combinations were collected.

Sour taste, watery character and tomato flavour were the dominant sensations during the first part of the evaluation of tomato without oil (Figure 4.6a). Sourness was the dominant sensation from the beginning of the evaluation and remained one of the dominant sensations up to 40 s. The watery sensation was dominant for 10 to 20 s and the tomato sensation started to be dominant after the first 10 s. Tomato flavour remained the only dominant sensation after 30 s and lasted until the end of the evaluation. Reduction in the sour taste dominance rate maximum value, suppression of the watery character dominance and onset of the bitter taste dominance were the main effects of combining tomato with oil A (Figure 4.6b). Similar effects on sourness and watery dominance were induced by adding oil B to tomato. Moreover, oil B contributed a typical unripe fruit flavour among the sensations, dominating the first part of the tomato/oil B evaluation. Results from consumer testing showed a significant increase in liking for tomato combined with either oil A or oil B compared to tomato without oil. Several factors relating to the general effects of adding oil rather than specific modifications of the sensory profile could account for the increased liking of tomato plus oil samples compared to that of tomato without oil. In fact, no significant differences in liking were found when comparing tomato plus oil A with tomato plus oil B, despite the differences in flavours and taste. Modifications of the sensory properties of tomato combinations induced by both oils, that is, reduction in both the watery attribute and the sour taste, might account for the increased liking for these samples. It is interesting to note that when oils are tested alone, there is a clear and significant difference in the liking mean scores. Tomato combined with oil B was perceived as fresher than both tomato without oil and tomato with oil A. The unripe fruit flavour, together with the other sensory properties specifically contributed by oil B to the tomato profile, accounted for the increased perceived freshness in tomato plus oil B compared with tomato plus oil A and tomato without oil. Thus the two oils have a varied sensory function. Again it is interesting to note that perceived freshness in tomato-oil pairings is not predicted at all by the evaluation of the oils alone.

Both the panel and consumer data highlight the risk of using sensory data from extra-virgin olive oil evaluated as such to build up quality categories based on flavour, taste and mouth-feel sensation intensities. Key sensations marking the sensory style of a given oil may not be experienced in conditions of normal use.

It appears evident that, in order to add value to the sensory differences among extra-virgin olive oils, the study of the sensory functionality of different styles in oil-food pairings is needed.

Annex 4.1: The method for evaluating extra-virgin olive oil sensory profiles

Subjects

Usually, a descriptive panel involves between 10 and 15 trained tasters, recruited according to their ability to detect differences in important product attributes. Subjects must be qualified prior to their participation. They must be users of extra-virgin olive oil. Previous exposure to varied extra-virgin olive oils is an important qualification. It is also important that individuals participate in a series of difference tests organized to represent a range of difficulties and to include relevant modalities (olfaction, taste, etc.).

Subject training and reproducibility

The number of training sessions depends on the number of products, but in general, 6–10 one-hour training sessions are needed. The initial list of attributes is normally reduced to achieve a list that comprehensively and accurately describes the product space: redundant and/or less-cited terms are grouped on a semantic basis and/or eliminated according to the subjects' consensual decisions. To facilitate the consensus and to calibrate the subjects, reference standards are presented to the panel, discussed and modified if need be. Refining descriptive terms, reference standards and definitions continues until the panelists reach a consensus.

During language development, subjects practise scoring products in order to familiarize themselves with products and the scale rating system. Different scaling methods are used by different researchers, and their efficiency seems to be similar. The unstructured, 10 cm linear scale anchored 'not perceived' at the left end and 'very intense' at the right end (other anchors are frequently used) is very common, but the nine-point category and 15-point unlabelled box scales are used, too.

In conventional descriptive analysis it is important to determine taster reproducibility. Normally, panel performance is evaluated at the end of the training period by having the panel evaluate, in triplicate, a subset of samples to be used for the real study. Data are frequently analysed for each attribute by means of a two- (sample and assessor) or three-way analysis of variance (sample, assessor, replication) to determine whether there are significant taster × sample interactions. The significant effect of this interaction implies that the panel leader will determine which taster should be further trained in the use of which attribute. The need for solid and validated information from sensory data in decision-making processes is extremely clear to sensory scientists. Panel Check software, a free statistical package developed at Nofima (Norway), allows the panel leader to control the quality of sensory profile data using both univariate and multivariate approaches (see Naes *et al.* 2010).

Experimental design

The sample evaluation in descriptive analysis is run after defining an appropriate experimental design. The term indicates a series of experimental procedures that have been developed to provide as much information as possible in the most efficient way. In descriptive analysis the objective is to collect a sensory profile of a heterogeneous group of products using a defined number of assessors and replicates. In general, the design should take into account taster variation, presentation order effect, first order and carry over effects, and any specific limitation associated with samples and assessors.

It is possible to say that when planning a profile study there are two important aspects to consider, the first being the initial choice of products. Before running a descriptive analysis, samples should be evaluated to test for the absence of defects. Secondly, another important aspect of an experimental design is how to present the samples to tasters. In sensory olive oil research the use of a complete design with replicates involving three factors (sample, replicate and taster) is extremely frequent. Randomization is a key principle in experimental design. Proper randomization ensures that the effect of extraneous factors is averaged out in the long run. The order of presentation represents a source of variation of sensory data in itself and a balanced design for first order and carry over effects is needed. This can be obtained by adopting modified Latin-square designs reported by MacFie (MacFie *et al* 1989). Software for collecting sensory data and running tests normally allows the experimenter to easily design the order of presentation with respect to this important requirement.

In descriptive analysis each taster is asked to replicate the evaluation of samples. The number of replications depends on the size of the differences that the experimenter is required to detect. Small expected differences require a higher number of replicates. However, the number of replications in olive oil sensory studies frequently ranges from 2 to 4 with 10–12 panelists involved in the test.

Sensory procedure

Bitterness, astringency, pungency and pepperiness are common descriptors of extra-virgin olive oils due to their phenolics content and profile. These sensations tend to persist for a rather long time after swallowing, showing a clear after-effect that can vary strongly among olive oils in intensity and duration and might affect consumer acceptance (Caporale *et al.* 2006). Thus, they are important sensory characteristics of oils. In a study that explored the dynamic perception of bitterness and pungency (Sinesio *et al.*, 2005), the attribute variation over time showed that each sensation acts according to a regular temporal sequence. A difference between the maxima of the two attributes of approximately 10 s (in the order bitterness and pungency) is independent of intensity. In agreement with these observations, Dinnella *et al.* (2012) showed that the dynamic changes of sensory dominances when tasting oils for 90 s follow the temporal sequence: bitterness, pungency, astringency.

A good procedure to describe the sensory characteristics of olive oils should consider the main sensory properties of oils in relation to the following points:

- Conditions of constant stimulation determine a decrease in responsiveness to bitterness (adaptation).
- Astringency is a tactile sensation perceived as a diffuse stimulus in the mouth and commonly described as a puckering, roughing and drying of the oral surface (Lee and Lawless 1991). The perceived intensity of an astringent stimulus increases with repeated ingestion. Because of this well-known 'carry-over' effect, the evaluation of an astringent product such as olive oil with a very high phenolics content cannot be made as a typical side-by-side comparison.
- Pungency and pepperiness are burning sensations and in general have a long-lasting nature. They are defined as chemesthetic sensations (chemical responsiveness mediated by trigeminal nerves). When the rest period between the evaluation of samples is omitted (or is too short) the perceived strength of these sensations continues to build to higher levels.

Considering all these aspects, it seems appropriate to suggest the sensory procedure for profiling extra-virgin oils as described by Monteleone *et al.* (2012).

Tasters are presented with up to four samples per session (served monadically). Each sample, identified by a three-digit code, is presented in a 100 ml amber glass containing 30 ml of oil, covered with a plastic Petri dish. The order of presentation of samples should be balanced for first order and carry-over effects. Following the order of presentation, subjects are asked to smell a sample and score the intensity of aroma (odour by nose) descriptors. Then they are asked to pour part of the sample into a teaspoon (around 3.5 ml), put it into their mouth and rate the perceived viscosity. Tasters are instructed to hold the sample in their mouth for up to 8 s, spit it out and, after a further 12 s, rate the perceived intensity of bitterness, pungency and astringency. Finally, subjects are asked again to pour the sample into the spoon, put it into the mouth and rate the intensity of odours perceived retronasally. Specific rinsing procedures between the evaluation of two samples are required to control possible carry-over effects. For this purpose, after each sample, subjects can be instructed to rinse their mouths with distilled water for 30 s, eat some plain crackers (or plain unsalted white bread) for 30 s and finally rinse their mouths with water for a further 30 s. Tests should be conducted in isolated booths, under red light (in order to limit visual bias). Scores are frequently recorded directly on a computer system using dedicated software. When more than four samples are evaluated, it is possible to run more than two sessions per day. However a break of at least 1 h between each session is recommended. In these cases the order of presentation of samples should be balanced within each replicate rather than each session.

Other general rules:

- Before evaluation, oils should be kept at a temperature ranging from 14 and 15 °C in containers of inert material, impermeable to light and closed tightly.
- The presence of air in the headspace of storage containers should be avoided.
- Oils should be presented at room temperature (around 25 °C).
- Oils should be evaluated within 20 minutes of sample preparation.

References

Bertuccioli, M. (1994) A study of sensory and nutritional quality of virgin olive oil in relation to variety, ripeness and extraction technology. Overview of three year study and conclusion. *Grasas y Aceites* **45**(1–2), 55–59.

Blake, A.A. (2004) Flavour perception and the learning of food preference, in *Flavour Perception* (eds A.J. Taylor and D.D. Roberts), Blackwell, Oxford, pp. 172–202.

Caporale, G., Policastro, S. and Monteleone, E. (2004) Bitterness enhancement induced by cut grass odourant (cis-3-hexen-1-ol) in a model olive oil. *Food Quality and Preference* **15**, 219–227.

Caporale, G., Policastro, S., Carlucci, A. and Monteleone, E. (2006) Consumers expectations for sensory properties in virgin olive oils. *Food Quality and Preference* **17**, 116–125.

Delgado, C. and Guinard, J. (2011a) How do consumer hedonic ratings for extra-virgin olive oil relate to quality ratings by experts and descriptive analysis ratings? *Food Quality and Preference* **22**, 213–225.

Delgado, C. and Guinard, J. (2011b) Sensory properties of Californian and imported extra virgin olive oils. *Journal of Food Science* **76**, 3.

Dinnella, C., Masi, C., Zoboli, G. and Monteleone, E. (2012) Sensory functionality of extra-virgin olive oil in vegetable foods assessed by temporal dominance of sensations and descriptive analysis. *Food Quality and Preference* **26**, 141–150.

Drescher, D. (2010), The super-premium olive oil experience: developing a culture of flavour discovery. *Beyond Extra-Virgin, The Fourth International Conference on Olive Oil Excellence*, organized by Association 3E (Milan, Italy), the Academy of Georgofili (Florence, Italy), The Culinary Institute of America (St Helena, California) and the Olive Center of the University of California Davis, Verona, 22 September 2010.

Gawel, R., (2007) Olive Oil Tasting Wheel, www.aromadictionary.com/oliveoilwheel.html (accessed 22 September 2013).

Hollowood, T.A., Linforth, R.S.T. and Taylor, A.J. (2000) The relationship between volatile release and perception, in *Flavour Release: Linking Experiments, Theory and Reality* (eds D. Roberts and A. Taylor), American Chemical Society, Washington DC, pp. 370–380.

International Olive Oil Council (2005) International Olive Oil Council, Organoleptic assessment of DO extra-virgin olive oil, T20/Doc No 22, IOOC, Madrid.

International Olive Oil Council (2011) Sensory analysis of olive oil. Method for the organoleptic assessment of virgin olive oil, IOOC/T20/Doc. No 15/Rev. 4, IOOC, Madrid.

Keast, R.S.J. and Breslin P.A.S. (2002) An overview of binary taste-taste interactions. *Food Quality and Preference* **14**, 111–124.

Labbe, D., Schlich, P., Pineau, N. *et al.* (2009) Temporal dominance of sensations and sensory profiling: a comparative study. *Food Quality and Preference* **20**, 216–221.

Lawless, H.T. (2000) Sensory combinations in meals, in *Dimensions of the Meal – The Science, Culture, Business and Art of Eating* (ed. H.L. Meiselman), Aspen Publishers, Gaithersburg, MD.

Lawless, H.T. and Heymann, H. (1998) *Sensory Evaluation of Foods: Principles and Practices*, Chapman & Hall, New York.

Lee, C.B. and Lawless, H.T. (1991) Time-course of astringent materials. *Chemical Senses* **16**, 225–238.

Lenfant, F., Loret, C., Pineau, N. *et al.* (2009) Perception of oral food breakdown. The concept of sensory trajectory. *Appetite* **52**, 659–667.

Lyon, D. H. and Watson, M.P. (1994) Sensory profiling: a method for describing the sensory characteristics of virgin olive oil. *Grasas y Aceites* **45**, 20–25.

MacFie, H.J., Bratchell, N., Greenhoff, K. and Vallis, LV. (1989) Designs to balance the effect of order of presentation and first-order-carry-over effects in hall tests. *Journal of Sensory Studies* **4** (2), 129.

Martens, H. and Martens, M. (2001) *Multivariate Analysis of Quality: An Introduction*, John Wiley & Sons Ltd, Chichester.

Meillon, S., Viala, D., Urbano, C. *et al.* (2010) Impact of partial alcohol reduction in Syrah wine on perceived complexity and temporality of sensations and link with preference. *Food Quality and Preference* **21**, 732–740.

Mojet, J. and de Jong, S. (1994) The sensory wheel of virgin olive oil. *Grasas y Aceites* **45**(1–2), 42–47.

Monteleone, E. (2010) From sensory characteristics to sensory functionality of super-premium olive oil. *Beyond Extra Virgin, The Fourth International Conference on Olive Oil Excellence*, organized by Association 3E (Milan, Italy), the Academy of Georgofili (Florence, Italy), The Culinary Institute of America (St Helena, California) and the Olive Center of the University of California Davis, Verona, 22 September 2010.

Monteleone E, Bendini A, Dinnella C. *et al.* (2012) L'olio extra vergine di olive, in *Atlante Sensoriale dei Prodotti Alimentari* (ed. Società Italiana di Scienze Sensoriali), Tecniche Nuove, Milano, pp. 114–128.

Naes, T., Brockhoff, P.B. and Tomic, O. (2010) *Statistics for Sensory and Consumer Science*, John Wiley & Sons, Ltd., Chichester.

O'Mahony M. (1986) *Sensory Evaluation of Food, Statistical Methods and Procedures*, Marcel Dekker, New York.

Pineau, N., Schlich, P., Cordelle, S. *et al.* (2009) Temporal dominance of sensations: construction of the TDS curves and comparison with time-intensity. *Food Quality and Preference* **20**, 450–455.

Sinesio, F., Moneta. E. and Esti, M. (2005) The dynamic sensory evaluation of bitterness and pungency in virgin olive oil. *Food Quality and Preference* **16**, 557–564.

Stevenson, R.J., Prescott, J. and Boakes, R.A. (1999) Confusing tastes and smell: how odours can influence the perception of sweet and sour tastes. *Chemical Senses* **24**, 627–635.

Tuorila, H. and Recchia, A. (2013) Sensory perception and other factors affecting consumer choice of olive oil, in *Olive Oil Sensory Science* (eds E. Monteleone and S. Langstaff), Wiley-Blackwell, Chichester.

5

Olive tree cultivars

Luana Ilarioni and Primo Proietti

Department of Agricultural, Food and Environmental Sciences, University of Perugia, Perugia, Italy

Abstract

The most widely grown cultivars worldwide are listed, accompanied by the major sources of information about them. Factors influencing cultivar productivity are briefly discussed, especially those related to self-sterility and cross-pollination. Common-sense recommendations are given concerning cultivar choice and plant certification. A clear-cut distinction is presented between new super-intensive and traditional semi-intensive olive groves.

5.1 Introduction

The aim of this chapter is to point out some critical relationships between olive tree cultivars and extra-virgin olive oil yield and quality. The names of some cultivars are becoming popular among experts and consumers alike as an indication of origin and particular sensory characteristics.

5.2 Cultivars

A cultivar (abbreviation: cv) is a group of similar plants that have been selected for one or more interesting characters; it is distinct, uniform and stable in these characters and it retains them by vegetative propagation. The word 'cultivar' probably derives from a combination of the words '**culti**vated' and '**var**iety'. The full name of an olive tree consists in the scientific Latin botanical name followed by a cultivar epithet, which is usually in the language of the country where the olive tree was first selected and cultivated. For example (note the standard use of capitals and italics): *Olea europaea sativa* cv Moraiolo (an Italian cultivar), or cv Arbequina (a Spanish cultivar), or cv Mission (a Californian cultivar), and so on. *Olea* is the genus, *europaea* is the species, and *sativa* is the subspecies. Wild subspecies are called *oleaster* and do not have further cultivar specification.

The Extra-Virgin Olive Oil Handbook, First Edition. Edited by Claudio Peri.

The *Olea europaea* species is predominantly self-sterile with a high degree of cross-pollination that leads to high levels of heterozygosity. For these reasons, the olive cultivars cannot be propagated by seed. Sometimes, propagation by seed is used to produce rootstocks to be grafted with the desired cultivar. Propagation is based on asexual techniques such as cuttings or grafting, through which individuals genetically identical to the mother plant can be propagated.

Selection carried out over centuries by olive growers or due to environment-induced genetic changes has given rise to more than 2000 cultivars, which are spread throughout the various olive oil producing regions.

Furthermore, the existing olive genetic heritage is rather stable because it is difficult to obtain new olive varieties through classical genetic methods (crossing, use of mutagenic agents, etc.). Consequently, in each olive-growing country, a large number of ancient cultivars are still cultivated. However, only a few cultivars are being cultivated for commercial production, due to their higher adaptability and productivity.

Cultivar identification and classification

There are still unclassified cultivars and some are totally unknown. Cultivars known in remote times are probably still being cultivated today, although because of the development of languages, the original names have changed, thus hindering identification in the descriptions given by ancient authors. The existence of homonyms (different cultivars labelled with the same name) and synonyms (different names for the same cultivar) is also very common.

In recent years, the use of advanced molecular methods has made evaluation of the genetic variability more reliable and precise and has allowed many cases of synonymy and homonymy to be solved.

General information on the main olive cultivars in the world can be found in the *World Catalogue of Olive Varieties* by the International Olive Council (IOC 2013) and other sources (Bartolini 2008).

In Table 5.1, a list of the main cultivars in different countries is given together with some information about their relative share of the country's overall olive-growing surface area (data processed from FAO sources http://faostat.fao.org/site/339/default.aspx, and Pannelli 2005). Italy, in particular, has an extraordinary genetic heritage, with more than 500 cultivars, some of which are present in only a single municipality, probably their area of origin.

5.3 The cultivar's relationship to productivity

'Olive-grove productivity' is defined as the average quantity (kg) of olives that are produced per unit surface area (usually the hectare).

The problem of olive grove productivity entails a discussion about marketing policies, which is beyond the scope of this handbook. To put it briefly, there are two different ways of competing in the marketing of extra-virgin olive oil: 'competitiveness on cost' and 'competitiveness on sensory style'. The first aims

Table 5.1 Olive tree cultivars. The nine major olive-producing countries in the table are listed according to the quantities of olives produced per year (values are in kton – thousands of tons – and are calculated as the average of the five-year period 2008–11) (data processed from FAO sources http://faostat.fao.org/site/339/default.aspx, and Pannelli 2005).

Country and total olive production, kton/year	Total number of cultivars and list of the most cultivated ones in order of decreasing olive-growing surface area
1. Spain **6223 kton/year**	Total number of cultivars in Spain: 262 The most important cultivars: *Picual, Cornicabra, Hojiblanca, Manzanilla de Sevilla, Arbequina, Morisca de Badajoz, Empeltre, Manzanilla Cacereña, Lechin de Sevilla, Picudo, Lechin de Granada, Verdial de Badajoz, Morrut, Sevillenca, Villalonga, Castellana, Farga, Verdial de Huevar, Blanqueta, Gordal Sevillana, Verdial de Velez-Malaga, Aloreña, Changlot Real, Alfafara* The 24 cultivars listed account for 96% of the total olive-growing surface area in Spain. The three cultivars at the top of the list account for 63% of the total olive-growing surface area.
2. Italy **3277 kton/year**	Total number of cultivars in Italy: 538 The most important cultivars: *Coratina, Ogliarola salentina, Cellina di Nardò, Carolea, Frantoio, Leccino, Ogliarola barese, Moraiolo, Bosana, Cima di Mola, Dolce di Rossano, Ogliarola messinese, Ottobratica, Sinopolese, Nocellara del Belice, Canino, Carboncella, Itrana, Moresca, Rotondella, Taggiasca, Tondina, Grossa di Gerace, Nocellara etnea* The 24 cultivars listed account for 58% of the total olive-growing surface area in Italy. More than 80 account for about 90% of the total olive-growing surface area in Italy.
3. Greece **2167 kton/year**	Total number of cultivars in Greece: 52 The most important cultivars: *Koroneiki, Kalamata, Mastoidis* The three cultivars listed account for about 90% of the total olive-growing surface area in Greece.
4. Turkey **1479 kton/year**	Total number of cultivars in Turkey: 80 The most important cultivars: *Ayvalik, Memecik, Gemlik* The three cultivars listed account for about 75% of the total olive-growing surface area in Turkey.
5. Morocco **1115 kton/year**	Total number of cultivars in Morocco: 6 The most important cultivars: *Picholine Marocaine* This cultivar accounts for about 97% of the total olive-growing surface area in Morocco.
6. Syria **941 kton/year**	Total number of cultivars in Syria: 75 The most important cultivars: *Zaity, Sorani, Doebli, Kaissy, Khodieri* The five cultivars listed account for about 90% of the total olive-growing surface area in Syria.

(*continued overleaf*)

Table 5.1 *(continued)*

Country and total olive production, kton/year	Total number of cultivars and list of the most cultivated ones in order of decreasing olive-growing surface area
7. Tunisia **899 kton/year**	Total number of cultivars in Tunisia: 44 The most important cultivars: *Chetoui, Chemali Sfax* The two cultivars listed account for about 95% of the total olive-growing surface area in Tunisia.
8. Portugal **413 kton/year**	Total number of cultivars in Portugal: 24 The most important cultivars: *Galega Vulgar, Cobrançosa, Cordovil de Serpa* The three cultivars listed account for about 96% of the total olive-growing surface area in Portugal.
9. Algeria **389 kton/year**	Total number of cultivars in Algeria: 36 The most important cultivars: *Chemlal of Kabylie , Sigoise, Limli, Azerradj* The four cultivars listed account for about 70% of the total olive-growing surface area in Algeria.

Olive tree biodiversity is a worldwide heritage. Other varieties can be listed in other countries that are of increasing importance in olive oil production, both inside and outside the Mediterranean area. For example:

Albania: *Kalinjot, Ulliri Bardhë i Tiranës, Kokërrmadhi i Beratit, KM Elbasani, Mixan*
Argentina: *Aranco*
California: *Mission*
China: *EZ-8, JF-6*
Croatia: *Oblica, Lastovka, Buža, PuntožaLevantinka,*
Cyprus: *Ladoelia, Flasou, Lythrodontas, Athalassa*
Egypt: *Toffahi, Aggeizi Shami, Marraki*
France: *Aglandau, Bouteillan, Cailletier, Picholine du Languedoc, Grossane,Salonenque, Lucques, Tanche*
Iran: *Zard, Roghani, Mari, Tokhm-e-Kabki*
Israel: *Souri, Nabali, Muhasan, Barnea, Askal*
Jordan: *Nabali Baladi, Rasei (Nabali Muhasan), Souri, Nasouhi Jaba, Kanabisi, Shami*
Lebanon: *Baladi, Souri, Ayrouni*
Libya: *Endory, Raghiani, Rasli, Hammudi*
Montenegro: *Žutica, Sitnica, Crnica*
Palestine: *Souri, Nabali (Baladi)*
Slovenia: *Buga, Štorta, Istrska belica*

at offering consumers a standard quality extra-virgin olive oil at the lowest price; the second aims at offering consumers a variety of extra-virgin olive oils with high quality standards and different sensory profiles. The first approach deals with production geared towards super-intensive cultivation and highly mechanized methods, while the second is based on highly specialized, niche and traditional methods.

This discussion has a great deal to do with cultivars because the introduction of super-intensive cultivation, which is possible only with very few olive cultivars, has become the most important breakthrough in olive oil production in economic terms.

Table 5.2 A comparison of super-intensive and intensive systems of olive production.

Terms of comparison	Super-intensive olive grove	Intensive olive grove
Tree density	1600 plants/ha	300 plants/ha
Cultivars	In orchards with a very high tree density it is important to choose cultivars such as the Spanish Arbequina and Arbosana and the Greek Koroneiki and their clones, which have very low vigour and a compact growth habit. It also has been observed that less vigorous genotypes come into production earlier than the more vigorous ones. They also have shorter life cycles.	All cultivars may be suitable provided that they contribute to sensory excellence and variety of the oils. Valuable links have been established between local culinary traditions and the oils from local cultivars. This is the inexhaustible richness of the link between food culture and biodiversity.
Productivity	8–10 tons/ha	6–8 tons/ha
Main advantage	Low manpower – standard quality	Specific sensory profile, best culinary combinations, preservation of olive biodiversity
Economy	Good at very high production and investments	Sustainable also for small size production depending on marketing ability in fostering sensory specificity and variety

The data and comments in Table 5.2 give an approximate, but significant, comparison of the two approaches. One point should be made clear: the two systems compared in Table 5.2 have a different marketing target and therefore involve very different marketing strategies.

The oil obtained with the first, super-intensive, system can be classified as a commodity, whereas the oil obtained with the second system is a food specialty. From an ethical point of view, the first system has a social mission: to make extra-virgin olive oil available to everybody at a sustainable price for mass consumption. The second system has a cultural mission, which is to preserve olive tree biodiversity and the extraordinary heritage of food traditions associated with the Mediterranean area and with the use of different extra-virgin olive oils in different culinary preparations.

The oils obtained with the two systems cannot be considered as different in quality: they serve different purposes and they can be both poor or excellent in quality depending on the ability of the producer. Other cultivar determinants of productivity are:

- The genetic predisposition for abundance and constancy of production that mainly depends on large amounts of fruit, high oil yield and low alternate bearing.
- Resistance to adverse climatic conditions, especially drought and frost.

- Resistance to pathogens (peacock spot, olive knot, etc.) and pests (olive fly, black scale, etc.). If olive cultivation is carried out according to organic agriculture rules, the resistance of olives to pathogens and pests is particularly important. Small fruit size and high phenolics content reduce the risk of fly attack. Early-ripening cultivars may escape late attacks because flies prefer green olives for ovideposition. In general, in olive groves using organic agricultural techniques, the presence of different cultivars ripening at different times should be avoided, as this could facilitate the succession of several generations of the fly.
- Suitability for mechanized harvesting. Fruit weight should not be too low and resistance to detachment should not be too high (see Chapter 8).
- Self-compatibility and cross-pollination.

Self-incompatibility (self-sterility) characterizes most of the cultivars. When using self-incompatible cultivars, in order to facilitate cross-pollination, at least 10–15% of the pollinator varieties should be planted and/or more inter-fertile cultivars. The best pollinators have abundant blooms and produce a lot of pollen in coincidence with the blooming of the cultivar to be pollinated. Even self-compatible cultivars take advantage of cross-pollination, so it is recommended that pollinators be planted even when self-compatible cultivars are chosen.

When the production target allows it, use 3-4 main intercompatible cultivars to avoid that if one cultivar is in an 'off-year', the other cannot be pollinated. This also takes into account 'discrepancies' in the blooming of different cultivars, which can occur because of seasonal conditions.

Some cultivars are listed in Table 5.3, with indication of self-compatibility or incompatibility and a corresponding list of pollinators. Pollinators are also listed in the case of self-compatible cultivars as, for example, Frantoio, Picual and Kalamata, pointing out the fact that pollinators that are essential for self-incompatible cultivars, are also useful to increase the productivity of self-compatible cultivars.

5.4 The cultivar's relationship to oil quality

The nutritional and sensory quality of extra-virgin olive oil is a complex issue, as was discussed in Chapters 3 and 4. On the other hand, cultivar and cultivar performance are also complex issues, as pointed out in this chapter. Presenting a relationship between the cultivar and the quality of oil risks being inconsistent because the composition and sensory profile of an extra-virgin olive oil depends, to a similar degree, on the cultivar, the environmental conditions, the degree of maturity of the olives at harvest and the conditions of the milling process. Results may change dramatically with changes in any of these main determinants of olive oil quality.

As an indicative reference only, Table 5.4 shows some ranges in variation of the oleic acid and phenolic compounds concentrations in extra-virgin olive oils from different cultivars.

Table 5.3 Self-incompatibility and pollinator cultivars.

Cultivar	Flower fertility	Pollinators
Coratina, Italy	Self-incompatible	Cellina di Nardò, Ogliarola, Frantoio, Moraiolo
Ogliarola salentina, Italy	Self-incompatible	Cellina di Nardò
Cellina di Nardò, Italy	Partially-self-compatible	Nociara, Ogliarola salentina
Carolea, Italy	Self-incompatible	Nocellara messinese, Cassanese, Pidicuddara, Itrana, Frantoio, Leccino, Moraiolo
Frantoio, Italy	Self-compatible	Moraiolo, Maurino, Leccino, Morchiaio, Pendolino
Leccino, Italy	Self-incompatible	Frantoio, Moraiolo, Pendolino, Razzo, Trillo
Ogliarola barese, Italy	Partially-self-compatible	Coratina
Moraiolo, Italy	Self-incompatible	Leccino, Frantoio, Carolea, Maurino, Pendolino, Morchiaio
Picual, Spain	Self-compatible	Arbequina and Hojiblanca
Cornicabra, Spain	Partially-self-compatible	
Hojiblanca, Spain	Self-compatible	Arbequina, Manzanillo, Picual
Manzanilla de Sevilla, Spain	Self-incompatible	Sevillano, Frantoio, Picual, Arbequina
Arbequina, Spain	Self-compatible	
Empeltre, Spain	Partially-self-compatible	Arbequina, Koroneiki, Bouteillan, Pendolino
Manzanilla Cacereña, Spain	Self-compatible	Barouni, Sevillano
Koroneiki, Greece	Partially-self-compatible	Mastoides
Kalamata, Greece	Self-compatible	Koroneiki, Mastoides, Frantoio
Picholine Marocaine, Morocco	Partially-self-compatible	Picudo

5.5 Common-sense recommendations

1. In the absence of specific knowledge or experience, cultivars traditionally cultivated in the area should be chosen. Due to the selection made by olive growers over centuries, a synergism exists between these cultivars and the environment, with increased resistance to climatic conditions and pests and diseases. When planting a cultivar native to another area, significant and unexpected changes in the vegetative and productive tree behaviour, oil quality and resistance to adversities could take place. As an example, Table 5.5 shows how widely the fatty acid composition of a single cultivar (Arbequina) may vary when changing the region of cultivation data from the authors and from the olive oil processing course, http://cesonoma.ucanr.edu/files/27761.pdf).

   Considering that the analysis of fatty acids is one of the most reliable and reproducible analyses and that the fatty acid composition is a relatively stable

Table 5.4 An indicative comparison of cultivar relationship with composition (and quality) parameters.

Cultivar	Oleic acid[a] (% total FA)	Total phenolic compounds[b] mg/kg as gallic acid
Arbequina	Medium	Low
Carolea	Medium-high	Medium
Cellina di Nardò	Medium	Low-medium
Coratina	High	Very high
Cornicabra	Medium-high	Medium-high
Empeltre	Medium	Medium
Frantoio	High	Medium
Hojiblanca	Medium-high	Low-medium
Kalamata	Medium-high	Low
Koroneiki	High	High
Leccino	High	Low-medium
Manzanilla Cacereña	High	Low-medium
Manzanilla de Sevilla	Medium	Medium
Moraiolo	High	High
Ogliarola barese	Medium-high	Medium
Ogliarola salentina	Medium	Medium
Picholine Marocaine	High	High
Picual	High	High

Notes: [a]The concentration of oleic acid has been conventionally graded as low if it comprises less than 75% of total fatty acids, medium if it comprises 75–80% of total fatty acids, and high if greater than 80% of total fatty acids. [b]The concentration of total phenolic compounds has been conventionally graded as low if it comprises less than 200 mg/kg of total phenolic compounds expressed as gallic acid equivalent; medium if it comprises 200–400 mg/kg; high if 400–600 mg/kg and very high if it comprises greater than 600 mg/kg.

Table 5.5 Concentration of some fatty acids in extra-virgin olive oil from cultivar Arbequina cultivated in different regions data from the authors and from the olive oil processing course, http://cesonoma.ucanr.edu/files/27761.pdf).

Region	Palmitic acid (C 16:0)	Oleic acid (C 18:1)	Linoleic acid (C 18:2)
La Rioja (Argentina)	20	52	21
Andalusia (Spain)	16	65	13
Catalonia (Spain)	13	72	10
Sicily (Italy)	18	64	12
Apulia (Italy)	18	66	10
Umbria (Italy)	14	74	8

and genetically determined character, these results are surprising. At the two extremes, the Arbequina extra-virgin olive oil from Argentina and that from Umbria look substantially different and suggest very different possible behaviour in chemical, physical and nutritional terms.
The common sense recommendation is to carry out preliminary studies and a few years of acclimation tests to verify the adaptability of a cultivar to a new environment.

2. Avoid monocultivar groves. With different cultivars, the risk due to meteorological adversities, pest attacks and variations in production from year to year is reduced. If cultivars ripen at different times, the harvesting of each cultivar at optimum ripeness is facilitated. Consequently, machinery and labour can be used over longer periods of time, with substantial economic saving.
3. In order to simplify differentiation of cultural practices and to optimize harvesting time according to olive ripeness, the different cultivars should be planted in blocks of rows.
4. Distance between cultivar and pollinator should be less than 30 m in order to guarantee effective transport of pollen by wind and facilitate cross-pollination.
5. Use certified trees, in which the genetic characteristics and the health status are guaranteed. In the EU all olive trees are grown according to CAC (*Conformitas Agraria Communitatis*) standards (Council Directive 2008/90/EC and Council Directive 92/34/EEC). Clonal propagation from parent stock (foundation stock) and phytosanitary suitability are guaranteed. Furthermore, nurseries applying a specific certification scheme VT (virus tested) and VF (virus free) can obtain plant material from certified 'propagation fields' as the sole source of supplies for their work of grafting and cuttings. 'Certified' olive plants are sold under specific labels.

References

Bartolini, G. (2008) *Olive Germplasm – Cultivar Synonyms, Cultivation Areas, Descriptors*, www.oleadb.eu/ (site accessed 23 September 2013).

International Olive Council (2013) *The World Catalogue of Olive Varieties – Olive Germplasm, Cultivars and World-Wide Collections*, IOC, Madrid.

Pannelli, G. (2005), Olivicoltura italiana e mondiale a confronto, *Olivo e Olio*, **7–8**, 4–8.

6

The role of oxygen and water in the extra-virgin olive oil process

Bruno Zanoni

Department of Agricultural, Food and Forestry System Management, University of Florence, Florence, Italy

Abstract

Oxygen and water play critical roles in the extra-virgin olive oil process. Some oxidation is essential for the development of positive sensory notes such as grassy and fruity flavours (the LOX pathway). At the same time, excessive oxidation leads to oil spoilage and the formation of sensory defects like rancidity. Similarly, the presence of water is necessary for enzymatic reactions that modify the polarity of phenolic compounds, allowing them to be transferred to the oil phase. However, excess water in the oil can spur enzymatic or even microbial degradation during oil storage with irreversible loss of sensory and nutritional quality. Suitable adjusting of the balance of oxygen and water is one of the most important control tools in the extra-virgin olive oil process.

6.1 The conflicting roles of oxygen

In the box titled 'Balance is everything' concluding Chapter 3, it was pointed out how water and oxygen play conflicting roles in the extra-virgin olive oil process, being, at the same time, the main factors of desirable and undesirable changes. For a good control of the extra-virgin olive oil process, experts need to understand the conditions favouring / inhibiting the wanted / unwanted outcome.

Figure 6.1 represents a simplified scheme of oxidative reactions in the olive oil process. Unsaturated fatty acids, both free and esterified in triglycerides, react with oxygen under the action of lipoxygenase, forming hydroperoxides that are the initiators of either desirable or undesirable changes.

The Extra-Virgin Olive Oil Handbook, First Edition. Edited by Claudio Peri.

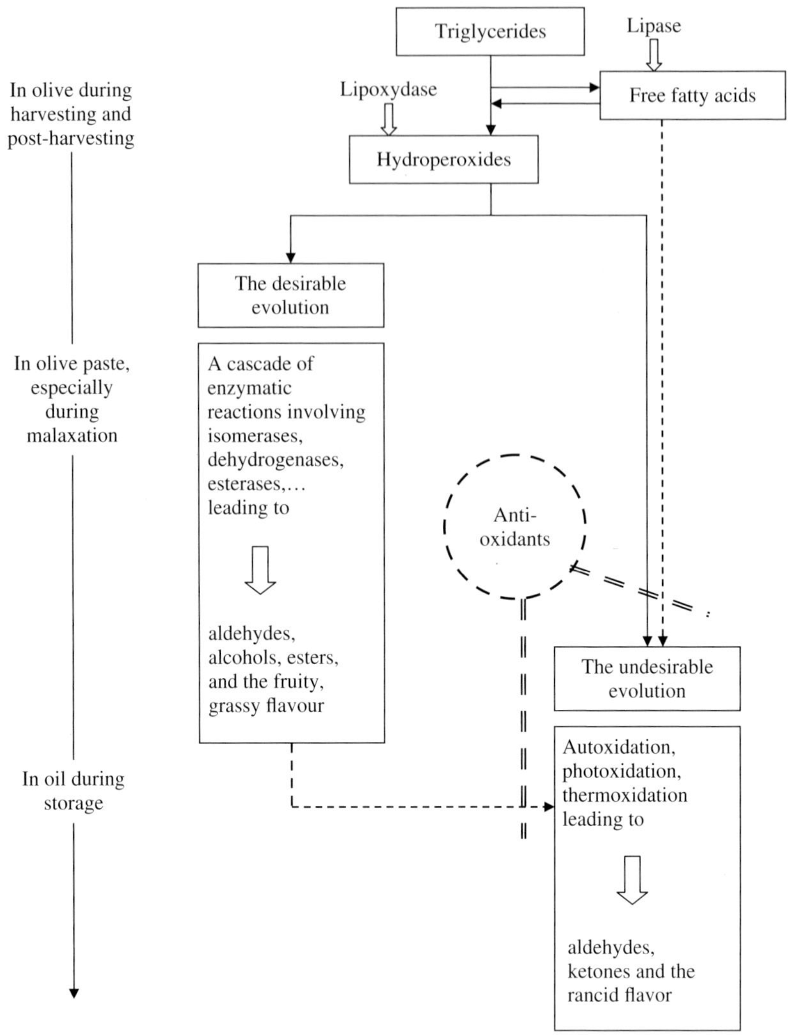

Figure 6.1 The complex role of oxygen in the extra-virgin olive oil process.

The preferred reactant of the desirable changes is α-linolenic acid and the overall cascade reaction is called the lipoxygenase (LOX) pathway.

Lyases, enzymes that catalyse the breaking of chemical bonds by means other than hydrolysis and oxidation, break fatty acid molecules into shorter chain pieces. Alcohol-dehydrogenases cause the aldehydes to be reduced to alcohols. Acetyl-transferase catalyses the esterification of alcohols with acetic acid.

Hexenyl-acetate or, more precisely, a series of C_6 compounds such as *trans*-2-hexenal, *trans*-2-hexen-1-ol and *cis*-3-hexenyl acetate, which are the possible products of these reactions, are among the compounds of the most appreciated grassy and fruity flavours of extra-virgin olive oil.

All the reactions of the desirable transformations are enzyme-catalysed, so they can only take place in the olive paste, during the milling process. Malaxing is the critical operation for the formation of these flavour compounds and their transfer from the aqueous phase to the oil phase. In fact, these molecules are soluble enough in lipids for favourable oil-water partitioning.

If the initial oxidation of α-linolenic acid is inhibited, for instance, by operating under strictly inert atmospheric conditions, the LOX pathway is not initiated and important flavour notes will not develop.

The undesirable changes also start from hydroperoxides but continue with a mechanism of autoxidation accelerated by light exposure (photoxidation) or high temperature (thermoxidation), if enough oxygen is available.

Such an autoxidative mechanism continues with formation of a mixture of highly reactive OH* and O* radicals. The molecules break at the double bond with formation of oxidized molecules, mainly aldehydes and ketones, with a perceivable rancid flavour.

The undesirable changes can start from olives and olive paste if the olives have been damaged mechanically or by pests. In this case, the good flavour components are overpowered by the bad ones and quality is irreversibly lost. The dotted lines suggest that reactants of the undesirable evolution can also come from products of the desirable evolution in an accelerated autocatalytic degradation path.

Autoxidation is inhibited by the absence of oxygen; photoxidation is inhibited by the absence of light; thermoxidation is inhibited by storing the oil at optimum temperature and enzymatic reactions are inhibited by elimination of enzymes and water. These considerations are the basis of suitable storage conditions, as is thoroughly discussed in Chapter 15. Antioxidants, especially phenolic compounds, intercept and neutralize reactive radicals, thus avoiding their involvement in oxidative degradation.

6.2 The role of water in the transformation of phenolic compounds

About 40 structurally different phenolic compounds have been identified in virgin olive oils (Cicerale *et al.* 2009). The various olive cultivars have very different amounts of phenolic compounds, but the distribution of the different phenolic categories in the overall phenolic fraction is rather similar. Figure 6.2 presents a simplified scheme of a possible evolution of phenolic compounds in the extra-virgin olive oil process. The figure emphasizes the role of water in the process.

The formulas in Figure 6.2 are drawn with slight differences from the formulas in Chapter 3. They indicate the three-dimensional arrangement of atoms. Solid wedged bonds point above the plane, while dashed wedged bonds point below the plane of the paper. The wavy symbol indicates unknown or unspecified stereochemistry. These are called "stereochemical" or "spacial" formulas.

The first molecule in Figure 6.2 represents the most abundant phenolic compounds in extra-virgin olive oil: oleorupein and ligstroside. They are a combination

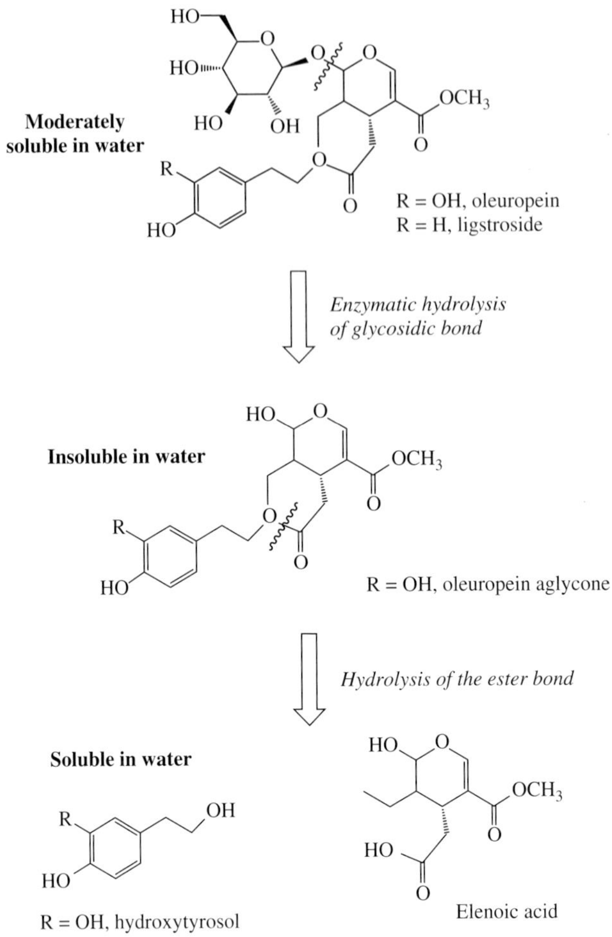

Figure 6.2 A simplified picture of the transformation of phenolic compounds in the extra-virgin olive oil process.

of elenoic acid esterified with a β-pyranose molecule on one side and tyrosol (in the case of ligstroside) or hydroxytyrosol (in the case of oleuropein) on the other side.

Oleorupein and ligstroside are moderately soluble in water and almost insoluble in oil. Therefore, their concentration in oil is always very low (a few mg/kg).

Oleuropein and ligstroside are found in the olive fruit but also in a very high amount in the olive leaves. They confer resistance to disease and insect infestation. They are present in relatively high concentrations in olive mill wastewater, with a significant antimicrobial and phytotoxic effect, which makes waste disposal more difficult.

An endogenous β-glucosidase causes the formation of oleuropein and ligstroside aglycones; they are esters of elenoic acid with tyrosol or hydroxytyrosol. The second molecule in Figure 6.2 represents oleuropein aglycone.

In this first transformation, the elimination of the glucopyranose moiety causes a change in the polarity of the molecule, which becomes almost insoluble in water but moderately soluble in oil. This change in polarity allows some oleuropein aglycone to be transferred to the oil phase, mainly during the olive paste malaxation. Aglycones of oleorupein and ligstroside and their various derivatives are the most abundant phenolic compounds in extra-virgin olive oil with concentrations in the range of 100 to 300 ppm or more. They are important in determining the bitterness and pungency sensations and the antioxidant power of the extra-virgin oil, which is essential for oil stability and health-promoting properties.

This point is worthy of a comment. The enzymatic reaction transforming oil-insoluble oleuropein into a slightly oil-soluble aglycone has a decisive impact on extra-virgin olive oil quality. If the transformation of oleuropein to the aglycone or the solubility of the aglycone in the oil were excessive, the antioxidant potential would be enhanced, but the bitter taste would be excessive. If, on the other hand, the transformation of oleuropein and the oil-solubility of the aglycone were excessively low, the oil would be less bitter, but also less stable. Only a very precise tuning of the enzymatic reaction and of the polarity of the molecules can assure a suitable balance of oil taste, stability, and health-promoting properties.

The final step of phenolic compounds transformation consists in the hydrolysis of the ester bond linking elenoic acid to tyrosol or hydroxytyrosol. Hydroxytyrosol is one of the most powerful *in vivo* antioxidants. It has immunostimulant, antibiotic and neuroprotective effects.

Again, the critical point is the change in polarity, hydroxytyrosol being much more polar than the oleuropein aglycones. Therefore, hydroxytyrosol can be formed only in the presence of water because it derives from an enzymatic reaction and also because it is soluble in water. This is a real problem because excessive water is a risk factor for oil stability, possibly favouring negative enzymatic activities such as, for example, lipolysis, or microbial growth. In conclusion, despite its antioxidant potential, hydroxytyrosol is considered as an indication of oil degradation instead of quality.

It can also be observed that in heating, cooking or frying, the antioxidants in extra-virgin olive oil are for the most part preserved due to the protection from oleuropein and ligstroside aglycones. Free tyrosol and hydroxytyrosol have much lower stability to heat.

A further comment may illustrate the extraordinary complexity and balance in extra-virgin olive oil. In ancient times, as well as up to the 1940s, the production of oil was carried out in very poor operating conditions: long processing times, much contact with air, ineffective separation of the oil from the water and the solid components of the olive paste. Nowadays, these factors are considered as the worst possible conditions for oil quality. Oils were frequently rancid and very few of them would have met the standards of extra-virgin olive oil. However, even at that time, olive oils showed striking health benefits (Colomer *et al.* 2008). Besides oleic acid and other particular components of the triglyceride fraction, was this due to a high concentration of hydroxytyrosol? Very probably so.

References

Cicerale, S., Conlan X.A., Sinclair A.J. and Keast, R.S.J. (2009) Chemistry and health of olive oil phenolics. *Critical Reviews in Food Science and Nutrition* **49** (3), 218–236.

Colomer, R., Lupu, R., Papadimitropoulou, A. *et al.* (2008) Giacomo Castelvetro's salads. Anti-HER2 oncogene nutraceuticals since the 17th century? *Clinical and Translational Oncology* **10**, 30–34.

Further reading

García-Rodríguez, R., Romero-Segura, C., Sanz, C. *et al.* (2011) Role of polyphenol oxidase and peroxidase in shaping the phenolic profile of virgin olive oil. *Food Research International* **44** (2), 629–635.

Servili, M., Selvaggini, R., Esposto, S. *et al.* (2004) Health and sensory properties of virgin olive oil hydrophilic phenols: agronomic and technological aspects of production that affect their occurrence in the oil. *Journal of Chromatography* **1054**, 113–127.

7

Extra-virgin olive oil contaminants

Cristina Alamprese
Department of Food, Environmental and Nutritional Sciences, University of Milan, Milan, Italy

Abstract

This chapter gives some basic information about extra-virgin olive oil contaminants and the associated risks. The list includes: mycotoxins, micro-organisms, pesticide residues, polychlorinated biphenyls (PCBs), polycyclic aromatic hydrocarbons (PAHs), organic solvents, phthalates, lubricants, and particulate materials. Maximum levels set for contaminants by Codex Alimentarius, IOC and EU regulations are reported. Conditions that determine or favour contamination and precautions for preventing them are discussed. In general, the health risk associated with the consumption of olive oils is low.

7.1 Introduction

Contaminants include micro-organisms, chemicals or particulate material that should not be present in extra-virgin olive oil at concentrations that could be harmful to human health.

Biological hazards derive from pathogenic and toxinogenic bacteria, fungi and viruses. They may grow and multiply in the olives before processing. Also, the presence of excess water and organic material may provide suitable conditions for microbial growth in unfiltered oil, giving rise to sensory defects.

Chemical hazards are the most serious hazards in extra-virgin olive oil. Toxic chemical contaminants may derive from:

- pesticide residues due to substances used in the treatment of olive trees or to combat pests and rodents in olive oil factories and stores;

The Extra-Virgin Olive Oil Handbook, First Edition. Edited by Claudio Peri.

- local contamination of groundwater or air or soil, and industrial wastes;
- smoke components from the environment surrounding the olive oil factory. They may contain soluble lipophilic constituents and fine particles that can contaminate the oil during processing, handling and storage;
- lipophilic volatiles (solvents) arising from industrial pollution in the area of olive cultivation or oil production and storage;
- plastic bottles, bags, containers, piping or conveyors containing, for instance, phthalates, used as plasticizers of polymeric materials (especially PVC), toxic monomers and oligomers.

Physical hazards include a variety of foreign bodies that can derive from:

- olive harvesting and handling (stones, metal, insects, undesirable plant material, soil, etc.);
- oil processing and handling operations (glass, metal, wood, bolts, screening and wire, cloth, grease, paint chips, insects, soil, and so on).

Various potential contaminants of extra-virgin olive oil are listed in Table 7.1 along with a qualitative evaluation of the likelihood of risk.

In general, the health risk associated with the consumption of extra-virgin olive oil is low. This reassuring observation is reinforced by two circumstances:

- The risks associated with degradation of the olives or the oil determine downgrading of the oil to the 'lampante' level. Contaminants are therefore removed in the refining process;
- Only lipid-soluble contaminants represent a risk because of their affinity and solubility in oil. The water-soluble contaminants, which may be present in the olive or may come into contact with the oil during the olive milling process, are eliminated with the wastewater or the pomace.

For all the other hygienic hazards, suitable conditions for safety are guaranteed by:

- location of the oil factory in a low polluted site;
- good hygienic control of the process and the environment (see Chapter 21);
- filtration of the oil;
- oil storage in hermetically sealed containers.

Table 7.1 Potential contaminants of extra-virgin olive oil (Meloni *et al.* 2005).

Type of contamination	Contaminant	Main causes of contamination			Likelihood of risk (see Chapter 21)
		Inadequate control of the agronomic and olive handling operations	*Inadequate control of the milling and storage operations*	*Inadequate control of environmental hygiene inside and outside the factory*	
Biological	Mycotoxins	+	+		Extremely unlikely
	Microbial growth		+		Extremely unlikely
Chemical	Pesticide residues	+			Unlikely
	Fungicide residues	+			Very unlikely
	Herbicide residues	+			Extremely unlikely
	Polycyclic aromatic hydrocarbons (PAHs)			+	Very unlikely
	Polychlorinated biphenyls (PCBs), dioxins			+	Very unlikely
	Smoke components			+	Unlikely
	Organic solvents			+	Unlikely
	Phthalates		+		Very unlikely
	Lubricants		+		Unlikely
	Detergents		+		Very unlikely
Physical	Fragments and residues from pests			+	Unlikely
	Stones, metal, plastic, wire, cloth, wood, soil, etc.			+	Unlikely
	Glass fragments		+		Unlikely

7.2 Contaminants of virgin olive oil

Legal limits for various olive oil contaminants are given in Table 7.2.

Table 7.2 Legal limits for some olive oil contaminants (Codex Stan 33-1981 1981–2013).

Contaminant	Virgin olive oils	Olive oil	Refined olive oil	Olive-pomace oil
Moisture and volatile material (%)	≤ 0.2	≤ 0.1	≤ 0.1	≤ 0.1
Insoluble impurities (%)	≤ 0.1	≤ 0.05	≤ 0.05	≤ 0.05
Trace metals (mg/kg):				
Iron (Fe)	≤ 3	≤ 3	≤ 3	≤ 3
Copper (Cu)	≤ 0.1	≤ 0.1	≤ 0.1	≤ 0.1
Heavy metals (mg/kg):				
Lead (Pb)	≤ 0.1	≤ 0.1	≤ 0.1	≤ 0.1
Arsenic (As)	≤ 0.1	≤ 0.1	≤ 0.1	≤ 0.1
Halogenated solvents, max. content:				
of each solvent (mg/kg)	0.1	0.1	0.1	0.1
of the sum of all solvents (mg/kg)	0.2	0.2	0.2	0.2

7.2.1 Mycotoxins

Mycotoxins are water-soluble, relatively low-molecular weight, secondary metabolites of fungal origin that are harmful to animals and humans. The amount of mycotoxins needed to produce adverse health effects varies widely among toxins, as well as for each animal or person's immune system. Mycotoxins can be acutely or chronically toxic, or both, depending on the kind of toxin and the dose. The European Community sets the tolerance level for some food products but edible oils are not specifically addressed (Commission Regulation (EC) No 1881/2006).

If olives are stored for several weeks under conditions that contribute to the growth of moulds, mycotoxin contamination may occur. Ochratoxin A and aflatoxin B, both mutagenic, have rarely been found in extra-virgin olive oil and, when found, they were at extremely low concentrations with no toxic effect. In some cases, higher concentrations have been found in 'lampante' oils derived from unhealthy, rotten olives. In this case, however, mycotoxins are eliminated by the refining process for the production of food-grade refined olive oil. In conclusion, it may be said that the presence of mycotoxins in extra-virgin olive oil is extremely rare and therefore the overall gravity of risk should be considered as negligible.

7.2.2 Micro-organisms

Microbial growth is very unlikely because extra-virgin olive oil does not contain sugars and nitrogenous compounds that are essential nutrients for growth of micro-organisms.

Some bacterial growth and fermentation can take place only in the presence of water and organic material. This is quite common especially when oils are not filtered. It is more common to find yeasts and moulds than bacteria. In any case, microbial growth does not represent a health hazard, but can cause quality degradation with increase in free acidity, peroxide and spectrophotometric values and formation of sensory defects, such as fusty and muddy sediment (moulds) or winey and vinegary (yeasts).

The presence of water is a necessary condition for bacterial growth and activity, so excess water should be carefully avoided. It is interesting to note that the Codex Alimentarius has set a limit of 0.2% water in extra-virgin olive oil (Codex Stan 33-1981 1981–2013). This is one of the least checked and one of the most often transgressed standards of extra-virgin olive oil.

7.2.3 Pesticide, fungicide and herbicide residues

Pesticides are chemical substances that are used to control or prevent pest attack. In the case of olive oil, control treatments against the olive fly (*Bactrocera oleae*) are the most frequently applied and therefore their residues are the most frequently found pesticide in the oil. Several pesticides are used, such as synthetic pyrethroids, organochlorine and organophosphorous insecticides. In regions where there is a low rate of attack, a single pesticide treatment at the start of the second generation of the fly is sufficient. In areas with high attack rates (such as coastal areas and valleys) treatment for the second and third generation is normally required. In this case, a careful control of residues in the oil may be advisable. In any case, pesticide treatments must be carried out with authorized products and active ingredients and must strictly conform to the required withholding periods in order to avoid residues in the fruit at the time of harvest. Conformity to these prescriptions should be duly documented.

Fungicides based on copper products and copper mixes have a relatively low toxicity, thus they are to be considered as a negligible risk. The legal limit for copper compounds in olives for oil production is set at 30 mg/kg by the European Community (Commission Regulation (EC) No 149/2008). Even the use of organic fungicides should be normally considered as safe in terms of oil contamination and residues. The use of herbicides should be considered as safe, especially in the case of harvesting from the tree (the only acceptable method in extra-virgin olive oil production) and not by picking up the olives from the soil.

A more subtle risk may derive from pest traps and pesticide treatments in olive mills and oil storage facilities. These practices, which are part of any good Hygiene Management System (Chapter 21), should be carried out by skilled operators under carefully controlled conditions.

Due to an increasing awareness of the possible risks involved with the use of pesticides, strict regulations for maximum residue limits (MRLs) for these contaminants have been fixed by the Codex Alimentarius (Table 7.3) and by the European Community (Codex Stan 33-1981 1981–2013; Commission Regulation (EC) No 149/2008).

Table 7.3 Maximum residue limits of pesticides in virgin olive oil (CAC/MRL 1 2009).

Pesticide	MRL (mg/kg)	GHS category
Cypermethrins (including alpha- and zeta- cypermethrin)	0.5	3
Kresoxim-methyl	0.7	3
Fenthion	1	3
Carbaryl	25	3

MRL: maximum residue limit; GHS: globally harmonized system of classification and labelling of chemicals (GHS 2003).

Table 7.4 A simplified summary of the Globally Harmonized System (GHS) of category classification of toxic substances (GHS 2003).

	Category 1	Category 2	Category 3	Category 4	Category 5
Exposure:					
Oral (LD_{50}, mg/kg bodyweight)	5	50	300	2000	5000
Dermal (LD_{50}, mg/kg bodyweight)	50	200	1000	2000	5000
Hazard statement:					
Oral	Fatal if swallowed	Fatal if swallowed	Toxic if swallowed	Harmful if swallowed	May be harmful if swallowed
Dermal	Fatal in contact with skin	Fatal in contact with skin	Toxic in contact with skin	Harmful in contact with skin	May be harmful in contact with skin

On the basis of the acute toxicity estimate value, chemicals are classified in four categories of the Globally Harmonized System of Classification and Labelling of Chemicals (UN 2003; Environmental Protection Agency 2012) (Table 7.4). Acute toxicity refers to adverse effects occurring after oral or dermal administration of a single dose of a substance, or multiple dose exposure within 24 hours. Most pesticides are generally included in Category 3.

Further considerations have to be made about pesticides and fungicides. First of all, in the Mediterranean area, there is a vast assortment of olive tree varieties. In these conditions a subtle balance between olive trees and their pathogens or other damaging agents has been established. Only when such balance is altered do severe attacks and considerable losses occur. Unfortunately, significant changes and variability in the climate are altering these equilibriums, as well as the new intensive olive growing practices (irrigation, fertilization, high density). Indiscriminate or excessive pesticide, fungicide and herbicide treatments contribute further to destabilizing the ecosystem equilibriums by destroying useful insect communities, microbial colonies in the soil and wild plant species.

Finally, it must be emphasized that the use of pesticides, fungicides and herbicides may represent a serious risk to workers' health in the handling of highly toxic substances.

Any guarantee or certification scheme of extra-virgin olive oil quality and safety must include:

- monitoring of pest and disease evolution in order to minimize the number of treatments;
- education, training and protection of workers for the most effective and careful use of toxic substances;
- registration of treatments including date, products, quantities, concentration and mode.

Use of traditional olive cultivars and sustainable agricultural practices, including integrated and organic agriculture, should be encouraged.

7.2.4 Environmental pollutants

Many environmental pollutants result from local contamination of groundwater or air or soil with industrial wastes. Among them the most feared are organochlorine compounds, known for their persistence and toxicity. Olive oils may contain residues of hexachlorobenzene (HCB), dichlorodiphenyltrichloroethane (DDT) and polychlorinated biphenyls (PCBs). Small amounts of PCBs may be formed during combustion of materials containing carbon and chlorine. Like PCBs, dioxins are ubiquitous environmental contaminants; they have been found in soil, surface water, sediment, plant and animal tissue worldwide. They have low water solubility and therefore they are persistent in the soil for years. On the other hand, they are soluble in lipids and tend to accumulate in the fat of animals and animal products (meat, fish, egg, milk) as well as in oil-bearing seeds and fruits. The European Community established maximum levels of 0.75 pg/g for the sum of dioxins and 1.5 pg/g for the sum of dioxins and dioxin-like PCBs in vegetable oils and fats (Commission Regulation (EC) No 1881/2006).

Similarly dangerous and widespread in industrial areas are the polycyclic aromatic hydrocarbons (PAHs). They are considered to be carcinogens and, due to their lipophilic characteristics, they can contaminate oils (Mafra *et al.* 2012).

The main sources are petroleum (petrogenic PAHs) and pyrolytic PAHs that are formed by incomplete combustion of organic materials and by combustion engines; they are widely found in industrial and municipal wastes and run-off. Electrogenic systems, sometimes used to provide electric power to the mill, can also produce PAHs. Moreover, the particulates generated by combustion can represent an effective system for diffusion of PAHs adsorbed on the solid nanoparticles. The International Olive Council and the European Community (RES-1/93-IV/05 2005; Commission Regulation (EC) No 1881/2006) has established a maximum concentration of 2 μg/kg for benzo(a)pyrene, considered as a marker of the presence and the effects of cancer-producing PAHs in edible foods. An opportune remark is that all these environmental contaminants should be considered as very uncommon in extra-virgin olive oil. If the olive grove or the olive oil processing facilities are in industrially polluted areas, periodical analyses should be carried out.

7.2.5 Smoke

Smoke contains lipophilic constituents and extremely fine particles that can pollute the oil during processing, handling and storage. No fires should be allowed in the olive milling and storage facilities or in the surrounding areas. Cars and trucks with the engine running should not be allowed to enter the closed space where the oil is stored or handled. Combustion of plastic materials in the surroundings of the olive oil factory must be carefully avoided. Smoke usually causes a detectable sensory defect and this may cause the oil to be downgraded to virgin or 'lampante'.

7.2.6 Solvents

Solvents deriving from industrial pollution in the area of olive cultivation or oil production and storage may concentrate in the oil.

The origin of solvent pollution is sometimes difficult to trace. A case has been reported of solvent contamination, deriving from the lab in which oil analyses were carried out. The solvent used in very small amounts for the analysis of free acidity of the oil (a useful routine determination in many olive mills) was wrongly and illegally discarded in the sewage system and from there solvent vapours spread into the atmosphere and were finally concentrated in the oil inside the storage tanks.

Other possible sources of contamination are petroleum-derived solvents used in agriculture to dissolve and dispense agronomic products in fields, or by solvents coming from printed paper, from air conditioning equipment, from painting equipment, and so forth.

In Table 7.1, solvent contamination is considered as unlikely. However, due to the high volatility and high penetrating ability of solvents, careful control of solvent contamination should be carried out in industrially polluted areas.

7.2.7 Phthalates

Phthalic acid esters (known as phthalates) are added to plastics, primarily vinyl, to make the material softer and to increase flexibility. Due to concerns raised over the potential effect of chronic phthalate exposure on human health, the European

Community has limited the use of phthalates in plastics for food use and, where their use is permitted, the specific migration limits of these chemicals into foods are set (Commission Directive 2007/19/EC). United States regulations treat phthalates that migrate into foodstuffs from food contact materials as indirect additives, that is additives 'that become part of the food in trace amounts due to its packaging, storage or other handling'. The Code of Federal Regulations of the FDA reports the list of substances that may be safely added as plasticizers in polymeric substances used in the manufacture of articles or components of articles intended for use in producing, manufacturing, packing, processing, preparing, treating, packaging, transporting, or holding food (21CFR178.3740 2013).

Although phthalates are widespread in the environment, levels tend to be low because they do not generally persist for extended periods when exposed to photochemical and biological breakdown. Phthalate contamination has been found in extra-virgin olive oils at very low concentrations, without any significant relevance in terms of safety. However, it may be advisable to evaluate them in cases of extensive use of plastics in bottles, bags, containers, piping or conveyors.

7.2.8 Lubricants

Contamination from the lubricants (oils and greases) used for olive oil mill equipment may be underestimated. A spill of gear oil caused by breaking of a seal may contaminate the oil. Careful attention should be paid to the integrity of sealings and to oil and grease consumption. A list of lubricants with accidental food contact that may be used in producing, manufacturing, packing, processing, preparing, treating, packaging, transporting, or holding food is provided by the FDA (21CFR178.3570 2013).

7.2.9 Detergents

Detergent contamination may derive from improper cleaning and washing procedures of the olive milling plant or of tanks, containers and bottles. Only food-grade detergents must be used, thus possible residues may not be a significant health hazard. Careful rinsing with potable water is the only effective preventive measure for this contamination.

7.2.10 Physical contaminants

Contamination by particulate materials should be considered as a 'normal' event in any olive oil factory. Particulates can derive from the soil, equipment, packaging material, people and the atmosphere. Circumstances for contamination are countless especially when the oil is handled in an open space for transfer from one container to another or filtering or bottling. Many animals, especially insects, rodents, and birds, are attracted to oil, and fragments of their body or droppings may contaminate containers or the oil itself, if they are allowed access to the areas where the oil is stored or handled. Other particulates may be accidentally dropped into the oil by the workers if their personal hygiene or clothes or behaviour are not carefully watched. The only preventive measure to avoid particulate contamination is care in the maintenance of buildings and equipment, in personal hygiene of people that are in contact with the oil, in preventing access of pests and in combating their proliferation in the internal and external environment of the oil factory. Filtration of the oil is the most effective operation to control any possible particulate contamination.

Among particulate contaminants, glass fragments may cause serious health damage with micro-lesions to the digestive tract. This risk is particularly high when fragments are so small that they are suspended in the oil and cannot be identified or perceived by the consumer. Glass fragment contamination can take place during bottling of the oil and can only be prevented by very close attention to the integrity of the bottles and by carefully cleaning them before filling.

Determination of the content of insoluble impurities in animal and vegetable fats and oils is generally carried out using the ISO method (ISO 663:2007 2007). Insoluble impurities are defined as the amount of dirt and other foreign insoluble material in n-hexane or light petroleum. A sample is treated with an excess of n-hexane or light petroleum, and then the solution is filtered. Insoluble impurities are gravimetrically determined, after washing the filter and residue with the same solvent and drying at 103 °C. Maximum limits set by the Codex Alimentarius are reported in Table 7.2.

References

21CFR178.3570 (2013) *Lubricants with Incidental Food Contact*, Code of Federal Regulations, Title 21, Vol. 3, Part 3570, US Food and Drug Administration, Silver Spring, MD.

21CFR178.3740 (2013) *Plasticizers in Polymeric Substances*, Code of Federal Regulations, Title 21, Vol. 3, Part 3740, U.S. Food and Drug Administration, Silver Spring, MD.

CAC/MRL1 (2009) *Maximum Residue Limits (MRLs) for Pesticides. OC 0305 – Olive Oil, Virgin*, Codex Alimentarius Commission, Rome.

Codex Stan 33-1981 (1981–2013) *Codex standard for olive oil and olive pomace oil*, Codex Alimentarius Commission, Rome. Revised 1989, 2003, amended 2013.

GHS (2003) *The Globally Harmonized System of Classification and Labelling of Chemicals (GHS)*, United Nations, New York and Geneva, http://www.consorzioextravergine.com/pubblicazioni/Catalogo_Contaminanti.pdf (accessed 11 October 2013).

ISO 663:2007 (2007) *Animal and Vegetable Fats and Oils – Determination of Insoluble Impurities Content*, International Standard Organization, Geneva.

Mafra I., Amaral J.S., Oliveira M.B.P.P. (2012) Polycyclic aromatic hydrocarbons (PAH) in olive oils and other vegetable oils; potential for carcinogenesis, in *Olives and Olive Oil in Health and Disease Prevention*, (eds V.R. Preedy and R. Ross Watson), Academic Press, Oxford, pp. 489–498.

Meloni M., Coni E., Conte L., *et al.* (2005) La contaminazione dell'olio extra-vergine di olive. Consorzio di Garanzia dell'Olio Extra Vergine di Oliva di Qualità, Rome.

RES-1/93-IV/05 (2005) *Detection of Polycyclic Aromatic Hydrocarbons in Edible Olive Oil*, International Olive Oil Council, Madrid.

Part II

The process

8

Olive harvesting

Luigi Nasini and Primo Proietti
Department of Agricultural, Food and Environmental Sciences, University of Perugia, Perugia, Italy

Abstract

Monitoring of the ripening of olives in order to decide the best harvesting period is a critical control step in extra-virgin olive oil production. The first part of this chapter presents the main phenomena of olive ripening. The second part presents the main harvesting systems: hand-held harvesting machines for traditional olive groves, trunk shakers for semi-intensive and intensive olive groves and straddle harvesters for super-intensive olive groves. Systems are compared in terms of harvesting efficiency and labour productivity. In Annex 8.1 a system of maturity assessment and harvesting decision is described in detail.

8.1 Introduction

The olive harvesting operation critically influences oil yield and quality as well as the cost of oil production. Optimizing olive harvesting entails obtaining the highest amount of oil of a predefined level of quality. In quality-oriented companies, oil yield is a dependent variable of oil quality. In fact, the harvesting decision is determined by the need to meet suitable sensory and analytical requirements and the yield of olives per tree and per hectare follows as a consequence.

The influence of olive harvesting on production cost is very important (see Chapter 23). Mechanization plays a strategic role in the planning of olive groves and in choosing the cultivars and the tree training system. Developments towards a high degree of harvest mechanization consist in choosing the right machinery but also adapting the trees to machinery use.

The Extra-Virgin Olive Oil Handbook, First Edition. Edited by Claudio Peri.

8.2 Olive ripening

The ripening process is the result of the combination of genetic, environmental and cultural conditions (Beltran *et al.* 2010). The most apparent changes are the increase in size of the olives and some typical change in the skin colour which, in many cases, follows a four-stage sequence (northern hemisphere):

- The 'green' stage. From August to mid-September, the olive skin is uniformly green.
- The 'light-green' stage. From mid-September to the first week of October, greenness progressively fades into a pale green colour.
- The 'purple' stage. From the first week of October to the last week of November, significant changes in the colour of the olive take place. The pigmentation starts to become reddish-purple often affecting only the apex of the fruit, then, with progressing maturity, extending over the entire fruit surface, with increasingly darker tones, progressively tending to black. The first part of the purple stage is often defined as 'veraison' a word of French origin meaning 'the onset of ripening' ("invaiatura" in Italian).
- The 'black' stage is when all of the skin is uniformly black and the olive can be considered as overripe. In the final stages of ripening, pigmentation may also affect the pulp from the outermost part towards the pit.

These changes in colour, however, cannot be considered as a general rule because the pigmentation process is influenced by several factors, such as climate, fruit load and, above all, the cultivar. For example, the fruit of Frantoio remains partially green in proximity to harvest time, while those of Moraiolo turn black as harvesting time approaches.

In general, harvesting should be carried out at the purple stage. The first question to be answered is: being that the purple stage is quite long (4–6 weeks), should harvesting take place at the beginning or at the end or in the middle of this period?

Secondly, it must be understood that olive ripening is a scalar process. Not all the fruits on an olive tree ripen at the same time. Ripening differs on branches that are high or low or on different sides of the tree. Olives in the more shaded parts of the canopy have a lower content of total phenols and flavouring compounds. Not all the trees in an olive grove follow the same pattern of ripening, depending on the size of the tree, the position in the canopy, and the fruit-load. Therefore, the description of ripening given above should be considered as representing an average condition of ripening patterns that each tree in the orchard and each fruit on a tree follows in its own way.

In the third place, the trend of the ripening process may change depending on the environmental conditions. The same cultivar in orchards with different exposures or altitudes or water availability may follow a similar trend but with some shifting of the ripening phenomena. In other words, for the same cultivar but in a different orchard, the optimal harvesting time may change by days or weeks. The same would obviously be true in comparing the ripening process of the same cultivar in the same orchard but in different years.

Some rules-of-thumb can be considered as, for example:

- a hot autumn and low rainfall can cause fruit to ripen quickly, resulting in a narrow window for optimum harvesting;
- a cool autumn may result in delayed ripening;
- low fruit load accelerates the ripening process, whereas the opposite happens with high fruit load;
- in general, it is suitable to anticipate harvesting when: (i) the olives are under risk of pest attack; (ii) the olives are damaged by hail; or (iii) there is a risk of early autumn frosts.

Monitoring of olive ripening with reliable systems is needed for deciding about the best harvesting period. Annex 8.1 presents a detailed discussion of methods for assessing olive maturity and for effectively linking the maturity to the desired quality characteristics of the oil.

8.3 Harvesting systems

An optimal harvesting operation can be defined as 'the ability to harvest more than 90% of the olives on a tree – in the shortest period of time and with the lowest number of workers – with minimum mechanical damage to the olives and trees and minimum risk for workers' safety and health.'

The continuous evolution of harvesting systems, which has been taking place at an accelerated pace since the 1970s, is driven by the need to increase harvesting efficiency by increasing labour productivity in order to decrease harvesting cost (Gil Ribes *et al.* 2010; Tous 2012).

The harvesting operation includes olive detachment from the tree and intercep tion/collection of the detached olives. Olive detachment from the tree is the most critical and costly of the harvesting operations. It can be carried out with different systems depending on the scale of operation and the training system of the olive trees (Table 8.1).

8.3.1 Hand picking

Hand picking is the most expensive harvesting system due to high labour costs. It is carried out with the aid of simple tools, for example small plastic rakes that are pulled along the fruit bearing branches. The detached fruit falls onto plastic nets spread on the ground below the crown of the olive tree; then, they are collected by hand and put into suitable crates or bins.

Table 8.1 The three main harvesting systems.

Harvesting systems	Suitable scale of application
Hand picking and hand-held harvesting machines	Suitable for small-scale production, any type of training system and in case of steep slopes
Trunk shakers	Suitable for high-scale production in semi-intensive or intensive olive groves
Straddle harvester	Suitable for high-scale production in super-intensive olive groves

Hand picking can be used with any type of tree training system and olive grove characteristics (size, tree spacing, slope of the ground and so forth); it does not require large investment or special skills. A serious limiting condition of hand picking is the high number of workers and the frequent need to use ladders, which decreases labour productivity and poses worker safety risks.

Under the best conditions, the productivity of hand picking is 10–20 kg of olives per hour per operator. Picking efficiency increases with increasing fruit load and when olive groves are on level areas, the trees are low and well pruned, so that the crown is easily accessible to the operator.

It has been calculated that hand picking may represent about 80% of the total labour time required for producing olives in an olive grove and 50–70% of the cost of the harvested olives. This is the reason why hand picking is disappearing and is being replaced by hand-held harvesting machines or mechanical systems.

8.3.2 Hand-held harvesting machines

Olive harvesting can be greatly facilitated by the use of harvesting machines held and carried by the operator. Hand-held harvesting machines are available in a great variety of models and the design and performance is continuously being improved. The harvest labour productivity varies from 30 to 50 kg per hour per operator, but under optimal conditions, a well-trained operator can harvest up to 150 kg per hour. The best results are obtained with medium or high weight fruit, low resistance to detachment and low and well-pruned trees, so that the crown is easily accessible to the operating device of the harvesting machines.

Hand-held harvesting machines are suitable for all types of tree training systems, but it is important that tree height does not exceed 4 m. Therefore, it is necessary that pruning be done to limit the growth in height and promote crown diameter development.

A hand-held harvesting machine consists of three parts: an operating device, a telescopic pole and a motor to provide the needed driving power.

The operating device

The most common operating devices are of two types: oscillating or vibrating or turning combs, and shaking hooks.

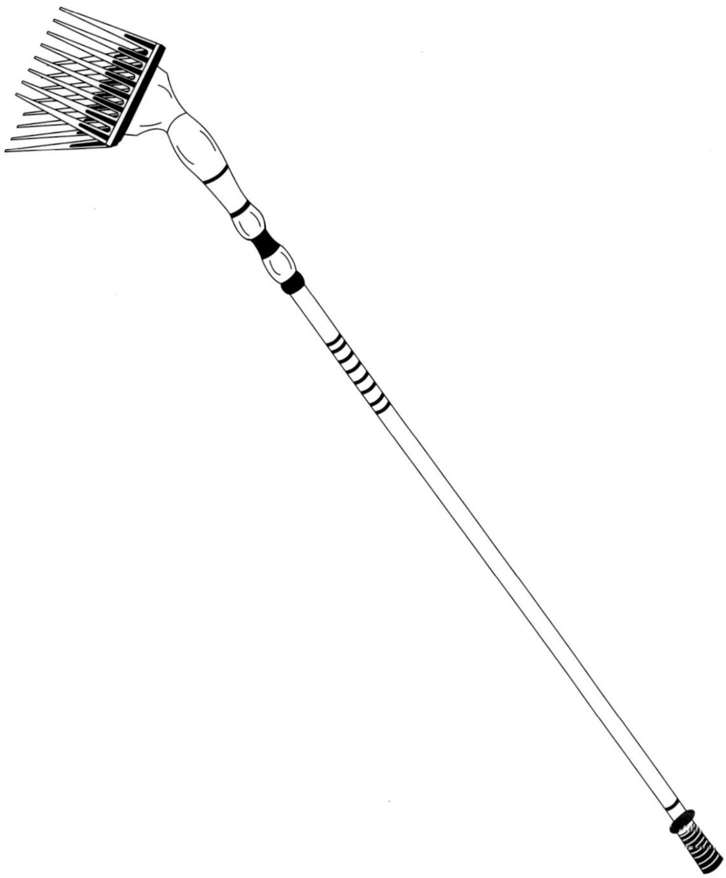

Figure 8.1 Hand-held harvesting machines: oscillating combs and the telescopic pole.

The most common operating modes of combs are:

- Oscillating combs mounted on pairs and swinging against one another (Figure 8.1). The teeth may vary in number and length. The combs with long and widely spaced teeth are suitable for harvesting trees that have a dense crown (e.g. cultivars with dense foliage and/or poorly pruned trees), while combs with a large number of teeth are particularly useful in early harvesting of small olives with a high resistance to detachment. Combs with teeth of two sizes or a decreasing thickness from the base to the tip penetrate more easily into the vegetation.
- Combs (beaters) with vibrating teeth on a rotating base. They have slightly lower labour productivity than oscillating combs, but cause less operator fatigue. They are more efficient when the vegetation is not dense and in harvesting the upper parts of the crown.

- Combs with undulated rotating teeth. The rotational movement of the teeth combines with the combing action on the crown by the operator. The labour productivity is lower than the other combs because their area of action is more limited. In general, they can cause greater operator fatigue because of the unbalanced weight.

With shaking hooks, the hooks are hooked to small olive branches (5 cm diameter maximum) to which they transmit a vigorous vibration of 1000–1500 strokes per minute.

Shaking hooks must be equipped with antivibration systems on the handles in order to minimize vibrations transmitted to the operator.

With combs, the fruit is detached by a beating action and by the effect of vibration on the shoots and branches. In the case of shaking hooks the fruit is detached only by the effect of vibrations.

Beating may cause bruising of the fruit and therefore it should be minimized, whereas vibration is effective only if the detachment force is not too high, preferably around 3 N. With shaker hooks harvest can reach 90–95% yield only in late harvest with ripe fruit.

The telescopic pole

All operating devices in hand-held harvesting machines are mounted on telescopic poles up to 2–4 m long made of lightweight yet durable materials, such as aluminium, fiberglass, nylon or carbon fibre. The weight of the pole plus the combs varies from 2 to 4 kg. In some cases the comb can be disconnected from the pole and more easily used in the lower portion of the crown. In other cases, the operating devices are connected to the pole by means of a swivel joint allowing the working angle to be adjusted, depending on the characteristics and shape of the vegetation.

With the advent of telescopic poles, ladders have been abandoned, thus decreasing both the risk to workers and working time.

The driving power

Hand-held harvesting machines are classified as pneumatic or electric. In pneumatic machines, compressed air generated by a compressor (self-propelled or hauled by a tractor) is fed through a connecting tube to the harvesting machine at a pressure of 6–8 bars and a flow-rate of about 200 litre per minute. The use of connecting tubes of considerable length allows movement of the compressor to be minimized. It is not advisable, however, to exceed 100 meters in length to ensure good operating conditions. Connecting tubes can be wrapped with automatic spring wrappers, which facilitate the operation.

Electric machines are powered by batteries (12 or 24 V), connected by a cable (maximum length about 20 m). The batteries, which can also be carried by the operator using an ergonomic vest, can operate for the whole day (autonomy is about 7 hours of work) and then they are recharged during the night to be ready for the next day. Electric machines have a low noise level. They are generally equipped with a safety electronic device to prevent motor damage in case the operating teeth get blocked in the vegetation. The weight of the electric motor plus that of the pole and the operating device is 2–3 kg.

Some shaking hooks are powered by a small endothermic engine (1.5 to 2.2 kW) carried by the operator by means of a harness. The endothermic engine is relatively heavy (9–15 kg) and has a high noise level. The breathing of combustion gases may also be a further threat to workers' health.

Two disadvantages of hand-held harvesting machines

In the first place, operating them may cause a relatively high level of worker fatigue, especially with heavy equipment or unbalanced distribution of the weight or intense vibrations or ergonomically inappropriate holding. In order to reduce fatigue, the harvesting teams should take turns. Periodically (about every 2 h) it is recommended that operators be alternated by switching between the tasks of harvesting and moving the nets.

In the second place, hand-held harvesting machines may cause some damage to the trees (bruising the bark and leaf fall), especially in early harvesting when it is necessary to dwell at length for effective fruit detachment. Bacterial and fungal growth may affect the damaged area of the bark, so a disinfectant treatment with copper products immediately after harvest is recommended. Mechanical damage to the fruit is generally low if the beating action is minimized.

8.3.3 Trunk shakers

Olives are harvested by means of a vibrating grip head attached to the trunk or, in the case of very large trunks, to the main branches. Trunk shakers can be self-propelled or mounted on tractors. The cost of self-propelled shakers is higher than that of tractor-mounted shakers and therefore they are mainly used by large olive groves or service companies. For trunk shakers a tractor with power greater than 60–80 CV is required, depending on the size of the grip head and the combination of the collecting frame for the olives.

The vibrating grip head consists of a jaw with a cushioned system to avoid damage to the bark of the trunk or branches. Vibrations are generated by two eccentric rotating masses turning in opposite directions or by one mass in an orbital movement.

The arm supporting the vibrating grip head may be telescopic, thus allowing greater versatility in movements, especially when it is necessary to apply the grip head to the main branches.

The vibration time is 10–15 s, depending on the olive cultivar, the ripening stage and the tree shape, but most of the olives drop in the first few seconds. In general, to avoid tree damage, it is preferable to apply two short vibrations than only one longer one. The falling olives are intercepted by nets manually spread on the ground or by upside-down umbrella-shaped mechanical looms (diameter from 5 to 10 m), which close below the crown (Figure 8.2).

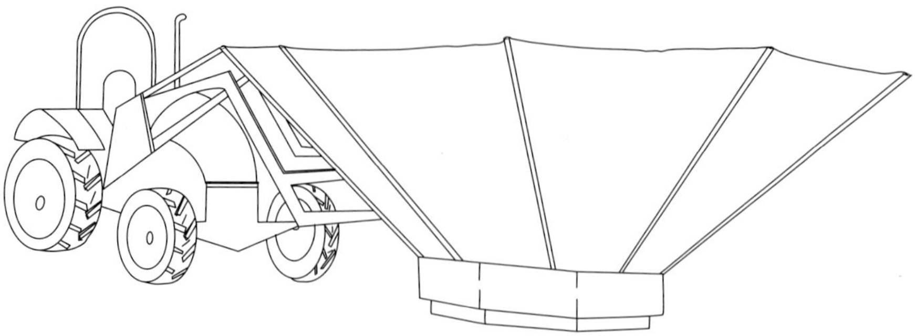

Figure 8.2 Trunk shacker with wrap-around umbrella for olives detachment-interception-collection.

When trunk shakers operate in combination with mechanical looms for fruit collection, labour productivity may reach a value of 200–400 kg per hour per operator. As an average, in one month 20–25 hectares of olive trees can be harvested. Service companies may offer trunk-shaking service to small olive growers.

The productivity of trunk shakers decreases greatly when it is necessary to attach the shaker to the branches and when nets have to be moved manually. The best conditions for effectively using trunk shakers are summarized in Table 8.2.

8.3.4 Interception and collection of the olives

Usually, this step is not taken into due consideration, even though it represents a substantial part of the harvesting labour and cost. The manual movement of nets from tree to tree and emptying them into crates or bins or trailers require several operators and the time needed for these operations is often longer than the time needed for removing the olives from the tree, either with hand-held harvesting machines or trunk shakers. Consequently, manual management of the nets reduces labour productivity. Furthermore, moving the nets is tiring, especially when working on sloping or wet soil. With wet soil without a green cover, the nets and consequently the olives can become dirty with mud and this may cause defects in the oil if the olives are not sufficiently cleaned and washed before processing.

These considerations have stimulated progress in harvesting technology towards the partial or total mechanization of interception of the detached olives. Two types of mechanical systems are currently applied: the inverted umbrella wrap around and the reel systems.

The umbrella type is used with trunk shakers. It consists of parts arranged to form a reverse cone with a centre wrapped around the tree trunk and a hopper for temporary storage of the olives (average capacity of 200 kg or more). When the hopper is full, it is unloaded into a trailer or bins. The umbrella system can substantially increase the labour productivity of trunk shakers achieving values of about 200–400 kg of olives per hour per operator. The layout of the olive grove has to be suited to this method and the canopy of the trees cannot be too large.

Table 8.2 The best conditions for the use of trunk shakers.

Best conditions	Notes and comments
Olive tree age: 8 to 60 years	Trunk shakers can be used when the trunk reaches a diameter of 8–10 cm. In old trees, trunks are usually too large (diameter greater than 50–60 cm) for effective trunk shaking and the shaker should therefore be attached to branches, which reduces labour productivity. Trunk shakers have the best efficiency with crown volumes up to 40–50 m^3.
Average density of plants: about 300 trees/ha	The optimal condition for trunk shaker operation is with a planting distance of 6 × 6 m or greater (distance between rows of at least 5 m). With trunk shakers with an interceptor loom, a distance of 1 m between two adjacent crowns along the row is necessary.
Training system: trunk should be single, regular, straight and at least 100 cm high. The highest yields are obtained with open centre (vase) and monocone training system	The best results are obtained with perfect coupling of the vibrating grip head to the trunk and branches (3 or 4) at a relatively narrow vertical angle (35–40°) as acute-angled branches transmit vibrations more efficiently than horizontal ones. This is possible in the open centre, but not in the monocone. The crown should not have long, low pendulous branching, especially if an inverted umbrella interceptor is used. The main and secondary branches should be without forks and not too long.
Ratio between fruit weight and detachment force: less than 3	A fruit detachment force is too high when it is higher than 6 N. The use of abscission products to encourage premature dropping has not proven to be a reliable method and has often caused marked defoliation. Considering the ratio between detachment force (N) and fruit weight (g), good harvesting yields are obtained with ratio values around 2, whereas harvesting yields are low with ratios greater than 3.
Best soil conditions: when protected by a green cover	Without a green cover and especially with wet soil, trunk shakers cause soil compaction.
Best application of the shaking jaw	Mechanical damage to the trees and the fruit is very limited if the shaker is correctly operated. Damage to the tree may be caused by using improper vibrations, inadequate tightening of the jaw, imperfect orthogonality between the plane of solicitation and the trunk, or by coupling too close to the main branches (which reduces the effect of the vibrations and can cause breakage of branches) or too close to the soil (this can cause breakage of some small roots located near the trunk).
In case of bark damage: disinfectant treatment	The part of the tree most frequently damaged is the bark of the trunk in the gripping area. This damage can occur, above all, when, due to favourable weather conditions and/or abundance of irrigation, the trees are still in vegetative activity during the harvest period. When this damage occurs, a disinfectant treatment with copper products immediately after harvest is necessary.

In the reel system, the nets are spread under the crown and rewound by a reel attached to a tractor. The system consists of a rectangular frame equipped with two or four wheels, hauled by a tractor; it has two longitudinal rollers, around which two nets are wound. The nets are manually unrolled by four operators and rewound mechanically after the shaking operation. When nets are rewound, the olives are conveyed onto a conveyor belt that pours them into a bin; the frames may be equipped with a fan for removing the leaves. In addition to increasing the labour productivity, the mechanization of olive interception and recovery greatly improves the working conditions of the operators by reducing the fatigue connected with moving the nets and managing the olives.

8.3.5 Straddle harvesters

In recent years, there has been increased interest in super high-density olive groves mainly because of the possibility of using modified mechanical grapevine harvesters (straddle harvesters) to greatly increase the harvesting labour productivity. It is a case in which tree training and all of the concepts of cultivation are changing in order to adapt them to an already available harvesting machine. Grapevine harvesters have been adapted to super high-density olive groves by simply increasing the number of shaker bars.

Straddle harvesters are equipped with auto-levelling and antiskid systems to ensure stability even on sloping terrain. In a single machine, straddle harvesters combine both the olive detachment and interception operations.

The harvested olives are cleaned of leaves and twigs by means of a fan located above the two containers (about 1700 l each) for the temporary storage. The containers are unloaded by tilting into a trailer.

Straddle harvesters require super high-density olive groves using dwarf cultivars in the form of hedgerows with a planting density between 1500 and 2100 trees per hectare. The cultivars that have thus far given the best results in super high-density olive groves are Arbosana, Koroneiki and especially Arbequina, which have low vigour and high fruit-bearing capacity.

Tree size is important because grape harvesters can handle trees with a maximum of 2.5–3 m in height and 1.5 m in width, otherwise the trees could be seriously damaged. The fruit-bearing portion of the trees must be about 50 cm from the ground in order to suitably intercept the harvested fruit.

The system consisting of superintensive cultivation plus straddle harvester has two main advantages compared to traditional systems:

- A very high yield or productivity considering either the olive production per hectare (soil productivity) or the amount in kilograms of harvested olives per hour per operator (Table 8.3). Also, straddle harvesters have a high harvesting efficiency, allowing 90–95% of the fruit to be removed even when they are small and have a high detachment force.

Table 8.3 A comparison of labour productivity of different harvesting systems.

Harvesting and interception systems	kg of olives per person per hour
Hand pick – nets	10–20
Combs and shaking hooks – nets	40–50
Trunk shakers – net reel system – intensive olive grove	100–150
Trunk shakers – wrap round – intensive olive grove	200–400
Straddle harvesters – super intensive olive grove	1000–1500

- The second advantage is standardization of oil quality. The possibility of harvesting at such a high rate allows the harvesting of very large quantities of olives to be concentrated in a short period of optimal maturity.

The system consisting of super-intensive cultivation plus straddle harvester has two main disadvantages compared to traditional systems:

- The impossibility of enhancing the olive tree biodiversity. Super-intensive cultivation is interesting for massive production of a standard extra-virgin olive oil. The excellence due to variety, tradition and territorial link is lost.
- The high investment needed to implement the system, making it suitable only for large-scale companies. This point becomes even more evident considering the need for very high capacity olive mills operating in an almost-direct connection with the harvester, which also requires very large investments.

Comparing the cost of the choice between super-intensive cultivation/straddle harvesting and intensive cultivation/trunk shaker harvesting is the object of extensive discussion and research and is beyond the scope of this handbook. Many factors should be considered, such as the length of time to reach full production after planting, the length of the useful productive life of the trees and their resistance to parasites. It can be concluded that both systems have advantages and disadvantages depending on critical conditions, such as, the company's marketing policy and the labour availability and cost.

Giant straddle harvesters

Very large straddle harvesters are used to harvest the olive trees in super-intensive olive groves by passing over the rows. The vibrations are transmitted through batter-vibrating reels mounted on the side that hit the crown of the tree. The olives are intercepted and discharged by conveyors into trailers. These machines are very heavy, about 38 t. A machine of this category is the Colossus, which has a 4 × 4 m shaking cage. The results are promising but, owing to their size and cost, they can only be used on very large tracks of land.

Compared to the grape harvesters, these machines are more versatile with respect to the vigour of the trees. Often they are a stopgap solution in super high-density olive groves no longer manageable with harvesting machines due to excessive vigour.

Giant straddle harvester machines are very expensive (and suitable only for large companies). Transporting them from one olive orchard to another is problematic and the harvesting efficiency is low compared to grape harvesters since the transmission of vibrations in the inner parts of the canopy is reduced.

Coffee harvesters

These are similar, but smaller, than straddle harvesters. They have two vertical, cylindrical heads made of a plastic shaft with radiating fingers, approximately 1 m in length, which move over the canopy. The heads are mounted above a self-propelled platform and a catch frame and conveyor system. The shafts are subjected to vibrations, which are transmitted to the fruit-bearing shoots and cause detachment of the olives. Coffee harvesters need one operator to drive the catch-frame and another to operate the rotating picking heads. The machine can pick a tree in about 60 s.

If the fingers have good contact with all the bearing zones, the harvester can remove 90% of the fruit. This harvester performs best with trees no more than 4 m high (with the fruit-bearing portion 0.9 m above the ground) and 3.5 m in width.

Annex 8.1: Methods for olive maturity assessment

The decision about olive harvesting critically influences oil yield and quality. Optimal olive harvesting consists in obtaining the highest amount of oil of the desired/predefined level of quality. At the same time, as is clearly shown in Chapter 23, the cost of harvesting is the major cost factor in the production of extra-virgin olive oil.

The cost of harvesting is the same whether the olives are harvested at optimal maturity or at a stage involving yield or quality losses. As a consequence, harvesting may be considered the single most important decision determining the cost/value balance. This is why harvesting should be the result of careful planning, based on knowledge of the maturity pattern and the relationship between the maturity stage and the oil yield and quality.

The experimental approach suggested in this annex consists in a first phase of initial set up of the maturity evaluation system and a second phase of system implementation and refining.

First phase: select a reliable system of olive maturity evaluation and establish its relationships with oil quality and yield

The maturity evolution of the olives should be correlated with at least one direct, quantitative index and one or more indirect semi-quantitative indices. The first one, usually determined by experts, validates the semi-quantitative index, which should be simple, fast and cheap enough to be routinely used by trained workers.

A preliminary condition

As mentioned above, olive ripening is a scalar process. Not all the fruit on an olive tree ripens at the same time. Ripening is different on branches that are high or low or on different sides of the tree. Olives in the more shaded parts of the canopy have a lower content of total phenols and flavouring compounds. Not all the trees in an olive grove follow the same pattern of ripening, depending on the size of the tree, the position in the canopy, and the fruit-load. Therefore, the condition for a suitable evaluation of maturity should be carried out on a representative sample of the production of the whole olive grove.

In general, it is suggested that a random sample of about 0.5 kg of fruit from several trees of the same orchard be taken. The fruit should be selected from high and low branches of the trees and from all sides. Collect all the fruit from a little branch here and there rather than individual fruits; this helps make the sample more random.

Select a suitable, quantitative index of olive maturity

The index most closely matching these requirements is the oil content of the olives. There are several reasons for choosing this index:

- In the first place, the oil content of the olives can be determined very precisely and reproducibly by applying the method based on the Soxhlet extraction or Automatic Soxhlet Extraction (Soxtec 2013) or other suitable instrumental methods, such as near infrared reflectance (NIR).
- In the second place, the evolution of the oil content follows a quite standard pattern, independent of the cultivar or the climatic conditions. In the northern hemisphere, the fruit begins accumulating oil at the beginning of August. During the month of September and the first half of October the amount of oil increases at an accelerated rate due to the simultaneous increase in the fruit weight and of the percentage oil in the pulp. In the second half of October, the increase in the amount of oil slows down, tending to plateau at about mid-November, when the maximum content of oil is reached (varying between 15 and 22% on a fresh weight basis). After this, the fruit enters the over-ripened stage when the total oil on the tree does not change or starts decreasing slightly due to overripe olives falling to the ground. Depending on the climatic conditions, this evolution may start earlier or later, but the pattern is similar. This circumstance allows interpolation to be done with some consistency between available experimental values.
- The evaluation of the oil content has a direct relationship with the oil yield, which is one of the critical points of process control.
- Knowledge of the oil content of the olives allows the effectiveness of the oil mill extraction to be verified.
- Unlike other chemical indices, evaluation of the oil content is not affected by interference due to other constituents.

In practice, a given number of olives (100 as a minimum) are randomly selected out of the same sampling bucket, weighed, finely ground (pulp and pits) and frozen or a chemical preservative is added until lab analyses. Without any additional analysis, the following information can be drawn from the above procedure:

- the unit average weight of the fruit;
- the oil content on a fresh weight basis;
- the water content and the oil content on a dry weight basis.

Select a semi-quantitative index of olive maturity

The method based on the oil content, which must be carried out by specialists in laboratories, may be progressively replaced by a simpler method that could be applied in the field by nonspecialists. Many different methods have been successfully used.

Method based on olive skin and pulp colour: The colour of each fruit is evaluated by first observing the skin colour of the whole fruit and then by observing the colour migration inside the pulp after cutting a portion of the pulp with a sharp knife. Maturity values of each fruit are classified in eight categories (see Table 8.4).

Table 8.4 The eight categories of skin and pulp colour of olives during the ripening process.

Maturity value	Colour description
Zero	Skin colour deep green – fruit hard
1	Skin colour yellow-green – fruit starting to soften
2	Skin colour with less than half the fruit surface turning red, purple or black
3	Skin colour with more than half the fruit surface turning red, purple or black
4	Skin colour all purple or black with all white or green pulp
5	Skin colour all purple or black with less than half the pulp turning purple
6	Skin colour all purple or black with more than half the pulp turning purple
7	Skin colour all purple or black with all the pulp purple to the pit

The steps of the evaluation procedure are:

1. Start the Maturity Index evaluation before the beginning of harvest (for example, mid-September) and repeat it twice a week until the harvesting decision is made.
2. Randomly select 100 fruits out of the same sampling bucket and evaluate the colour characteristics.
3. The maturity value of each of the one hundred olives is determined according to Table 8.4 and the fruits in each category are counted.
4. The Maturity Index (MI) is obtained by multiplying the number of fruits in each colour category by the number of the corresponding maturity value, adding all the numbers together and dividing by 100 as follows:

$$MI = [(0\ x\ n_0) + (1\ x\ n_1) + (2\ x\ n_2) + (3\ x\ n_3) + (4\ x\ n_4) + (5\ x\ n_5) + (6\ x\ n_6) + (7\ x\ n_7)]/100$$

where: $n_0, n_1, n_2, \ldots . n_7$ = number of olives in each of the eight categories of the maturity value.

The MI based on the evaluation of colour increases with the ripening process.

Method based on pulp firmness: During ripening, the olive pulp becomes progressively softer due to partial hydrolysis of protopectin. The MI based on flesh firmness is a measure of the force (in N) needed for the tip of a penetrometer to penetrate through the skin and pulp. The diameter of the tip varies from 1.0 to 1.5 or 2.0 mm according to the range of variation in the firmness (the riper the olives and the softer their pulp, the larger is the tip diameter to be used).

The preparation of the sample is the same as described in the method for colour evaluation.

The MI value is calculated as the average of 100 measurements (the N sum resulting from 100 measurements divided by 100). Simple calculations are suggested by the penetrometer manufacturer in order to standardize measurements made with tips of different diameters.

The MI value based on the evaluation of olive firmness decreases with the ripening process.

Method based on fruit detachment force: The force needed for fruit detachment, measured with a simple dynamometer, decreases with fruit ripening following specific patterns in different cultivars. It is approximately 6 N at the beginning of ripening, around 4–5 N in the intermediate stage and it drops below 3 N at an advanced ripening phase.

The fruit detachment test is carried out in the orchard by detaching 100 olives or more from the trees selected with the same criteria as for the colour or pulp firmness evaluation. The MI value is calculated as the average of at least 100 measurements.

The MI based on the fruit detachment force is particularly useful when the olives are to be harvested mechanically by a trunk shaker, whose effectiveness in harvesting reaches a maximum value when the fruit detachment force is lower than 4 N.

Other methods: More complex methods can be used to evaluate fruit maturity including bench-scale extraction of the oil from small (1–2 kg) olive samples. These methods, which are mostly applied in research, allow different oil components (such as phenolic compounds) to be evaluated. These methods, however, are complex and costly and the oil extracted with bench-scale methods is not comparable with the oil obtained in commercial milling processes.

Find the optimal harvesting-maturity combination

During the first years of method testing and set up, harvesting is carried out at purposely spaced intervals and various combinations of the oil content and MI value. The two most basic parameters to be compared with the maturity indices are: the extraction yield and the oil sensory profile. Such a procedure should allow evaluation of which conditions of the oil content and the MI values are compatible or incompatible with the company's production goals.

Second phase: system implementation, simplification and refining

During this phase, which becomes routine practice after the initial years, the oil content determination is suspended and maturity is evaluated by a regular running of MI tests based on colour or pulp firmness or fruit detachment force. Data on extraction yield and sensory profile allow the method to be progressively refined. From time to time (for example, every 5 years) the oil content monitoring is repeated as a system validation practice.

References

Beltran, G., Uceda, M., Hermoso, M. and Frias, L. (2010) Ripening, in *Olive Growing* (eds D. Barranco, R. Fenández Escobar and L. Rallo), RIRDC, Canberra, pp. 147–170.

Gil Ribes, J., Lopez Gimenez, J., Blanco Roldán, G.L. and Castro García, S. (2010) Mechanization, in *Olive Growing* (eds D. Barranco, R. Fenández Escobar and L. Rallo), RIRDC, Canberra, pp. 393–447.

Soxtec (2013) *Automatic Soxhlet Extraction*, http://www.cyberlipid.org/extract/extr0010.htm (accessed 25 September 2013).

Tous, J. (2012) Olive production systems and mechanization. *Acta Horticulturae* **924**, 169–184.

9

Olive handling, storage and transportation

Primo Proietti
Department of Agricultural, Foods and Environmental Sciences, University of Perugia, Perugia, Italy

Abstract

Avoiding mechanical damage and controlling time-temperature relationships are key factors for satisfactorily handling and storing olives during the period from harvesting to milling. A ten-point list of handling and storage practices is proposed as well as a semi-log diagram for the choice of suitable time-temperature relationship. The criteria for optimizing the harvesting-milling link are discussed.

9.1 The autocatalytic nature of olives and oil degradation

The handling and storage of olives in the period between harvesting and milling is critical for oil quality. Improper conditions may trigger a chain of degrading reactions whose autocatalytic mechanism is such that degradation increases at an accelerated pace. In fact, some products of oil degradation such as monoglycerides and peroxides become the catalysts for further degrading reactions. Figure 9.1 shows a flow-chart of olive degradation and its effect on oil quality.

The two critical conditions for satisfactory olive handling and storage before milling are:

- avoid mechanical damage to the olives, and
- control the time-temperature relationship.

9.2 Avoid mechanical damage to the olives

Crushing and bruising of olives cause cellular structures to break and hence mixing of the oil, which is originally stored in vesicles (cell organelles called spherosomes)

The Extra-Virgin Olive Oil Handbook, First Edition. Edited by Claudio Peri.

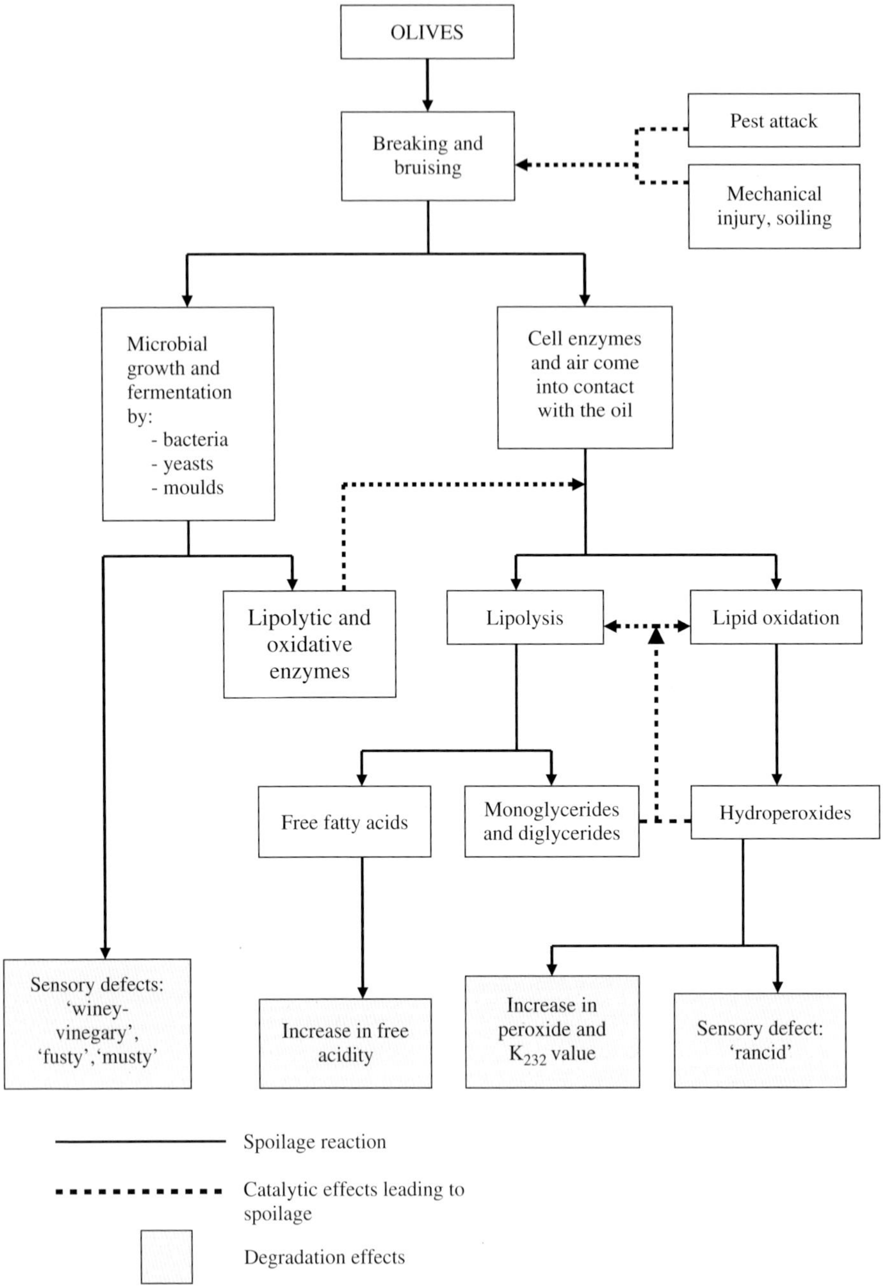

Figure 9.1 The mechanisms of olive spoilage and oil degradation.

with other cellular components, especially hydrolysing and oxidizing enzymes. In these conditions, in the presence of oxygen from the air, oil degradation readily starts. Due to the presence of sugars in the cellular juice, bacterial and yeast fermentation may also take place with formation of ethanol, acetic acid and ethyl-acetate causing the 'winey/vinegary' sensory defect.

Fermentation is an exothermic phenomenon, so it also causes an increase in temperature, which further favours microbial growth and fermentation. The overall result of olive fermentation and heating is the so-called 'fusty' defect.

When storage under improper conditions goes on for days, mould growth, generally *Penicillium* and *Aspergillus* species, causes a massive production of degrading enzymes and the appearance of the 'musty', 'humid', 'earthy' sensory defects.

For oil transportation and storage, rigid plastic containers should be used (15–20 kg crates or 200–300 kg bins) with olives in layers not thicker than 30 cm. Thick layers cause crushing of the olives in the bottom layers. Crates and bins for olive transportation and storage must have ventilation holes and should be equipped with a foothold for the overlay, to allow good air circulation (Figure 9.2). Trucks for olive transportation in bulk with mechanical or hydraulic discharge can be used only if the period of time from harvesting to milling is a matter of very few hours Figure 9.3.

Strictly avoid putting the olives on the floor or in sacks.

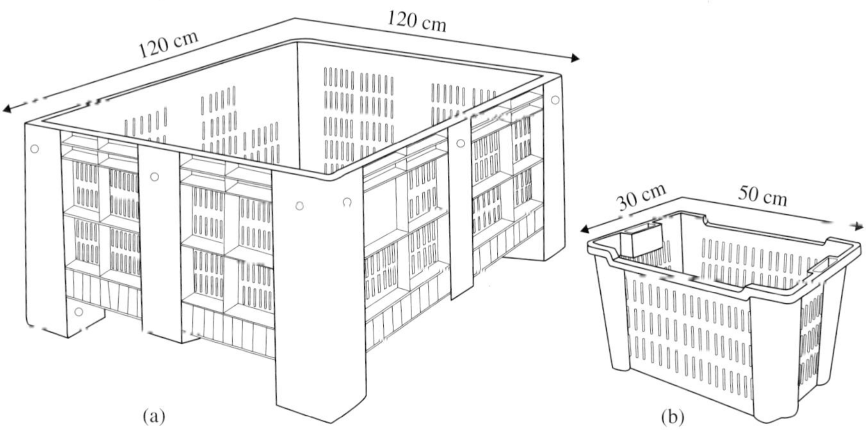

Figure 9.2 Plastic containers for olive handling, storage and transportation: (a) 200–300 kg bin to be handled by forklift trucks. (b) 15–20 kg crates for manual handling.

9.3 Control the time-temperature relationship

In general, as a rough estimate, we may say that the rate of degradation reactions increases with temperature according to an exponential relationship, while the overall degradation effect is proportional to time. A frequent way to represent

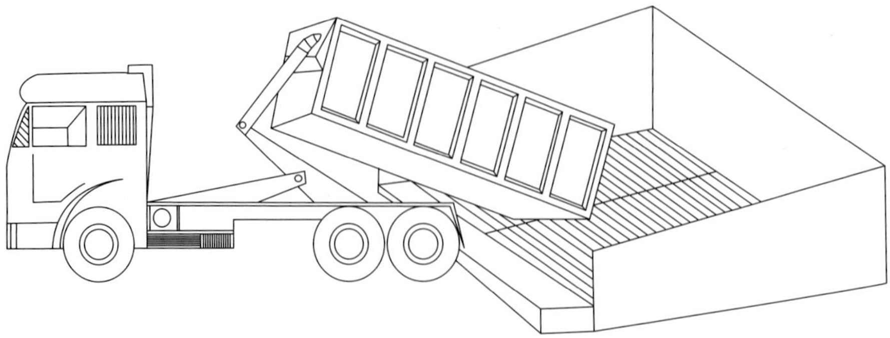

Figure 9.3 Truck for olive transportation in bulk with mechanical or hydraulic discharge at the mill's discharge hopper.

time/temperature relationships of technological operations is in a semi-logarithmic graph, with temperature on the abscissa in a linear scale and time on the ordinate in a logarithmic scale. In such a plot, time-temperature conditions that determine equal degrading effects are represented by oblique lines. Figure 9.4 shows the

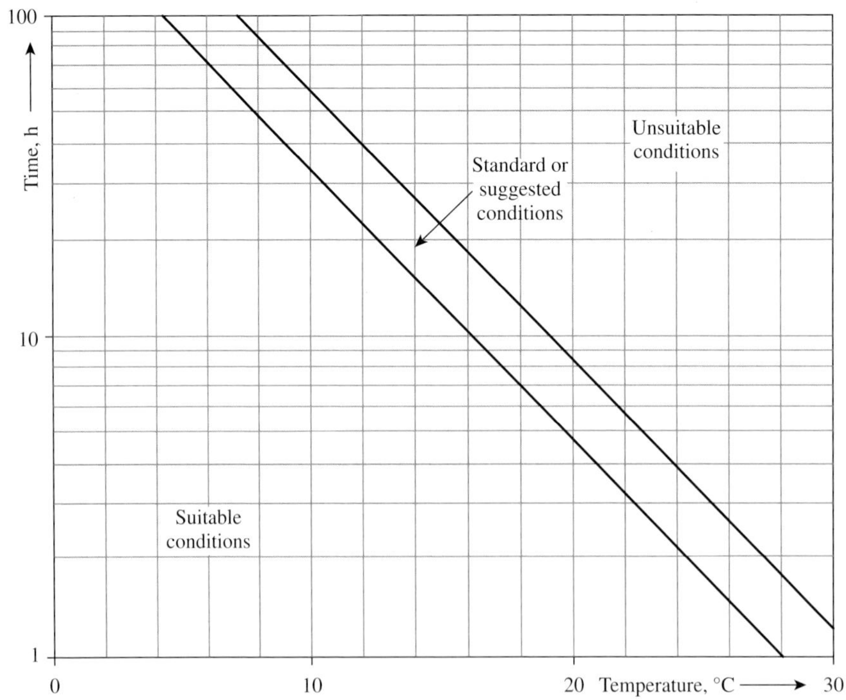

Figure 9.4 Time-temperature relationship of olive-storage conditions in the period from harvesting to milling. *Source*: Peri, C. (2013). Reproduced with permission from Wiley.

time-temperature relationship of the storage of healthy olives. The graph is the result of five years of field tests carried out by Association 3E for the production of super-premium olive oils (Peri *et al.* 2010; Peri 2013).

The temperature scale varies from zero to 30 °C, while the time scale varies from 1 to 100 hours. The two oblique lines represent the border area for satisfactory olive storage. They divide the diagram into two parts: the time-temperature conditions in the lower area are suitable, while those in the upper area are unsuitable for good olive storage. Between the two lines, standard conditions can vary in relation to the cultivar and the degree of maturity of the olives.

It can be seen, for example, that it is advisable not to exceed a holding time of 1 h at temperatures of 29–30 °C, while at 5–7 °C, storability can be extended up to 4 days. At 15 °C the olives can be safely stored up to about 20 h and at 10 °C up to 50 hours.

In order to meet the recommended conditions of olive handling and storage, the decalogue in the box should be considered.

Postharvest handling and storage of olives

1. Closely control temperature according to Figure 9.4.
2. Avoid olive breaking and bruising during harvesting by using appropriate harvesting machines and mechanical facilitators (see Chapter 8).
3. Avoid contact between olives and soil in order to avoid microbial contamination, especially moulds.
4. Avoid mixing good and healthy olives with broken or mouldy ones.
5. Avoid mixing fly-infested olives with healthy ones. The flies' larvae are a super-effective dispenser of degrading enzymes and micro-organisms in the olive pulp.
6. For olive transportation and storage use rigid plastic containers with ventilation holes and layers of olives not thicker than 30 cm.
7. Avoid exposure of harvested olives to the sun or rain. Rain on stored olives not only favours microbial growth but also acts to inoculate and propagate micro-organisms. Storage at the grove or at the mill should therefore not be in the open air but in a covered, well-ventilated space. Relative humidity in the storage environment should not exceed 80%.
8. All the above points become more critical as olives ripen because the mechanical resistance of the olive skin and pulp decreases.

9. A critical control step at the mill reception area should be visual evaluation of the integrity of the olives so that mixing of good and bad olives is carefully avoided.
10. The olive storage area should be clean and free of atmospheric contaminants especially solvent vapours, smoke, exhaust, fusty smell, and so forth.

9.4 Management of the harvesting-milling link

It is often recommended that olives be milled 'in the shortest time possible' after harvesting. This is scientifically inaccurate because the time-temperature relationship, as presented in Figure 9.4, is not taken into consideration. It is also misleading in process management terms as it frequently leads to irrational procedures if applied in a rigid or too restrictive way.

The problem that arises is that of rationalization of the harvesting-milling link. Trying to mill the olives immediately after harvesting creates a rigid connection between the harvesting operation and that of milling, with the consequence of organizational inconsistencies and diseconomies of scale. Micro-mills with a very low hourly working capacity are becoming common. They operate in direct connection with the harvesting by adapting the rate of milling to the rate of harvesting. The labour costs are high and it becomes impossible to carry out the milling process under steady, well-controlled conditions.

Instead, if the olives are handled and stored under proper conditions until suitably sized batches are obtained, they can be milled in a short time in larger-sized mills, in standardized conditions and with more reliable qualitative and quantitative results. Optimizing this step of olive handling and storing before milling is critical for optimal sizing of the mills.

References

Peri, C., Kicenik Devarenne, A. and Pinton, S. (2010) 3E Super-Premium selection for extra-virgin olive oil. *Beyond Extra-Virgin, The Fourth International Conference on Olive Oil Excellence*, organized by Association 3E (Milan, Italy), the Academy of Georgofili (Florence, Italy), The Culinary Institute of America (St Helena, California) and the Olive Center of the University of California Davis, Verona, 22 September 2010.

Peri, C. (2013) Quality excellence in extra-virgin olive oil, in *Olive Oil Sensory Science* (eds E. Monteleone and S. Langstaff), John Wiley & Sons, Ltd, Chichester.

10
Olive cleaning

Claudio Peri
University of Milan, Milan, Italy

Abstract

Olive cleaning is carried out in two steps. In the first 'separation' step, particulate foreign materials are removed by sifting, vibrating screens and air blowing of leaves. In the second 'washing' step olives are shaken into a washing basin and finally rinsed with clean water. Suggestions are given concerning the rinsing step, the daily plant cleaning, the frequency of water replacement and the control of water temperature.

10.1 Introduction

The aims of this operation are twofold: (i) removing leaves, as well as pieces of wood, small branches, stones, damaged olives and other particulate material that may have been accidentally collected with the olives, and (ii) washing dust and soil from the olives.

Particulate material should be separated from the olives by mechanical and pneumatic action, whereas removal of soil requires washing with water. The cleaning operation is therefore carried out in two sections: the separation and the washing sections.

10.2 The separation section

Figure 10.1 presents the functional scheme of the separation section.

The separation of olives from particulate foreign material is carried out in three steps:

1. In the first step, the olives are fed into a rotating cylindrical sifter with chinks or holes larger that the size of the olives. Two actions take place: (i) a shearing action causing the olives to be pulled off the small branches and (ii) separation

The Extra-Virgin Olive Oil Handbook, First Edition. Edited by Claudio Peri.

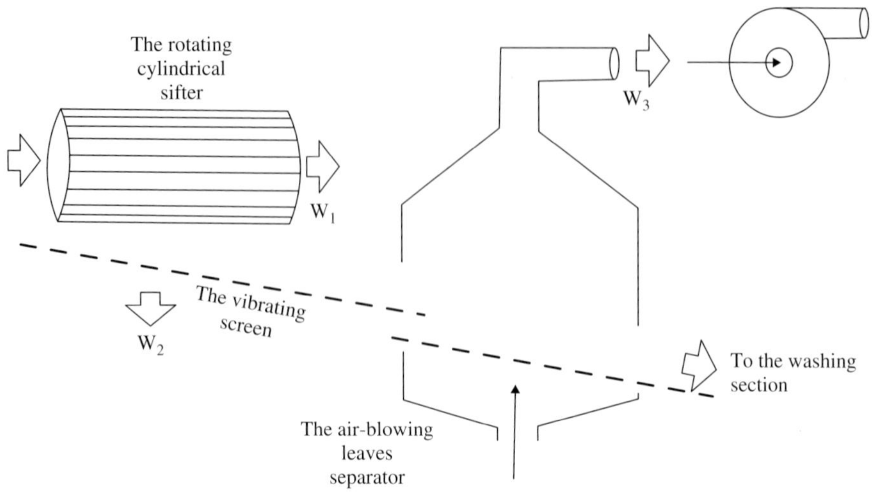

Figure 10.1 Olive cleaning: the separation section.

of the olives through the holes of the sifter, while larger material, such as small branches or stones, is retained inside the sifter and discharged as the first dry waste (W_1 in Figure 10.1).

2. In the second step, the olives pass over a vibrating screen with openings smaller than the olives. In this step, small particulates including soil, pebbles, broken olives, pass through the screen and are discarded as the second dry waste (W_2).
3. In the third step olives slide over an inclined screen while air is blown across them. Particles that are lighter than the olives, leaves in particular, are swept away and finally discarded as the third dry waste (W_3).

In some plants, the order of the three steps may be different, but the aim and the result are the same. There is an increasing interest in carrying out the separation step at the olive grove, as a pre-processing step. Such an arrangement allows the dry waste to remain in the field and do not accumulate at the olive mill site.

10.3 The washing section

Figure 10.2 presents the functional scheme of the washing section.

The washing of the olives is carried out in two steps:

1. In the first step, olives are dumped into a washing basin where they are vigorously shaken by bubbling air. This action facilitates the wetting and removal of dust and soil. The olives are pushed through the washing basin by the incoming

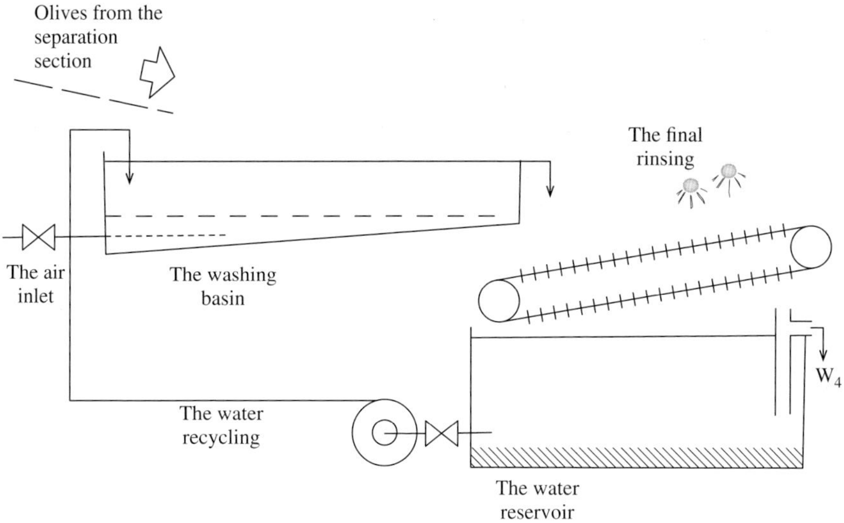

Figure 10.2 Add outlet W_4 including the arrow.

flow of the recycled water and overflow with the water at the opposite end of the basin.

2. In the second step, the washed olives, still wet with dirty water, are transferred through a draining belt under sprays of clear potable water for the final rinsing. The clean wet olives are finally transferred by a belt elevator to the milling section.

A reservoir underneath receives both the dirty water overflowing from the washing basin and the final rinsing water coming from the draining belt. The reservoir acts as a sedimentation tank in which the heavier soil particles settle to the bottom, while a centrifugal pump recycles the water back to the washing basin.

The addition of rinsing water (at a flow rate roughly 10% of the weight of the olives) causes a continuous overflow of dirty water from the sedimentation tank, which is discarded as waste water (W_4).

10.4 Control points

The critical points of this operation are:

- water replacement and plant cleaning
- rinsing
- water temperature.

Water replacement and plant cleaning

Recycling of the rinsing water causes partial replacement of the dirty water with relatively clean water. This is not sufficient to guarantee an acceptable level of water cleanliness. A periodical (daily at least) total replacement of the dirty water with clean potable water is needed. At the same time, the entire washing section should be cleaned: removal of sediment from the washing basin and the sedimentation tank, washing of the draining belt, the recycling pump and piping, and the spray rinsing nozzles. The separation section must also be cleaned at a similar frequency, in particular the vibrating screens and the air blowing circuit.

The rinsing step

One of the lesser understood points is the essential role of the final rinsing of the olives. Observing the olives coming out of the washing basin, they appear to be wet and brilliant as if they were perfectly clean. In fact, they are wet with dirty water and therefore far from being really clean. The dirty water often contains Fe (iron) and Cu (copper) ions, which are strong catalysts of oil oxidation. The purpose of the rinsing step is precisely to replace the dirty water wetting the olive surface with clean, uncontaminated water. This necessitates a suitable number of sprays over the entire mass of olives and on each single fruit while it is being transferred over the draining belt.

The water temperature

In the northern hemisphere, olive harvesting takes place in the winter, mostly from late October to early December, when the atmospheric temperature is often lower than 10 °C. Tap water in this period also has a low temperature (12–14 °C). If washing with cool water is carried out on even cooler olives, two problems may occur: lower effectiveness of the washing operation and, secondly, a low temperature at the milling operation, resulting in a low temperature of the olive paste at the beginning of the malaxing operation. The optimal range of temperature in the malaxing operation is 24–27 °C.

A good solution is to use water at a temperature of 20–24 °C for washing the olives. This allows more effective washing and better control of the temperature conditions during the following milling-malaxing-decanting sequence of the process.

Such a temperature can be easily obtained by circulating water at a suitable temperature in a coil heat exchanger in the washing basin or in the decantation tank.

11

Olive milling and pitting

Alessandro Leone
Department of Science of Agriculture, Food and Environment, University of Foggia, Foggia, Italy

Abstract

The operating mode and performance of various olive milling and pitting machines are presented. In particular: (i) single and double-grid hammer mills; (ii) disc mill; (iii) stone mill; (iv) total and partial pitter mills. Pressure (with impact, shock and percussion mechanisms) and shearing action reduce the olive pulp to a very fine size in order to obtain the most complete release of the oil from the vacuoles. At the same time, the pits are reduced to relatively coarse fragments, guaranteeing an effective draining network. The effects of milling on oil quality are discussed.

11.1 Introduction

The aim of milling is to reduce the olives to a homogeneous paste by breaking the pits, skin and pulp cells and the vacuoles containing tiny droplets of oil. The oil flowing freely from the vacuoles can then be separated from the water and the solid constituents. The olive paste is a semiliquid mixture of two different types of solids (rigid pit fragments and soft fleshy parts from the pulp and skin) and two types of immiscible liquids (water and oil). This operation is important for both oil quality and yield.

Assuming the size of a pulp cell to be about 20 μm thick, and the thickness of the pulp from the skin to the pit to be about 6 mm (6000 μm), about 300 layers of cells are present in the pulp, each of them containing oil vacuoles. This is why the pulp must be crushed very finely in order to facilitate release of the oil.

Milling also breaks up the hard woody pits. The pit fragments make a rigid framework in the olive paste, which facilitates draining and separation of the liquid from the solid components. During malaxation, pit fragments also contribute to friction and cutting effects on cells, thus facilitating oil release and coalescence. Conversely, milling the pits too finely may generate powder, causing clogging of the decanter and reducing the extraction yield.

The Extra-Virgin Olive Oil Handbook, First Edition. Edited by Claudio Peri.

A mass balance of olive and olive paste

Table 11.1 presents an average breakdown of the main structural components of the olive.

The oil in the kernel represents about 20% of its weight, corresponding to about 1 kg of oil per 100 kg of olives, which is not a negligible amount. Even considering that only 70 to 80% of the oil in the kernel is extracted, this may represent about 5% of the total extracted oil. This oil fraction is interesting not only in terms of yield, but also in terms of quality because the oil in the kernel is rich in linoleic acid, an essential fatty acid.

After the milling operation, a homogeneous paste is obtained with an average composition as reported in Table 11.2. The importance of rigid solids is evident. Fragments of woody pit shells make up more than 20% of the olive paste; they create a rigid network that facilitates draining and separation of the water and oil in the decanter.

Table 11.1 An average mass balance of the olive structural components.

Olive component	kg/100 kg of olives
Skin	2
Pulp	72
Pit	26
Woody shell	*21*
Kernel (the seed)	*5*

Table 11.2 An average mass balance of olive paste.

Average composition of olive paste	Percentage by weight
Water plus water-soluble solids (the vegetation water)	57
Oil (pulp plus kernel oil)	15
Insoluble solids, with:	28
Rigid solids from pit shells	*21*
Soft solids from pulp	*7*

The effects of pressure and shear forces

Milling must satisfy two contrasting requirements: on the one hand, reduction of the pulp to a very fine size in order to obtain the most complete release of the oil from the vacuoles; on the other hand, reduction of the size of the pits to relatively coarse fragments in order to guarantee an effective draining network.

From a mechanical point of view, milling is the result of two different actions: pressure and shearing. Pressure is a force that is applied to a body in an orthogonal direction, whereas shearing is a force that is applied in a tangential direction.

Shearing causes efficient size reduction of soft pulp tissues with a low consumption and dispersion of energy, so that overheating does not take place. Pressure, instead, is necessary for milling the olive pits through a combined impact, percussion and shock mechanism, with a high dispersion of energy and significant overheating of the olive paste. The relative intensity of pressure and shearing forces and their effects depend on the design and rotating speed of the crusher. The effects of milling on oil quality are briefly outlined in the box.

The milling–quality relationship

An intense milling action causes a significant size reduction of the tissues and disruption of cellular material. Phenolic compounds are released to a greater extent and enzyme activities are triggered with formation of oleuropein aglycones that are partially soluble in oil. As a consequence, a more intense milling action results in an oil with greater bitterness and pungent characteristics and a higher content of phenolic antioxidants (Inarejos-Garcia *et al.* 2011).

It is evident that an intense milling action should be applied to cultivars with a low phenolic content, whereas a milder milling action is suitable for cultivars that have a high phenolic content. Doing the opposite may result in an oil with a flat sensory profile in the first case and an oil with excessive bitterness and pungency in the second.

The intensity of the milling action should therefore be considered as an important control parameter of the extra-virgin olive oil process. In modern olive oil factories, milling intensity is controlled through variable rotating speed mills and sometimes also by alternating different types of milling mills (as, for example, hammer and disc mills) or alternating milling and pitting mills, depending on cultivar and ripeness of the olives.

11.2 Milling machines

The characteristics and performance of the three most common milling machines – hammer, disc, and stone mill – are compared in Table 11.3 (Amirante *et al.* 2010, Earle and Earle 2013).

11.2.1 Single-grid hammer mill (Figure 11.1)

A single-grid hammer mill consists of an outer casing enclosing a cylindrical body with grid walls. The diameter of the grid holes is variable depending on the desired

Table 11.3 Comparison of the most common olive mills.

Type of crusher	Operating mode, hourly capacity (kg/h)	Rotating speed (rpm)	Feature controlling the dimension of paste particles	Mechanism of action	Power of the electric motor (rotor) (kW)	Intensity of mechanical action	Risk for oil quality and yield
Hammer mill – single grid	Continuous, 2000–7000	2500–2900	Diameter of holes, 6–8 mm	Prevailing pressure	Up to 50	high	Heating, oil emulsion
Hammer mill – double grid	Continuous, 2000–4000	1300–1500	Diameter of holes: 9–11 mm 1st grid, 6–8 mm 2nd grid	Balanced pressure and shear	Up to 30	medium	negligible
Disk mill	Continuous, 2000–3000	1300–1500	Distance between discs, 2–4 mm	Balanced pressure and shear	Up to 20	medium	negligible
Stone mill	Discontinuous, 600–2000	10–15	Space between millstones and base, 1–5 mm	Prevailing pressure on pits and shear on pulp	Up to 10	low	Oxidative degradation

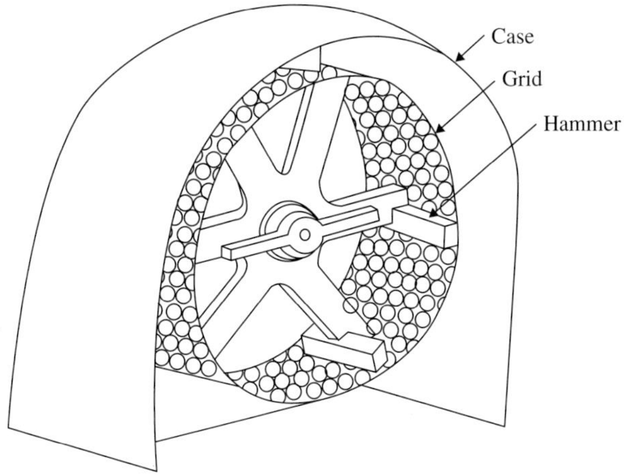

Figure 11.1 The single-grid hammer mill. The illustration shows the inside of the mill as it would appear when removing the case wall on the feeding side. In fact, olives are fed through a central opening in the missing wall by a screw conveyor, which operates in a direction perpendicular to the rotating hammers. The olive paste coming out of the grid drops by gravity under the mill into a screw conveyor that feeds the malaxing machine.

reduction in size: finer pastes are obtained from grids with smaller holes. Inside the grid there is a rotor bearing three to six spokes with steel plates (hammers) at the ends. The steel plates may be fixed or mobile on the arm. The rotor is mounted on an axis with an electric motor of variable power depending on the working capacity required. The perforated grid can be fixed or contra-rotating.

Olives enter via a small screw feeder operated by a low-powered electric motor. During the operation, as the olives enter in a central position, the hammers turn and collide violently with the fruit, breaking up the pits and partially tearing the pulp. Centrifugal force pushes the mass towards the walls of the grid, forcing it out through the holes with an intense shearing action that further reduces the size of the pulp and cellular components.

The external case collects the paste as it comes out over the entire circumference of the grid and conveys it to the transfer system that feeds the malaxer.

11.2.2 Double-grid hammer mill

A double-grid hammer mill consists of two concentric cylindrical grids with different hole diameters and a single rotor fitted with two series of spokes and hammers.

The olives are fed in centrally and pushed through the first grid with the first set of rotating hammers (first stage). Centrifugal forces push the paste out of the holes of the first grid, where it is further crushed by the second set of hammers and forced through the second grid with smaller holes (second stage). The pulp coming out of the second grid is conveyed to the transfer system that feeds the malaxer.

In summary, in the double-grid hammer mills, a progressive size reduction is obtained, with more homogeneous-sized particles and, at the same time, less energy dispersion and heating effects. In a double-grid hammer mill emulsification and oil heating are lower compared to the single-grid mill.

11.2.3 Disc mill (Figure 11.2)

The disc mill crushes olives by the action of a mobile-toothed disc mounted in axis with the electric motor and a fixed-toothed disc with a central hole. The teeth on each disc are in concentric circles at different distances from the centre of rotation, so that the concentric circles of teeth on one disc occupy the free space between the concentric circles of the teeth of the other disc when the discs are placed together. Both discs are enclosed in a steel casing that conveys the pulp to the malaxer. In the working position, the two discs are opposite each other so that the movable disc teeth skim those of the fixed ones. The olives are fed in through the central opening of the fixed disc, so that they occupy the free spaces between the teeth of the two discs. The centrifugal forces created by the rotation of the mobile disc push the olives across the discs in a radial direction towards the edges. The olives are forced into a narrow space between the teeth, thus milling the pits and lacerating the pulp. The paste is thrown off the edges of the discs, collected by the containment case and conveyed to the malaxer.

The fineness of the paste may be controlled by adjusting the distance between the fixed and mobile disc: the closer the distance between the two discs, the finer the paste.

A possible problem with the crusher is that teeth may break due to wear or to the accidental presence of pieces of rock or metal entering the machine with the olives. Broken teeth may then cause damage to other machinery in the production cycle.

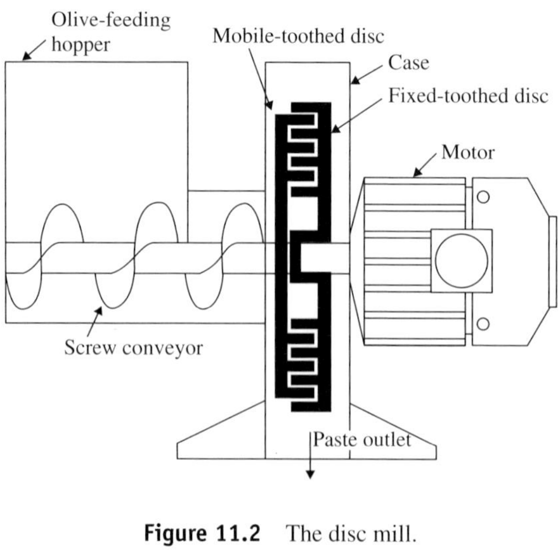

Figure 11.2 The disc mill.

Pits are broken up by the action of the shear teeth on the opposite teeth during rotation, while the cell tissues are torn by the dual shear action of the teeth and rubbing of pit fragments on the pulp tissue.

Shear forces release less energy than pressure and impact forces in a hammer mill, and this results in less overheating than the hammer mill, and a lower extraction of phenolic components. Oils obtained with disc mills are therefore less bitter and pungent than oils obtained with hammer mills. Oil emulsion is also minimized.

11.2.4 Stone mill (Figure 11.3)

The stone mill is the oldest type of milling machine. It consists of a circular granite base resting on 3 or 4 metal bearings and a steel casing. A vertical metal shaft at the centre of the granite base is connected to an electric motor by a gear reduction. Two to four circular granite millstones are connected to the vertical shaft by means of semi-axes.

Each millstone is at a different distance from the central rotating shaft, so that their combined rolling/milling action covers the entire radius of the base. The distance between the millstones and the base is 1 to 5 mm.

The operation is discontinuous, with loading of the olives, milling and subsequent discharge of the paste. The paste is discharged by the rotation of the system and the lateral thrust of blades designed for this purpose. Milling time ranges from 15 to 20 min.

The fineness of the paste can be varied by varying the distance between the millstones and the granite base; the shorter the distance, the finer the paste obtained.

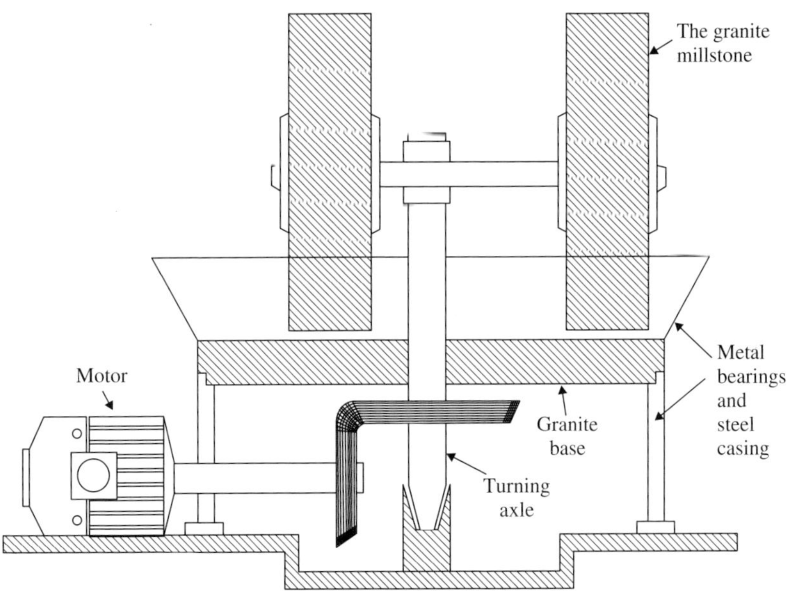

Figure 11.3 The stone mill.

The rolling action of the millstones over the stone base breaks down the pits, while the size reduction of the pulp is due to shear effects between the millstone and the base, aided by the presence of pit fragments.

Compared to the continuous mechanical crushers such as hammer and disc mills, stone mills have some advantages and disadvantages (Veillet *et al.* 2009).

The main advantage of stone mills is the perfect balance between pressure, which causes pits to break into coarse particles, and shear action due to the rotating stones and pit fragments, which causes reduction of pulp tissues and cells to a fine size. In these conditions and considering the low rotating speed of the millstones, oil homogenization and overheating do not take place. The mild milling action also results in lower extraction of phenolic compounds and hence in less bitter and pungent oils. Unlike continuous mechanical crushers, mixing and kneading help to start the coalescence of oil droplets, thus facilitating the following malaxation operation.

The main disadvantages of stone mills are: (i) long exposure of the olive paste to air and oxygen with possible oxidative degradation of the oil, and (ii) exposure to environmental contamination. Most of all, (iii) discontinuity is a serious limitation in terms of working capacity, control and standardization of operating conditions and manpower cost. These disadvantages surpass the advantages and therefore stone mills are rapidly disappearing from modern olive mills.

11.3 Pitting machines

The technique of pitting olives in order to produce oil from the pulp of the olives has been practiced since ancient times. In his agricultural treatise *De Re Rustica*, the Roman agronomist Columella writes that the most valuable oils for the king were made from depitted olives.

Pitting consists in removing the olive pits and in reducing the pulp into a paste of suitable fineness, fluidity and homogeneity. Pit removal is carried out with continuous pitter mills, which eliminate all pits (total pitter mill) or part of them (partial pitter mill) (Amirante *et al.* 2010).

In the absence of the rigid network due to the pit fragments, the soft solids from the pulp have a high tendency to clog, which may hinder the satisfactory separation of the oil. This problem is now almost completely overcome by closely controlling the fluidity and water content of the pitted paste so that the oil from the pulp can be better separated by malaxation and the centrifugal action of decanters.

11.3.1 Total pitter mill

The total pitter separates all pits from the pulp and skin tissues. It consists of a horizontal perforated drum with 4–6 mm holes, with an internal coaxial rotating shaft with spokes bearing rubber-coated metal rods. During rotation at about 700–800 rpm the rubber-coated rods touch the walls of the drum. The drum is surrounded by a casing and a tank below it for collecting the pulp.

The olives are fed by a screw feeder at one side of the cylinder, and are forced to rub against the outer drum by the thrust of the rubber-coated rods. The combined actions of acceleration and friction crush the pulp and break it gently away from the pit. This process involves negligible temperature increments.

The pulp exits through the holes in the drum and falls into the external collection tank, while the rotating shaft pushes the pits out of the side of the grill opposite to the feeder where they are extracted by a screw. The hourly capacity is about 2000 kg/h of olives.

11.3.2 Partial pitter mill

The partial pitter mill has two sections: one is for milling of both pits and pulp, while the second partially separates the pit fragments.

The milling section consists of two counter-rotating toothed rollers, one at about 70 rpm, and the other at about 140 rpm: pits are crushed and pulp tissues are partially torn. The paste obtained is fed into the second section by a screw. This section, which is similar to that of the total pitter mill, consists of a horizontal cylindrical perforated stationary drum, with small holes (2.5–3.5 mm) and a rotating reel with rubber-coated rods.

The paste is fed from the first to the second drum where the rotating shaft (700–800 rpm) forces it to rub on the perforated walls of the drum pushing it out of the holes. The size of the holes is such that some stone fragments exit with the paste, while others stay in the grill and are discharged from the opposite end of the feeder. The hourly capacity is 2000–6000 kg/h of olives.

This machine makes it possible to download pit fragments in a variable proportion from zero (the result is similar to that of a mill) to 100% (the result is similar to that of a total pitter). Generally 50–70% of pit fragments are extracted.

A common limitation of pitter mills (total or partial) is that the paste produced is not homogenous enough for an optimum malaxation. Therefore, it is sometimes necessary to use a mechanical crusher as a finisher in order to obtain a more homogeneous paste.

In pitting operations two problems should be avoided: first, incomplete separation of pulp from the pits. Vertical streaks of pulp in the pit are an evident sign of improper operation. The second problem, mainly with partial pitters and olive cultivars with pits with low mechanical resistance, is the formation of pit dust with clogging effects in the decanter and consequent reduction of the oil extraction yield. This is obviously also a problem with hammer mills.

Pits can also become more fragile at the more advanced state of olive maturity. Pitting only works properly with varieties of olives with hard pits and at an early stage of maturity.

Total pitting may cause problems in the first stage of malaxation and in decanter extraction. In malaxation, the absence of pit fragments reduces the effect of breaking cell walls and vacuoles. This problem can be partially overcome by increasing the malaxation time of pitted pastes.

In decanter centrifugation, the absence of pit fragments causes a reduction of draining effects, making separation of liquids and solids more difficult. In this case, it is necessary to reduce the flow rate of the decanter and the differential speed between the bowl and screw conveyor, leading to reduction in working capacity.

Despite possible adjustments of the decanter, the extraction yield of plants using totally pitted olive pastes is always slightly lower than that obtained using paste containing a network of pit fragments.

This problem is avoided with partial pitter mills. These have, in fact, a performance similar to traditional mills, but with a controllable proportion of rigid solids and a milder action on pulp tissues.

References

Amirante P., Clodoveo M.L., Tamborrino A. *et al.* (2010) Influence of the crushing system: phenol content in virgin olive oil produced from whole and de-stoned pastes, in *Olives and Olive Oil in Health and Disease Prevention* (eds V.R. Preedy and R.R. Watson), Academic Press, London, pp. 69–76.

Earle, R.L. and Earle, R.D. (2013) Size reduction, in: *Unit Operations in Food Processing*, www.nzifst.org.nz/unitoperations (accessed 26 September 2013).

Inarejos-García, A.M., Fregapane, G. and Salvador, M.D. (2011) Effect of crushing on olive paste and virgin olive oil minor components. *European Food Research and Technology* **232** (3), 441–451.

Veillet, S., Tomao, V., Bornard, I. *et al.* (2009) Chemical changes in virgin olive oils as a function of milling systems: Stone mill and hammer crusher. *Comptes Rendus Chimie* **12** (8), 895–904.

12

Olive paste malaxation

Antonia Tamborrino
Department of Agro Environmental and Territorial Sciences,
University of Bari, Bari, Italy

Abstract

Complex physical and biochemical phenomena take place during malaxation with critical effects on the extraction yield as well as on the nutritional and sensory quality of the oil. The optimal range of time-temperature relationships is shown in a semi-log graph. The subjects of temperature and residence time control are thoroughly discussed. It is suggested that the olive paste be heated in a scraped-surface heat exchanger (SSHE) feeding the malaxer so that the separation of the heating from the holding temperature functions allows a more precise control of time-temperature relationships. Control of residence times requires that malaxers are operated in parallel and that automatic washing takes place after each malaxing operation.

12.1 Basic phenomena in malaxation

The aim of malaxation is to make oil separation easier in the subsequent centrifugation steps. In modern extra-virgin olive oil processing, malaxation is the only discontinuous operation in a series of continuous operations. It is preceded by the continuous process of washing and milling and is followed by the continuous process of decanter and centrifuge separation. This discontinuity of malaxation derives from the fact that the transformations taking place in this operation are rate-limited and cannot be effectively carried out in too short a time. If we assume that the overall time for processing extra-virgin olive oil, from milling the olives to the final centrifugal finishing, is about one hour, two-thirds of it is required for malaxation.

The importance of malaxation for separating the oil from the olive paste was probably obvious from the very beginning of olive oil processing, millennia ago. Those ancient people must have observed that a slow mixing of the olive paste caused a

The Extra-Virgin Olive Oil Handbook, First Edition. Edited by Claudio Peri.

progressive formation of an oil phase and its rising to the surface of the paste slurry. It was also probably found that adding a little bit of water accelerated and improved the oil separation. This phenomenon, which is called 'coalescence', is essential for oil separation even today, when very effective separation techniques, such as decanter centrifugation, are applied.

At the same time, with advances in knowledge and science, it is understood that the effect of malaxation is not only coalescence of the oil but profound physical and chemical changes that are responsible for giving the oil its main nutritional and sensory characteristics (Servili *et al.* 2003, 2008; Kalua *et al.* 2006; Gómez-Rico *et al.* 2009; Tamborrino *et al.* 2010; Stefanoudaki *et al.* 2011; Taticchi *et al.* 2013). In summary, it can be asserted that the real transformation of olives into olive oil is the result of molecular changes that take place during malaxation. Table 12.1 presents the main phenomena associated with malaxation.

The connections among these various phenomena are very complex. Some of them are roughly schematized in Figure 12.1.

The paste fed to the malaxer is a solid-water medium in which small oil droplets, with a diameter less than 30 μm, are evenly dispersed.

Table 12.1 The nature and effects of transformation phenomena during olive paste malaxation

Nature of the transformation	Transformation phenomena
Mechanical	• Breaking of cells by the cutting and shearing action of pit fragments
	• Mixing
	• Rheological changes
Physical	• Heat exchange and temperature control
	• Control of the composition of headspace atmosphere
	• Diffusion and equilibrium phenomena between the aqueous and the oil phases
Physico-chemical	• Changes in hydrophobic – hydrophilic equilibria
	• Coalescence of oil droplets
Chemical and biochemical	• Enzymatic reactions
	• Cellular respiratory phenomena
	• Chemical oxidation of unsaturated fatty acids in the presence of Cu^{++}, Fe^{+++}

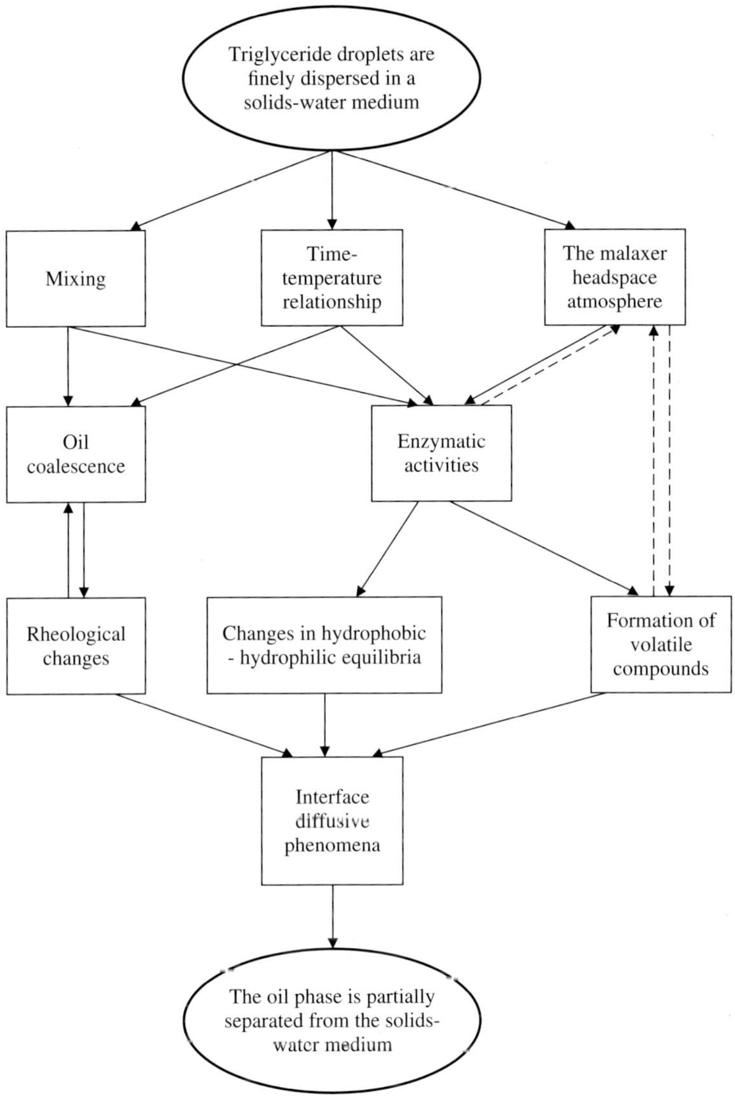

Figure 12.1 A schematic representation of rheological, physical, chemical and physical-chemical phenomena in olive paste malaxation.

The main phenomena during malaxation are:

- Oil coalescence.
- Triggering of complex enzymatic activities, mainly due to hydrolases, oxidases and to complex chain reactions such as the lipoxygenase (LOX) pathway, respiration and so on.

- As a consequence of oil coalescence, significant rheological changes take place in the paste. Curiously, contrary to the normal effect of a mixing operation, which consists in phases combining and increasing in viscosity, malaxing of olive paste causes the phases to separate and decrease in viscosity (Tamborrino *et al.* 2013). This is precisely what is needed to make oil separation easier in the following step.
- The enzymatic activities entail some critical changes, such as a change in the equilibrium of hydrophobic and hydrophilic compounds. This is due, in part, to the formation of monoglycerides and diglycerides that have amphiphilic properties and to other changes in solubility of important oil components. For example, the action of glucopyranose hydrolase reduces the hydrophilicity of oleuropein and allows some transfer of oleuropein aglycones into the oil phase (Chapter 6).
- Complex enzymatic reactions are responsible for the formation of short-chain carbonyl and aldehydic derivatives from polyunsaturated fatty acids, namely linoleic and α-linolenic acids (the lipoxygenase – LOX - pathway). These derivatives play a critical role in the flavour characteristics of the oil (Chapter 6).
- Respiratory metabolism is responsible for a decrease in oxygen and an increase in CO_2 in the malaxer headspace (dotted line in Figure 12.1).
- Due to the effects of the mixing action and the right temperature, all these newly-formed components undergo a diffusive transfer from the aqueous phase into the oil phase.

The final result of malaxation is a mixture that is very different from the incoming paste. More than 80% of the oil has coalesced into an oil phase that contains new polar and amphiphilic compounds, which play a critical role in the nutritional and sensory profile of the oil.

12.1.1 Coalescence

Coalescence is a phenomenon due to hydrophobic-hydrophilic equilibria in the olive paste. It is a consequence of the polarity of the water molecules. In fact, water consists of two hydrogen atoms joined to one oxygen atom. Oxygen is negatively charged, while hydrogen is positively charged. Thus, water molecules are dipoles and link to one another by hydrogen bonds.

Although it may appear that, in malaxation, the oil droplets actively gather into a continuous lipid phase, the reality is that water is attracted by other water molecules so that the oil molecules are squeezed out of the aqueous medium and forced to join into the lipid phase. In conclusion, water, with its hydrogen-bonding ability, provides the force for separating the oil. The separation of the oil is more precisely described as an *expulsion* rather than an *extraction* phenomenon.

12.1.2 The time-temperature relationship

Considering the temperature influence on the malaxing operation, it may be observed that:

- coalescence of the oil increases with the increasing in the temperature and decreasing in viscosity;
- diffusion of substances from the aqueous to the oil phase is also accelerated by an increase in temperature.

On the other hand, an excessively high temperature may have detrimental effects on the oil quality due to:

- enzymatic changes involving both lipolytic and oxidative degradation;
- increase in the vapour pressure of the volatile compounds causes some loss of aromas.

In the malaxing operation, time-temperature conditions should be chosen to maximize the desired and minimize the undesired effects.

In general, the extent of a transformation depends on the reaction rate and time. The reaction rate is accelerated by an increase in temperature according to an exponential relationship, while a linear relationship links the extent of the transformation to time.

A way to represent time/temperature relationships of technological operations is in a semi-log graph, with temperatures on the abscissa in a linear scale and times on the ordinate in a logarithmic scale. In this type of graph, the time-temperature conditions producing equivalent effects of transformations or reactions are represented by oblique lines.

Based on literature information and practical experience, Figure 12.2 shows the range of time-temperature conditions that are suitable for optimal malaxation (Peri 2013). The temperature interval on the abscissa is between 20 and 35 °C, whereas the times on the ordinate in logarithmic scale vary from 10 to 100 minutes. The two oblique lines indicate the conditions within which coalescence takes place at a suitable rate. The rectangular area traced in the centre of the diagram, between 25 and 30 °C and between 20 and 50 minutes, is the area in which the best compromise between yield and quality can be found.

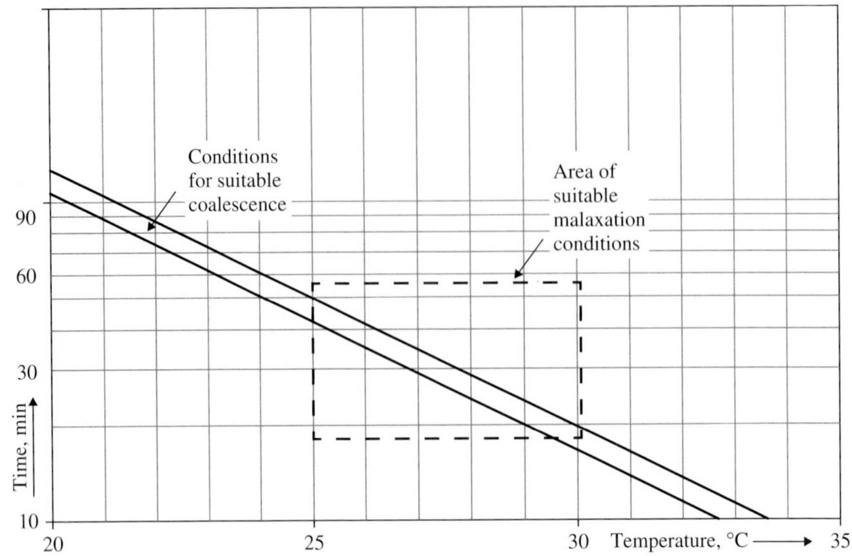

Figure 12.2 Time-temperature relationships in malaxation (*Source*: Peri, C. (2013). Reproduced with permission from Wiley).

Based on the above considerations, it should be noted that use of the term 'cold extraction', a process in which the malaxing temperature does not exceed 27 °C (the Commission Regulation (EC) No 1019/2002), is incorrect. First of all because there is nothing that demonstrates that 27 °C is a temperature tied to some particular change or phenomenon, and secondly because indicating a temperature without indicating a time is incomplete and misleading.

12.2 Malaxers

A malaxer consists of a stainless steel chamber fitted with a horizontal shaft bearing a set of stainless steel blades and an external coil. In order to guarantee the best mixing effect and avoid oil emulsification, the rotating speed should not exceed 20 RPM. While the blades assure the mixing of the paste, the coil continuously removes the layer of paste at the malaxer wall, in order to favour heat exchange and minimize overheating. The coil also acts as a conveyor during the discharging phase by pushing the paste towards the open gate valve of the discharge port.

The malaxing chamber is equipped with a jacket for hot water circulation for heating the paste to the desired temperature. Malaxers have a hermetical cover with safety inspection windows. Open-top malaxers have been abandoned for safety reasons.

Some malaxers are equipped with devices for injecting an inert gas in the headspace. So far, this device has proved useless for achieving better malaxing results. In normal operating conditions, the headspace contains enough oxygen to promote transformations that are essential for development of the odour and flavour of the oil (for example, the LOX pathway). At the same time, the residual respiratory activity of catabolic pathways results in further oxygen consumption and CO_2 formation. In conclusion, enzymatic and catabolic reactions in a hermetically closed malaxer operate as a self-controlled system of the atmosphere composition and guarantee a limited and optimal level of oxidative transformations.

12.2.1 The problem of temperature control

The malaxer is a very ineffective heat exchanger. In the first place the heating surface is small compared to the mass of paste to be heated. In the second place, the high viscosity and the slow motion of the paste result in very low heat transfer coefficients at the malaxer wall. As a consequence, the temperature gradient on the paste side is very high, with overheating effects upon the paste components.

If the temperature of the water is increased, in an attempt to accelerate heating of the paste to the desired optimal value, a double negative effect takes place: on one hand, the heat transfer rate does not increase much due to the resistance to heat transfer in the paste layer at the malaxer wall; on the other hand, the paste layer is at risk of overheating and damaging the quality of the oil.

Despite these obvious considerations, in many mill plants the paste is heated using water at excessively high temperatures. As a rule of thumb it can be considered that the temperature of the paste layer in contact with the malaxer wall is only 1–2 °C lower than the temperature of the water. Thus, if the temperature of the water is, for example, 40 °C, the paste layer at the wall has a temperature of 38–39 °C, while the measured temperature in the paste mass is much lower.

The first solution to the problem consists in implementing a control system consisting in:

- setting a fixed constant maximum temperature of the heating water (for instance, 30 °C). This limit should be guaranteed by automatic regulation of the heat exchanger producing warm water, and
- setting a flow-rate control of the water, which determines a suitable average temperature difference between the water in the jacket and the paste inside the malaxer.

The system is schematically presented in Figure 12.3. The water enters the water jacket at the chosen constant temperature (e.g. 30 °C). The temperature of the paste is measured with a thermometer. The thermometer is connected to a temperature controller, which compares the actual paste temperature with the desired temperature. The temperature controller acts upon a flow-rate regulating device in the water line (a valve); the greater the difference between the actual and the target temperature, the higher the flow rate. When the actual temperature approaches the target temperature, the flow rate of the water is reduced close to zero and the system operates only to maintain the temperature constant.

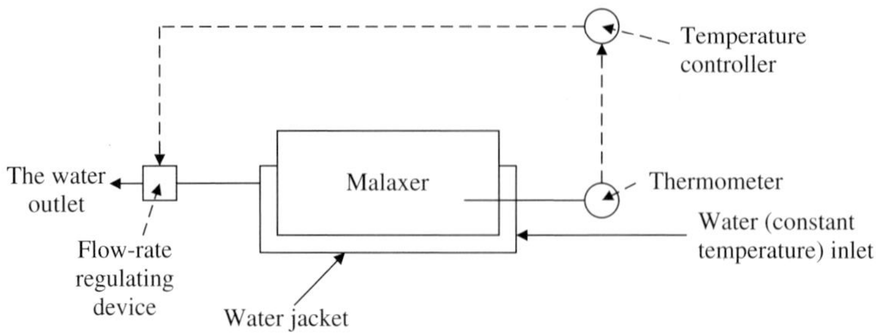

Figure 12.3 The scheme of temperature control in malaxers.

For the best monitoring of the operation, it is suggested that both the temperature of the paste and the water are recorded. In some plants, the heat exchange between the water and the paste is improved by the presence of turbulence promoters on the water side.

There is, however, a better system. It involves separating the heating from the malaxing function by introducing the use of a scraped-surface heat exchanger in series with the malaxer according to the scheme in Figure 12.4.

The scraped-surface heat exchanger is a very effective heat exchanger for semi-liquid, viscous products like the olive paste. It has a cylindrical shape, while rotating blades assure continuous and effective mixing of the product. This heat exchanger can assure high heat transfer rates thanks to a high specific surface area and a high heat transfer coefficient due to optimal mixing and shearing effects at the heat exchanger wall. The SSHE can be equipped with precise and effective automatic control of the product and the heating fluid temperature.

The olive paste coming from the crusher can be continuously fed into the SSHE through a positive displacement pump where it reaches the optimal temperature for malaxation (e.g. 27 °C). The paste is then fed into the malaxer, which can operate at constant temperature as a reaction vessel. Water in the jacket is maintained at optimum, constant temperature (e.g. 27 °C).

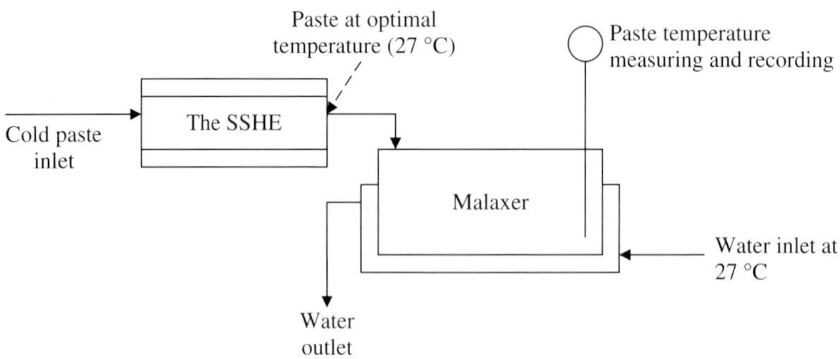

Figure 12.4 The separation of olive paste heating from olive paste malaxing by the introduction of a scraped-surface heat exchanger.

This solution, based on separation of the heating and temperature holding functions, allows more precise time-temperature relationships to be applied with beneficial effects for oil extraction yield and quality. Furthermore, this represents a fast and easy solution for upgrading the performance of pre-existing plants, with low investment (Amirante *et al.* 2006; Esposto *et al.* 2013).

12.2.2 The problem of residence-time control

One of the most frequent mistakes in malaxing operations is ineffective control of the residence time of the paste in the malaxer. In most malaxers, when a batch of paste is discharged at the end of a malaxing operation, a layer of variable thickness of paste sticks to the wall and bottom of the malaxer so that this paste undergoes malaxation again for a second cycle. It may be expected that, at the end of the second malaxation, a fraction of the previous residue is again part of the residue that sticks to the wall. In the small fraction that remains at the malaxing temperature for repeated cycles, degrading reactions can produce degradation and even sensory defects. The consequence is that, after an induction period of hours, the malaxer starts becoming a source of defects spreading into good oil. The problem of improper residence time becomes very serious when malaxers are connected in series. In this case a fraction of the paste may closely approach the theoretical condition of infinite residence time.

The solution to this problem, which is so often the cause of oil degradation, is very simple and consists of two points:

- never assemble malaxers in series, always in parallel
- carefully empty and wash the malaxer after each malaxing batch so that no paste residues remain in the following malaxing cycle.

Figure 12.5 compares three malaxers with different performances in terms of residence time control.

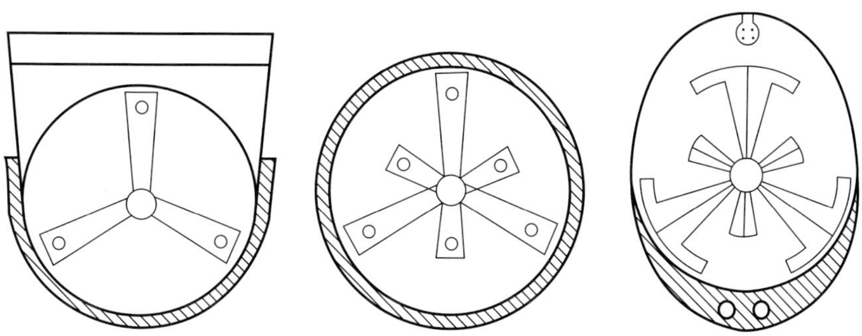

(a) (b) (c)

Figure 12.5 Different shapes of malaxer tanks: (a) traditional; (b) cylindrical; (c) ovoidal.

Figure 12.5(a) shows the shape of a traditional malaxer. The heating surface is in the lower part of the malaxing chamber and a large surface area is out of reach of the rotating blades and coil. In the absence of careful washing and cleaning, solid and liquid residues stick to the wall and the roof of the malaxing chamber and remain for hours in conditions that foster degradation and oxidation of the oil.

Figure 12.5(b) is an improved malaxer design. In the first place the heating surface is larger with a proportional improvement in heat transfer. Dead spots, unreachable by the blades and spiral motion, are absent, but the problem of paste residues sticking to the malaxer wall still remains to be solved.

In modern plants, malaxers are equipped with water nozzles for washing after each malaxing cycle. Two points should be carefully considered: (i) the water sprays should reach all possible angles and surfaces in the malaxer chamber, including its roof, and (ii) the washing should automatically follow emptying of the malaxer. This operation should not to be left to the operator's discretion and care.

Figure 12.5(c) shows a malaxer with an ovoidal shape. This may be a good solution because of the optimal positioning of the washing nozzles.

References

Amirante, P., Clodoveo, M.L., Dugo, G. *et al.* (2006). Advance technology in virgin olive oil production from traditional and de-stoned pastes: influence of the introduction of a heat exchanger on oil quality. *Food Chemistry* **98**, 797–805.

Esposto, S., Veneziani, G., Taticchi, A., Selvaggini, R., Urbani, S., Di Maio, I., Sordini, B., Minnocci, A., Sebastiani, L. and Servili, M. (2013), Flash thermal conditioning of olive pastes during the olive mill mechanical extraction process: impact on the structural modifications of pastes and oil quality, *J. Agric. Food Chem.* **61**(20), 4953–4960.

Gómez-Rico, A., Inarejos-García, A.M., Salvador, D.M. and Fregapane, G. (2009) Effect of malaxation conditions on phenol and volatile profiles in olive paste and the corresponding virgin olive oils (Olea europaea L. Cv. Cornicabra). *Journal of Agricultural and Food Chemistry* **57**(9), 3587–3595.

Kalua, C.M., Bedgood, D.R., Jr., Bishop, A.G. and Penzler, P.D. (2006) Changes in volatile and phenolic compounds with malaxation time and temperature during virgin olive oil production. *Journal of Agricultural and Food Chemistry* **54**(20), 7641–7651.

Peri, C. (2013) Quality excellence in extra virgin olive oil, in: *Olive Oil Sensory Science* (eds E. Monteleone and S. Langstaff), John Wiley & Sons, Ltd, Chichester.

Servili, M., Selvaggini, R., Taticchi, A. *et al.* (2003) Volatile compounds and phenolic composition of virgin olive oil: optimization of temperature and time of exposure of olive pastes to air contact during the mechanical extraction process. *Journal of Agricultural and Food Chemistry* **51**, 7980-7988.

Servili, M., Taticchi, A., Esposto, S. *et al.* (2008). Influence of the decrease in oxygen during malaxation of olive paste on the composition of volatiles and phenolic compounds in virgin olive oil. *Journal of Agricultural and Food Chemistry* **56**(21), 10048–10055.

Stefanoudaki, E., Koutsaftakis, A. and Harwood, J.L. (2011) Influence of malaxation conditions on characteristic qualities of olive oil. *Food Chemistry* **127**, 1481–1486.

Tamborrino, A., Clodoveo, M.L., Leone, A. *et al.* (2010) The malaxation process: influence on olive oil quality and the effect of the control of oxygen concentration in virgin olive oil, in *Olives and Olive Oil in Health and Disease Prevention*. Academic Press, London, pp. 77–83.

Tamborrino, A., Catalano, P. and Leone, A (2013), Using an in-line rotating torque trasducer to study the rheological aspects of malaxed olive paste, *J. Food Eng.* (DOI):10.1016/j.jfoodeng.2013.09.024

Taticchi A., Esposto S., Veneziani G. *et al.* (2013) The influence of the malaxation temperature on the activity of polyphenoloxidase and peroxidase and on the phenolic composition of virgin olive oil. *Food Chemistry* **136**, 975–983.

13

Centrifugal separation

Lamberto Baccioni[1] and Claudio Peri[2]

[1]*Agrivision, Florence, Italy*
[2]*University of Milan, Milan, Italy*

Abstract

Liquid-liquid and solid-liquid centrifugal separations have been the key technological breakthroughs in the extra-virgin olive oil process. They were introduced in the olive mills in the late 1930s and early 1960s, respectively, leading to a radical change in olive oil extraction technology. Decanters are thoroughly described, two-phase and three-phase operations are compared. Disc centrifuges for oil clarification and finishing are described, in particular continuous centrifuges with automatic solid discharge. Some critical points are discussed: (i) noise control of decanters; (ii) decanter and centrifuge calibration; (iii) cleaning-in-place of decanters; (iv) control of the oil stress at the centrifuge discharge.

13.1 Introduction

Olive paste, derived from the milling and malaxing operations, is the input material for the separation step in the production of extra-virgin olive oil. Olive paste is a heterogeneous mixture of three phases. In order of decreasing density, they are:

- The 'insoluble solids phase', consisting of organic semisolid components and the woody fragments from the pit shells. It is 25–30% by weight of the olive paste, with 75% pit fragments and 25% cell wall fragments.
- The 'aqueous phase', consisting of water and water-soluble components (salts, simple sugars, simple phenolics and so forth). It is 50–60% of the total paste weight, with 92–95% water and 5–8% soluble solids.
- The 'oil phase' consists of 97–99% triglycerides and 1–3% minor components, the latter being a complex mixture of lipophilic, hydrophilic and amphiphilic components with critical roles in sensory and nutritional quality.

The Extra-Virgin Olive Oil Handbook, First Edition. Edited by Claudio Peri.

It is estimated to be 10–20% by weight of the olive paste, depending on cultivar and fruit maturity.

- The insoluble solids fraction derives from the olive cell walls and the pit shells.
- The cell wall consists of fibres of cellulose acting as a kind of framework, and a semisolid mixture of water, carbohydrates (hemicellulose, pectic substances), minerals and proteins that fill the space between them.
- The pit shell fragments are essentially made of lignin, the main component of wood.
- An important observation from the separation point of view is that the solids from cell walls are highly deformable under pressure, causing clogging and hindering in liquid separation. The pit fragments, on the contrary, are rigid and ensure easier draining of the liquid through the solids cake.

Perfect separation of the three phases is impossible and some retention of oil by insoluble solids should be expected.

The most important objective of the separation process is to recover as much oil as possible from the olive paste. A recovery of 80–85% or more of the oil present in the paste indicates a good performance of the separation process. For example, if the oil content of the paste is 18% by weight, a recovery of 85% would result in an extraction yield of $18 \times 0.85 = 15.3$ kg of oil from 100 kg olives. In this case, 2.7 kg of oil would remain in the pomace, which is an acceptable result. A recovery of 75% would result in the separation of $18 \times 0.75 = 13.5$ kg of oil from 100 kg olives and the loss in the pomace of 4.5 kg of oil per 100 kg olives, which is a poor result.

In the late 1960s, the process of olive oil extraction underwent a radical change with the introduction of decanter centrifugation in place of pressing for separating the insoluble solids from the liquid fraction (water and oil) of the olive paste. This innovation is comparable in importance and impact to the introduction of centrifugation in place of gravity settling for separating oil from vegetation water. Thus, the introduction of centrifugation may be considered as the key innovation, radically changing a system that had remained the same for centuries – since very ancient times.

13.2 The three-phase process

Figure 13.1 presents the scheme of olive oil extraction based on a three-phase decanter separation.

The first centrifugation step is carried out with a three-phase decanter in which the three phases are separated. This is based on their different densities, in the order

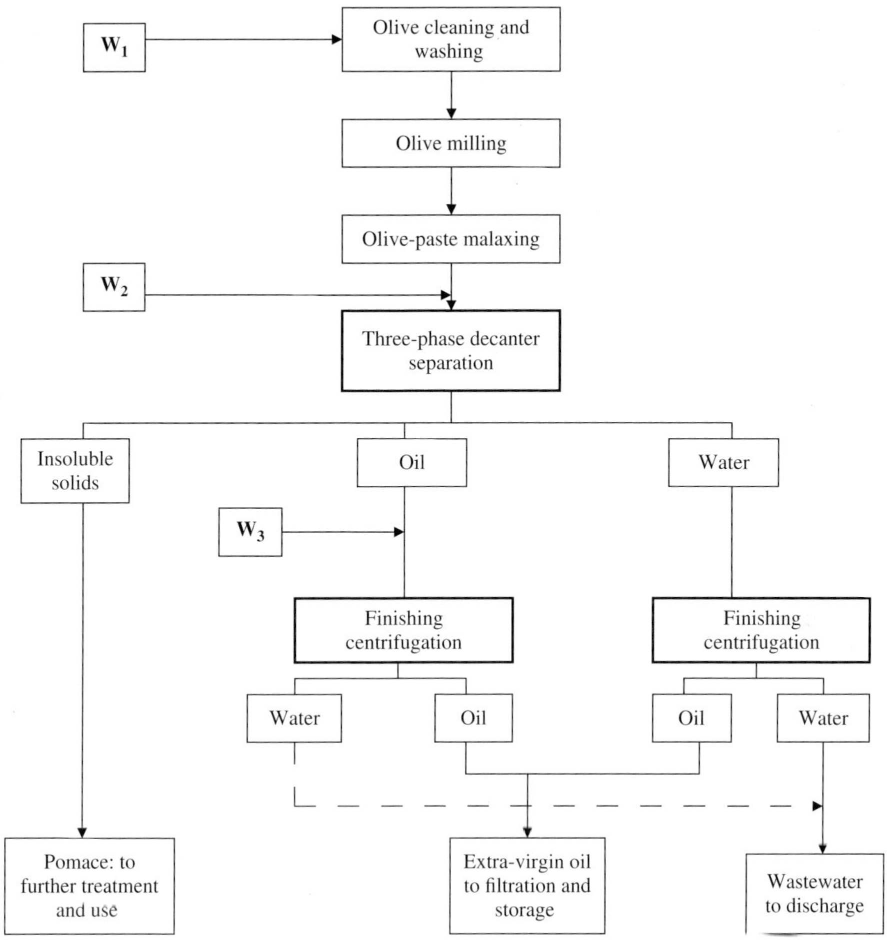

Figure 13.1 Scheme of the three-phase process.

of increasing density: the oil – the aqueous – and the insoluble solids phases. The insoluble solids are discharged as pomace and sent to further fractionation and utilization (Chapter 22).

Due to the relatively low rotating speed (usually from 3000 to 5000 RPM) and limited separation effectiveness of decanters, the oil fraction still contains 2–5% of water droplets in emulsion and insoluble solids impurities in dispersion. Therefore, the oil phase must undergo a second finishing centrifugation for the final removal of water and dispersed solids. Some water W_3 is added to improve the centrifuge performance in the oil separation and cleaning. The final product of this centrifugation step is clarified oil, while solid impurities and residual water are discharged as waste.

Similarly, the aqueous phase undergoes a finishing centrifugation treatment with separation of the vegetation water as the main product, recovery of a small quantity of oil and discharge of solid impurities. Due to the improved performance of the new-generation three-phase decanters in the separation of the oil from water, the finish centrifugation of water is often avoided.

The second centrifugation step is carried out in high-speed disc centrifuges, usually from 5000 to 7000 RPM, which are unsuitable for treating products with a high solids content but are very effective in separating impurities dispersed in a liquid phase (either oil or water).

In Figure 13.1, W_1, W_2 and W_3 indicate points where water is usually added. For most effective oil separation, in the first-generation three-phase decanters, a considerable quantity of water was added at W_2, accounting for 60–100% or more of the weight of the processed olives. This resulted in huge quantities of wastewater (100–120 kg per 100 kg of processed olives), creating serious waste-disposal problems.

A new generation of three-phase decanters is now available (the so-called LWC – Low Water Consumption – decanters), in which only 15–30% water is added, resulting in 50–70 kg wastewater per 100 kg of processed olives.

13.3 The two-phase process

The wastewater problem has spurred research on new decanter design and operation. Three-phase LWC and two-phase decanters are the alternative solutions now available. Figure 13.2 shows the two-phase process, in which the addition of water to the decanter is eliminated or reduced to 5–10% of the weight of the processed olives. The insoluble solids and vegetation water are discharged together from the decanter as a semi-liquid slurry.

The oil undergoes a finishing centrifugal treatment to eliminate solid impurities and some water (W_2), accounting for 15–25% of the oil weight is added to improve the oil cleaning and recovery. Some water (not shown in Figure 13.2) can be added either at the malaxation or the decanter stage in order to control the paste viscosity and improve the malaxing – decanter performance.

The two-phase system requires only one finishing centrifugation and does not require the handling of huge wastewater volumes. It is therefore less expensive in terms of investment and operational costs than the three-phase system.

The choice of a two or a three-phase decanter requires careful evaluation by the milling company, taking into account various technical, economic and environmental aspects, as outlined in Table 13.1.

13.4 Decanters

13.4.1 The classic two-phase decanter

Decanter centrifuges consist of a screw conveyor and a solid exterior bowl having both a combined cylindrical and conical shape. The bowl and the conveyor rotate at different speeds. The difference in speed between the two is responsible

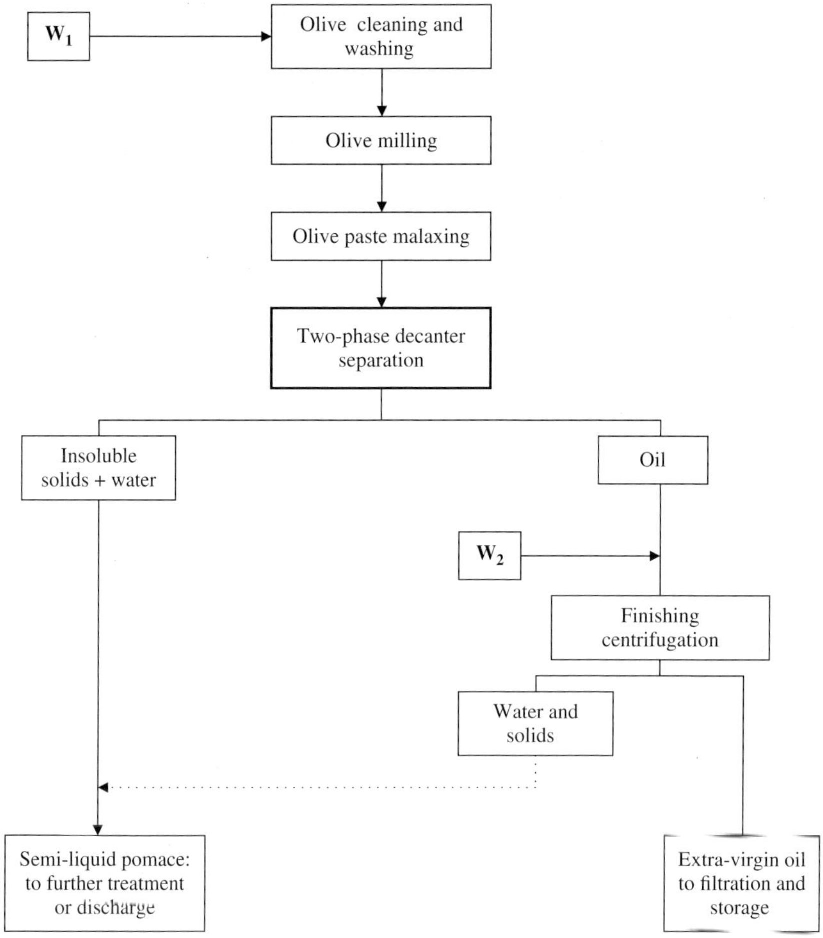

Figure 13.2 Scheme of the two-phase process.

for conveying the solid sediment from the cylindrical to the conical part of the bowl towards discharge (Peri and Zanoni 1994).

The bowl is mounted between fixed bearings anchored to a rigid frame. Gearboxes are cantilevered out board of these bearings, and a non-rotating feed pipe enters the rotating assembly through one end of the screw conveyor. The frame is isolated from the support structure by spring-type vibration isolators.

Figure 13.3 shows the main components of the rotating set of the decanter: (a) the rotating screw conveyor with detail of the stationary feeding pipe and the distribution port discharging the olive paste into the bowl; (b) the assembly of the rotating screw and bowl; (c) the whole rotating set with detail of the screw and bowl drive.

Figure 13.4 shows the functional scheme of a classic decanter for treating semisolid slurries and separating a concentrated solids phase (a cake) and a liquid phase.

Table 13.1 Comparison of pluses and minuses of the three- and two-phase decanter separation processes.

Points to be compared	Three-phase process	Two-phase process
Water content of pomace	The quantity of pomace is 55–60 kg per 100 kg of olives, with a water content of 48–54%. This value is sufficiently low to favour the further use of pomace, including residual oil extraction in refineries. Due to lower water content, three-phase pomace also has a higher calorific value than two-phase pomace	The quantity of pomace is 60–80 kg per 100 kg of olives, with a water content of 58–65%, too high for recovering valuable by-products. The pomace has a lower value.
Water addition in the process	A considerable addition of water is required before decanting. The sum of added water in may vary from 70 to 110 kg per 100 kg olives (first generation decanters) or 20 to 60 kg (LWC decanters). A higher content of water facilitates separation of the oil from the water layer, thus increasing the extraction yield	A small addition of water is required (5–10 kg per 100 kg olives). In this case the water and solids act as a unique phase and the separation of the oil is favoured by a water content that does not exceed the absorption capacity of the solids
Investment and operating cost (including wastewater disposal)	Higher cost	Lower cost
Production of wastewater	A considerable quantity of wastewater is produced at the mill with the problem of disposal or treatment at the milling site	There is no production of wastewater at the milling site and this greatly simplifies waste disposal
Oil quality	People who favour the two-phase decanter say that a lesser addition of water favours a greater retention of phenolics in the oil. This point is debatable. In the first place, a reduction in the total phenolic content of some oils that have a sensory note that is too bitter may be an advantage, not a disadvantage. Secondly, the transfer of phenolic compounds from the water to the oil phase is not simply a matter of partitioning equilibria between the two phases, but mostly and foremost a problem of change in solubility of phenolic compounds in the oil which is due to enzymatic reactions and therefore to the time and temperature of the malaxing operation. The washing effect of water on the oil is considerable in the finishing centrifugation of the oil, which is carried out in both the three-phase and the two-phase decanters. Finally, modern three-phase decanters recycle some vegetation water for diluting the paste. In this case, the diluting effect on phenolics is minimized.	

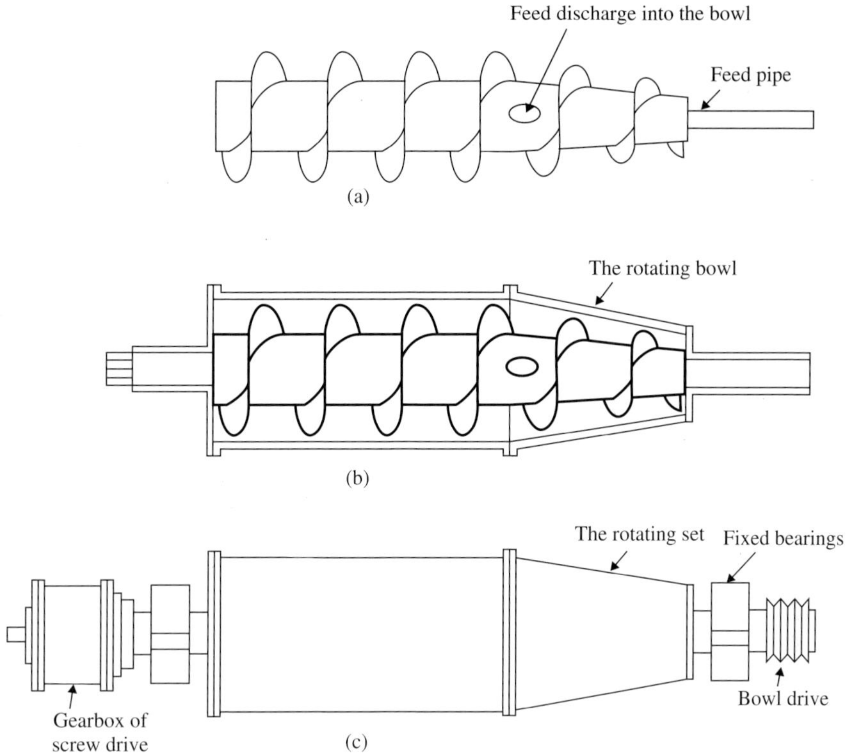

Figure 13.3 The decanter rotating set: (a) the rotating screw conveyor with detail of the stationary feeding pipe and the port feeding the olive paste into the bowl; (b) the assembly of the rotating screw and bowl; (c) the rotating set with detail of screw and bowl drive.

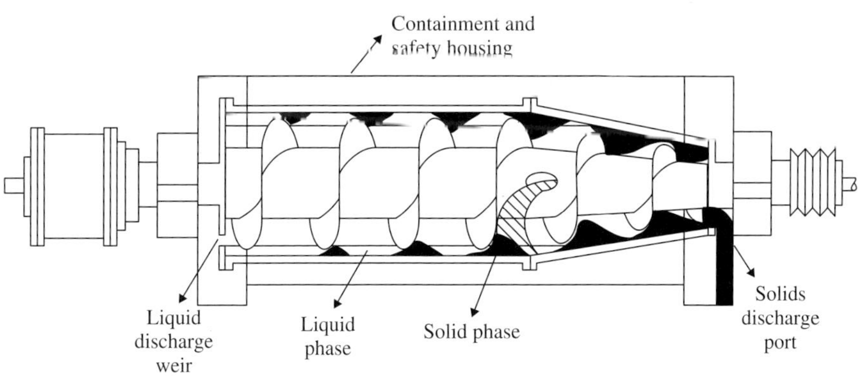

Figure 13.4 The functional scheme of the traditional two-phase decanter for solid-liquid separation of semi-solid slurries. The level of liquid, which is determined by the position of the liquid discharge weirs, determines the point of separation of the 'pool' from the 'beach' sections. The flow of the solid phase is in the opposite direction of the flow of the liquid phase.

The slurry is fed through a stationary tube near the centre of the screw conveyor at a point approximately corresponding to the junction of the cylindrical and the conical section. The slurry passes through distribution ports into the bowl. The bowl rotates at a speed in the range of 3000–5000 RPM. Under the action of centrifugal force, the slurry forms a concentric layer at the bowl wall. The section corresponding to the cylindrical part is called 'the pool', while the section corresponding to the conical part is called 'the beach'. In the separation pool, the solids, which are heavier than the liquid, settle toward the bowl wall, while the clarified liquid moves radially toward the pool surface. Subsequently, the liquid flows toward the bowl head, from which it discharges over the weirs.

In the dry part of the beach, the solid cake is dewatered with the expressed liquid returning back to the pool. The angle of the cone may vary from 5° to 20°, depending on the application and performance required.

The annular pool height can be changed by adjusting the radial position of the weir openings. This is an important regulating device. When the liquid layer is too thin, the finer particles may be entrained by the fast moving liquid stream, eventually ending up in the liquid phase. A deeper pool guarantees a thicker liquid layer to assure resettling of suspended solids. This can be at the expense of cake dryness due to the reduction of the dry beach. Consequently, there is a compromise between liquid clarity and cake dryness, which can be controlled by regulating the position of the weir.

The screw conveyor rotates at a slightly different speed than the bowl and conveys the deposited solids toward the dry beach.

The centrifugal force helps dewater the cake, yet at the same time hinders the transport of the cake to the dry beach. A balance in cake conveyance and cake dewatering is therefore the key to setting the pool depth and the centrifugal acceleration at the right value.

The solids are submerged in the pool when they are in the cylinder and at the beginning of the beach. In this region, liquid buoyancy helps reduce the effective weight of the cake under the centrifugal acceleration, resulting in lower conveyance torque. Further up the beach, the solids emerge above the pool and move along the dry beach where buoyancy force is absent, resulting in a higher torque.

The speed with which the solids are transported towards the discharge port is controlled by the differential speed. High differential speed facilitates high solids throughput where the cake thickness is kept to a minimum so as not to impair liquid clarity due to entrainment of fine solids. Cake dewatering is also improved due to reduction of the drainage path with lower cake height; however, this is offset by the fact that higher differential speed also reduces cake residence-time, especially in the dry beach. The opposite happens with low differential speed. An optimal differential speed is therefore required to balance liquid clarity and cake dryness.

Automatic systems for controlling the differential speed and the rotating speed of the screw conveyor are based on torque measurement.

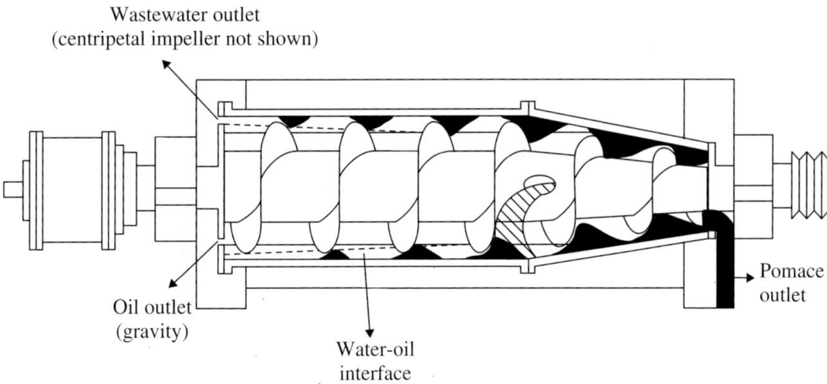

Figure 13.5 The functional scheme of a three-phase decanter for olive oil extraction. A dotted line indicates the oil-water interface. The radial position of the weir openings set the position of the interface and is a critical factor of separation effectiveness.

13.4.2 The three-phase decanter for olive oil extraction

Figure 13.5 represents the functional scheme of a three-phase decanter for olive oil extraction. The dotted line indicates the oil-water interface. The radial position of the weir openings set the position of the oil-water interface and is a critical factor of separation effectiveness.

Three-phase decanters are similar in principle to the classic decanter described above, but separate three phases instead of two. The two liquid phases are discharged either via gravity over two sets of adjustable weir plates or via a dual discharge system where the heavy liquid phase (water) is discharged via a stationary impeller under pressure and the light liquid phase is discharged via gravity over a ring dam. The advantage of the dual discharge system is that the liquid interface zone (and ultimately the pool height) is adjustable while the machine is operating at full speed.

In the three-phase decanters, a length of the dry beach is needed in order to allow expressing and to get higher oil recovery from the solids cake. On the other hand, in three-phase decanters, water must be added to the incoming slurry in order to get an oil-water interface as sharp as possible. In the last generation, LWC decanters, a baffle disc between the conical and the cylindrical part of the bowl separates the solid discharge from the liquid discharge levels and improves the oil recovery.

13.4.3 The two-phase decanter for olive oil extraction

Figure 13.6 represents the functional scheme of the two-phase decanter for olive oil extraction. The products of separation are oil and a semi-liquid pomace.

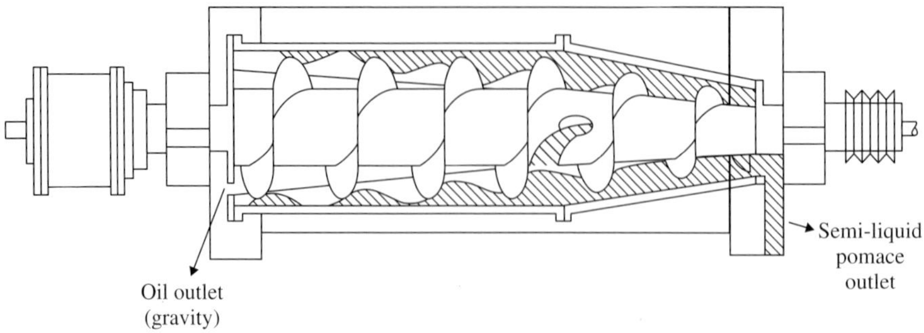

Figure 13.6 The functional scheme of two-phase decanter for olive oil extraction.

The two-phase decanters separate the oil phase from a denser phase consisting of the mixture of water and solids. A limited addition of water (less than 10% of the paste weight) may be needed in order to adjust the paste to an optimal semiliquid consistency. However, in general, no water is added. In fact, an excess of 'free water' may compromise a sharp separation of the semiliquid cake from the oil layer.

In two-phase decanters, buoyancy helps reduce the weight of the cake under centrifugal acceleration. This condition holds true all the way from solids-water entrainment up to the discharge port, thus resulting in lower power consumption of two-phase compared to three-phase decanters.

13.5 Disc centrifuges

The second step in centrifugation is carried out to make the oil as clear and stable as possible. Decanters can only roughly separate the three phases, due to a relatively low centrifugal acceleration. Dispersed solids less than 10 μm in diameter and droplets of water of a similar dimension can seldom be separated in a decanter. Disc centrifuges with a high rotating speed (5000–7000 RPM) are used for the purpose of oil finishing and clarification.

13.5.1 Manual-discharge disc centrifuges

Figure 13.7 shows the functional scheme of a disc centrifuge for oil clarification.

The turbid oil coming from the decanter, with the addition of some water in order to make an easier separation of the continuous water phase, is fed into the centrifuge proximate to the axis of the bowl. At the bottom of the bowl, the incoming turbid oil is accelerated by centrifugal force through a conical distributor, which evenly distributes the liquid to the appropriate disc stack channels. At a suitable radius of the bowl, depending on the oil-water proportion in the feed, the flow of the incoming liquid is diverted upwards into vertical channels formed by corresponding openings in the stack of closely spaced discs: 100 to 150 discs are assembled, spaced 0.8

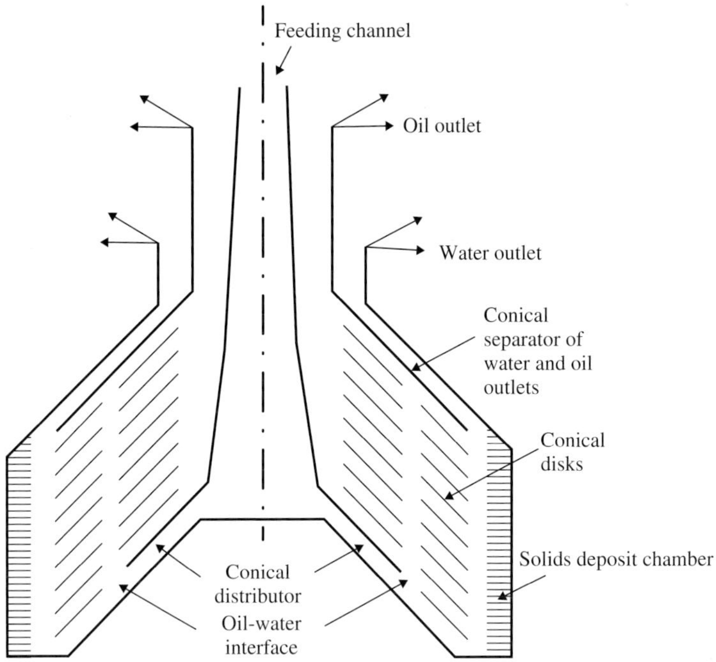

Figure 13.7 The functional scheme of a liquid-liquid and liquid-solid centrifugal separator with manual discharge of solid deposits.

to 2 mm apart. The angle made by the conical discs from the horizontal is 35–40 °. Under the effect of the centrifugal force, the difference in density and the push of the incoming liquid, the solids and water flow centrifugally in the thin layer between the discs, while the oil flows centripetally toward the axis of the centrifuge. The solids settle against the underside of the disc, move down to the large end of the conical disc and finally accumulate at the bowl wall. At optimal operating conditions, the upward channels formed by the disc holes are, at the same time, the feeding channels of the incoming liquid and the separation boundary of the inner oil layer and the outer water layer. A solid conical separator on top of the bowl divides the outlets of water and oil that discharge separately through overflow ports. In more advanced centrifuges, the rotating liquid is diverted to a stationary impeller (a centripetal pump) from which the kinetic energy of the stream is converted to hydrostatic pressure. In this case, the centrifuge acts as a pump for conveying the oil and the water to suitable destinations. In the olive oil application, the centripetal pump is used only for water discharge and not for the oil because of its friction effects with mechanical stress and increase in the oil temperature.

In the simplest design represented in Figure 13.7, the accumulated solids must be removed manually on a periodic basis. This requires stopping and disassembling the bowl and removing the disc stack. Manual removal of solids is economical only when the fraction of solids in the feed is very small. This was the case in older times, when oil and water separation was carried out by pressing. In that case, the draining materials used in the pressing stack for the separation of the liquid from the solid phases, retained most of the solids with a filtering mechanism and therefore the suspended solids content of the oil was quite low. With the use of decanters the solids content in the oil or in the water phase is relatively high and therefore manual discharge disc centrifuges are replaced by self-cleaning disc centrifuges.

13.5.2 Self-cleaning disc centrifuges

Figure 13.8 represents the functional scheme of a liquid-liquid and liquid-solid centrifugal separator with automatic solid discharge.

These centrifuges automatically discharge accumulated solids on a timed cycle while the bowl is at full speed. Solids accumulate in the sludge holding area just inside of the maximum diameter of the double cone-shaped bowl (Figure 13.8). When the solids chamber is full, the double bottom of the bowl, which is hydraulically held closed to the top portion of the bowl, drops by evacuating the hydraulic operating fluid. The solids are discharged under the pressure due to the centrifugal force in a very short time through a series of peripheral nozzles into an outer casing where they are diverted out of the machine. After the discharge of solids and a little quantity of entrainment water, the double bottom is automatically pushed up by the same hydraulic system and the solids accumulate again at the peripheral sludge holding area while the centrifuge keeps operating continuously.

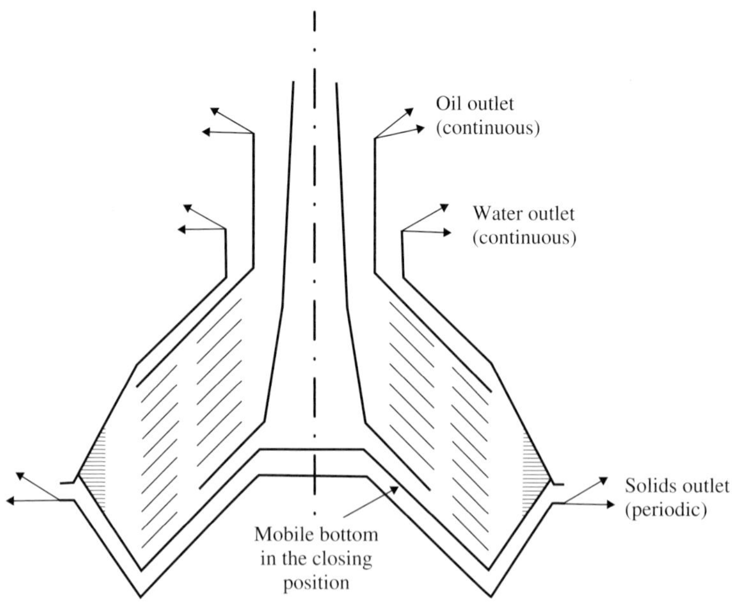

Figure 13.8 The functional scheme of a liquid-liquid and liquid-solid centrifugal separator with automatic solid discharge.

13.6 Final comments and remarks

Centrifugation is an essential part of a modern extra-virgin olive oil process. Old methods based on pressing or selective percolation cannot be considered as suitable in terms of effectiveness for oil recovery, hourly capacity, flexibility and cost.

Compared to decanter separation, pressing and selective percolation were labour intensive. Furthermore, they required conditions where there was extensive contact with the air, temperatures were inappropriate and there was a high risk of contamination.

Decanters and disc centrifuges are highly reliable equipment. They can operate for days without interruption and are available from small to large sizes.

A series of automatic controlling devices assure adaptability of the working cycles and conditions to variable operating requirements and physical characteristics of the processed products. Some points, however, should be considered very carefully as possible sources of risk.

Noise

Excessive noise of decanters is an underestimated risk for workers operating in an olive mill. This point is outlined in the box.

The noise of decanters as a risk to workers' health

Noise can cause permanent and disabling hearing damage. However, hearing loss is not the only problem. People may develop tinnitus (ringing, whistling, buzzing or humming in the ears), a distressing condition that can lead to disturbed sleep.

Noise can also reduce people's awareness of their surroundings and this can lead to safety risks, putting people at risk of injury.

The maximum limit allowed by the health and safety laws for daily personal noise exposure is 87 dB, while 85 dB is the upper exposure action value. Exposure action values are defined as the levels of noise exposure, which, if exceeded, require specific action to be taken.

These points should be considered very seriously in olive mills because a normal level of noise around a decanter is in the order of 90 dB. This level is easily and usually exceeded because of the addition of the considerable noise of hammer mills and disc centrifuges operating in the same room. Furthermore, neglecting this problem in the design of the building and space in a milling factory results in concentration or amplification of the noise level.

A simple rule for evaluating if there is a noise problem is to verify if people have to raise their voices to carry out a normal conversation when about 2 m apart.

What should be done?

In the first place, noise measurements can be taken in the workplace and information can be obtained from equipment suppliers on noise levels of equipment.

In case the results of this inquiry indicate the presence of a risk, many precautions can be taken to reduce noise and noise exposure:

- engineering/technical controls to reduce, at the source, the noise produced by a machine or process;
- using screens, barriers, enclosures and adsorbent materials to reduce the noise on its path to the people exposed;
- designing and laying out the workplace to create quiet workstations;
- limiting the time people spend in noisy areas;
- a low-noise purchasing policy for machinery and equipment;
- proper and regular maintenance of machinery and equipment that takes account of noise;
- conduct hearing checks especially if hearing problems are detected.

The following steps should be taken urgently:

- provide workers with hearing protection devices and make sure they use them fully and properly;
- identify hearing protection areas in the workplace where access is restricted, and where wearing protective hearing devices is compulsory;
- choose a suitable protection factor, sufficient to eliminate risks from noise but not so much protection that wearers become isolated.

Calibration

From the description of decanters it is understood that the optimal performance in terms of oil recovery, power consumption, efficiency in oil and water separation, depend on a suitable calibration of the rotating speed and the differential speed of the screw conveyor and bowl and positioning of the weirs at the outlet of the liquid phases. A number of controls and measurements must also be assured, such as the flow-rate and temperature of added water, the torque of the rotating screw conveyor, and so on. Before each harvesting campaign, the equipment must be carefully checked by experts under the responsibility and direction of the plant supplier.

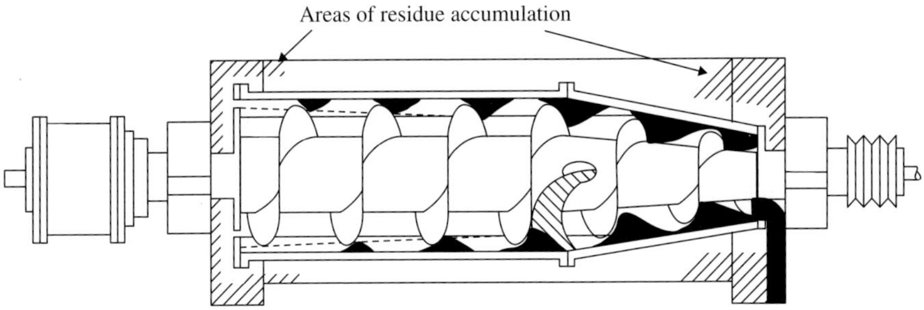

Figure 13.9 Residue accumulation in a decanter.

Written reports of this yearly operation should be considered as an important source of information for the best plant use and maintenance.

Cleaning

Figure 13.9 shows some critical points where solid and liquid residues can accumulate in decanters. These materials undergo rapid oil degradation, especially oxidation and rancidity, with the risk of contamination and development of sensory defects in the oil.

Periodical and careful cleaning of the soiled parts is needed, especially where accumulation occurs in the stationary casings of the machines.

Modern decanters have a built-in cleaning-in-place (CIP) system that allows complete and easy cleaning daily.

Mechanical stressing of the oil

At discharge from the high-speed rotating bowl, the jet of oil against a stationary casing produces a sudden loss of kinetic energy with shear-stress effects, mixing with air, and possible foaming or emulsifying of solid traces, oil and water.

These effects can be minimized by having the oil discharge by gravity as close as possible to the rotating axle of the bowl (minimal kinetic energy).

Further reading

Amirante, P., Baccioni, L., Catalano, P. and Montel, G.L. (1999) Nuove tecnologie per l'estrazione dell'olio di oliva: il decanter a cono corto a pressione dinamica variabile e controllo della velocità differenziale tamburo/coclea. *Rivista Italiana delle Sostanze Grasse* **76**, 129–140.

Amirante, P., Clodoveo, M.L., Dugo, G. *et al.* (2005) Virgin olive oil from de-stoned paste: introduction of a new decanter with short and variable dynamic pressure cone to increase oil yield. Proceedings of the Intrafood Congress on 'Innovations in Traditional Foods' held in Valencia 25–28 October 2005 (eds P. Fito and F. Todrà), Elsevier Science Inc., New York, Vol. II, pp. 1183–1186.

Peri, C. and Zanoni, B. (1994) *Manuale di Tecnologie Alimentari*, CUSL, Milano.

Records, A. and Sutherland, K. (eds) (2001) *Decanter Centrifuge Handbook*, Elsevier Science Inc., New York.

14

Filtration of extra-virgin olive oil

Claudio Peri
University of Milan, Milan, Italy

Abstract

A short presentation of filtration theory explains the role of the main factors determining filter performance: surface area, pressure difference across the filter and viscosity. The most suitable filter media and filtering equipment and two filtration systems are presented: a traditional one with a positive rotating pump providing pressure difference, and an alternative method exploiting hydraulic and nitrogen pressure. Best practices in extra-virgin olive oil filtration are summarized in ten points.

14.1 Introduction

Filtration is a solid–liquid separation technique. When a turbid suspension is forced by a suitable pressure difference to pass through a filter medium of suitable porosity, solid particles are retained / captured by the filter medium and the outgoing liquid is clearer than the incoming suspension. The liquid that passes through the filter medium is called the filtrate. Filter media with a coarse porosity eliminate coarse particles and filter media with a fine porosity also eliminate fine particles. Two widespread prejudices are to be clarified and overcome: the first is that filtration is detrimental to the quality of extra-virgin olive oil; the second claims that centrifugation can guarantee a perfect cleaning of the oil.

Filtration is necessary

Some producers maintain that extra-virgin olive oils do not need filtration but also that filtration is detrimental to oil quality. This point of view should be considered as erroneous and probably the result of improper implementation of this operation.

The Extra-Virgin Olive Oil Handbook, First Edition. Edited by Claudio Peri.

In fact, fine particles that are suspended in a virgin olive oil, even after the most effective centrifugal finishing, contain water and enzymes that may impair oil stability and ruin its sensory profile. Filtered oil has a better appearance and colour and does not form deposits in the bottles, which are not appreciated by the consumer. Filtration makes an extra-virgin olive oil more stable and also more attractive.

If the suspended particles are not removed they slowly agglomerate and flocculate, forming a deposit on the bottom of the storage containers. Such a deposit continues to be at risk of enzymatic spoilage and, in the worst case, of development of anaerobic micro-organisms with further spoilage and hygienic risk.

Carrying out a filtration process after a period of rest in the storage tanks obviously makes filtration easier because most of the coarse particles have decanted but this fails to prevent enzymatic spoilage. It is therefore recommended that filtration be carried out as soon as possible after centrifugal separation and finishing.

The difference between filtration and centrifugation

Some producers and even some milling-plant makers consider filtration as an alternative to centrifugation and stubbornly maintain that centrifugation can perfectly clarify extra-virgin olive oil. In reality, filtration and centrifugation are complementary, not alternative operations.

Separation by filtration is due to a sieving mechanism based on the difference in size between the suspended particles and the pore diameter of the filter medium. Centrifugal separation, instead, is based on the difference in density between the liquid and the suspended particles. In turbid oil the suspended particles consist of cell fragments containing hydrophilic components associated with water. They are therefore denser than the oil phase and can be separated by centrifugation. However, centrifugal separation is also influenced by the size of the particles, and the decanting velocity in centrifugation is proportional to the square of the diameter of the particles. As a consequence, separation by centrifugation of large particles from a cloudy oil is easy but becomes increasingly difficult as the diameter of the particle decreases. Particles with a diameter of 0.1 mm or more are effectively separated by centrifugation. For particles with a diameter in the range of 0.1 to 0.01 mm or lower, centrifugation cannot assure complete separation and filtration with fine filtration media is necessary.

14.2 Filtration principles

14.2.1 The filtration mechanisms

Two main mechanisms allow suspended solids to be retained or captured by a filter medium. When solids are retained on the surface of a filter medium due to the fact that their size is larger than the size of the filter pores, the filtration is called surface or cake filtration. When solids are trapped within the pores or body of the medium (different phenomena are responsible for retaining solids, including surface absorption and electrostatic attraction), it is called deep-bed filtration. Both mechanisms are active in the filtration of cloudy extra-virgin olive oil.

14.2.2 The filtration rate and the Hagen–Poiseuille equation

The filtration rate is measured as the volume of filtrate (***V***) which is obtained per unit time (θ). The fundamental relationship of the filtration rate is given by the Hagen–Poiseuille equation (Peri 1983):

$$\frac{1}{A}\frac{dV}{d\theta} = \frac{\Delta P}{\mu(\alpha w V/A + r)}$$

in which $dV/d\theta$ is the filtration rate (volume of filtrate per unit time), A is the filter area, ΔP is the pressure difference across the filter medium (the driving force of filtration), μ is the viscosity of the filtrate, α is the resistance to the flow of filtrate due to the solids that are retained by the medium, w is the weight of solids per unit volume of filtrate, r is the resistance to flow due to the (clean) filter medium.

Comments on the Hagen–Poiseuille equation may help understand some critical aspects of filtration.

The driving force

The driving force of filtration is the difference in pressure across the filter medium, ΔP. Such a pressure differential may be induced by a hydrostatic (gravity) pressure or by pressure applied upstream from the filter medium or by vacuum applied downstream from the filter medium.

According to the Hagen–Poiseuille equation, the filtration rate is proportional to ΔP: an increase in pressure should result in a proportional increase in the filtration rate. This is not true in our case because the solids that are retained at the surface or trapped inside the filter medium are highly compressible and deformable under pressure. Therefore, an increase in ΔP causes compaction of the solids with a more than proportional increase in α and, in the end, this is not an advantage but a disadvantage in terms of filtration rate. Due to the compressibility of the solids, the best ΔP in filtration of extra-virgin olive oil is 0.2–0.5 atm.

Temperature

The filtration rate is inversely proportional to μ, the oil viscosity. Therefore, an increase in temperature causes a decrease in viscosity and a proportional increase in the filtration rate.

In Table A4 of the Appendix, the viscosity of extra-virgin olive oil is given as a function of temperature. Temperatures higher than 25 °C should be avoided (risk for oil quality) and temperatures lower than 17 °C should also be avoided (low filtration rate).

The temperature of the oil at the outlet of the final centrifugal finishing is generally in the range of 22–24 °C, which is a good condition for filtration. This indicates that the best time to filter is immediately after centrifugation, followed by cooling to about 16 °C and transfer to the storage tanks.

14.2.3 Filtration cycles

In the Hagen–Poiseuille equation there are two variables that increase with time, they are: V, the volume of filtrate and thus the V/A ratio, the filter area A being constant, and α, the resistance due to the progressive retention of solids and the consequent reduction of the porosity of the medium.

As the V/A and α values increase, the filtrate rate decreases. When the filtrate rate drops below a level that is not compatible with the processing needs, filtration is stopped, the filter is disassembled and cleaned, the exhausted filter media is discarded and, finally, the filter is reassembled with new filtration media to start a new filtration cycle.

Two practices are possible:

- filtration is carried out until filtration rates become very low
- shorter filtration cycles are alternated with more frequent disassembling / changing of filter media / reassembling.

In the first case the average filtration rate is low, but the retention capacity of the filter is fully used: the amount of filter media used is therefore reduced. In the second case there is a higher amount (and cost) of filter media used, but the average filtration rate is higher. The choice about the two alternatives is a matter of cost and also of time of exposure of the oil to conditions that may be detrimental to its quality.

14.2.4 Filter surface area

As is evident from the Hagen–Poiseuille equation, an increase in the filter area, A, determines a proportional increase in the filtration rate $dV/d\theta$. With plate filters a variable number of filter sheets can be assembled, thus allowing the operator to adapt the filtration area to the filtering needs of each filtration cycle.

Increase in the filtration rate is more than proportional to increase in the filter area because of the simultaneous increase in A and decrease in the V/A ratio. In fact, an increase in A determines an increase of the filtering surface area and also a slower clogging effect due to the distribution of retained solids in a larger mass and surface of the filtration medium.

14.2.5 Exhausted filter sheets

At the end of each filtration cycle, the exhausted filter sheets are impregnated with the oil that fills up the pores of the medium. Pressing treatments can be used to recover a part of this oil to then be mixed with lower quality lots. When possible, however, exhausted filter sheets are mixed with pomace and sent to refineries for pomace oil extraction.

14.2.6 Oil turbidity

The turbidity of an oil can be visually evaluated or more precisely measured with a nephelometer. This is a portable, cheap and easy to use instrument for measuring the turbidity of liquids containing suspended particulates. It employs a light beam and a light detector set to one side (usually 90 °) of the source beam. Turbidity is a function of the light reflected into the detector from the particles.

A nephelometer can be used for controlling the effectiveness of filtration and for choosing the right filtration media. Using the nephelometer for evaluating the effectiveness of centrifuges and their cleaning performance may also be very useful.

14.3 The filter media

Filter media used for filtering extra-virgin olive oil are generally filter sheets of pure cellulose fibre, with or without diatomaceous earth. More rarely, synthetic inert materials are used. The average diameter of the pores is between 10 and 30 μm and the sheet thickness varies from 2 to 5 mm.

Filter sheets have an anisotropic structure with porosity being progressively finer in the direction of the liquid flow. This arrangement favours a differential solid retention as a function of the thickness of the sheet, with the coarser particles being retained first and the finer particles being retained in the last layers of the sheet before the filtrate exits. This condition slows down clogging and allows for a more complete utilization of the retention capacity of the filter. In mounting the filter sheets in the plate filter, operators should take care to put the filter sheet in the correct position with the coarser part at the inlet side of the turbid oil and the finer part at the outlet side of the limpid filtrate. Doing the opposite would result in very fast clogging of the filter surface and in the incomplete use of its retention capacity. The two sides of the filter sheet are easy to distinguish because the coarse inlet side has a rougher surface than the finer outlet side.

Other filtering materials have been proposed, especially rigid porous media made of sintered stainless steel or porcelain. These can be used for a second clarifying filtration step after the coarser particles have been eliminated by filter sheet filtration.

14.4 Filtration equipment

The most common filtering equipment used in the filtration of extra-virgin olive oil is the 'plate filter' (Figure 14.1).

Filter sheets alternate with filter plates, the latter being of two types: 'feeding plates' in which the turbid oil is fed under pressure, and 'collecting plates' through which the limpid filtrate is collected and discharged.

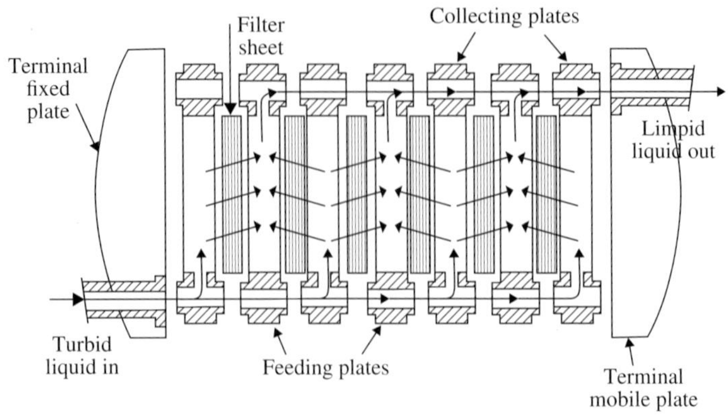

Figure 14.1 The functional scheme of a plate filter.

When the filter is assembled, the terminal mobile plate is pushed by some suitable mechanical system towards the terminal fixed plate so that the filter plates and filter sheets are tightly packed and the channels on the upper and lower sides of the plates join together and form the feeding and collecting ducts of the filter. The duct for feeding the turbid oil has openings to the feeding plates, while the duct for the discharge of the limpid filtrate has openings to the collecting plates.

In mounting the filter sheets, great attention should be paid to mount the rough side of each filter sheet to face the feeding plate. At each filtering cycle, a different number of filter sheets and filter plates can be mounted according to the needs (the total quantity of oil to be filtered and the average filtration rate to be assured).

14.5 Filtration systems

The traditional assembly of a filtration system

Figure 14.2 shows the functional scheme of a traditional filtering system.

The turbid oil arriving from the final centrifugal finish of the milling process is stored in a buffer tank, which is at room temperature. This tank is maintained at a pressure slightly higher than atmospheric pressure. The buffer tank has a conical bottom where the coarser suspended particles are decanted and can be discharged after each filtration cycle. At the beginning of the cycle, the tank valve is opened, the pump is switched on and the turbid oil is fed into the filter. A suitable pressure should be maintained at the filter outlet so that the limpid oil can be fed into the storage tank, which may already be under an inert gas. Two optional features are shown schematically in Figure 14.2:

- A recycling loop from the filter outlet to the pump inlet. This is a safety loop that may be activated in case of plant failure or breaking of filter sheets or

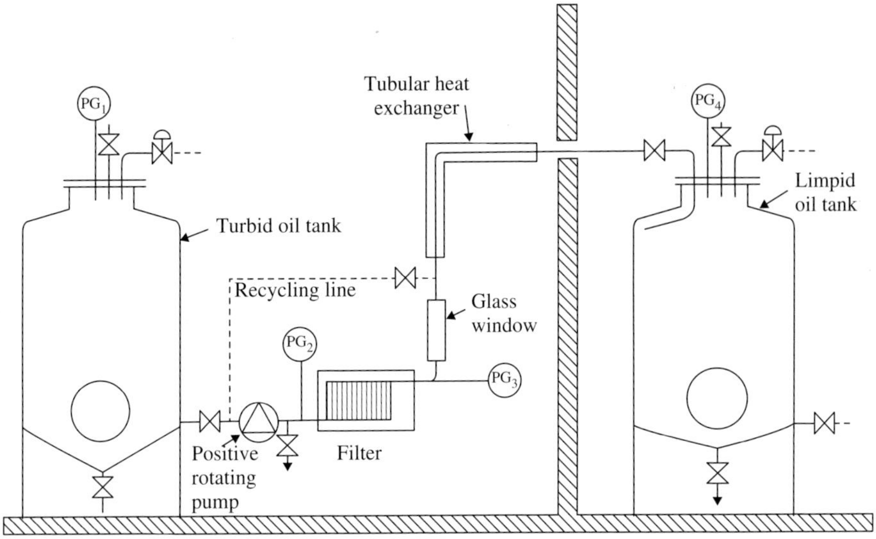

Figure 14.2 The traditional assembly of a filtration system.

improper assembly of the filter pack. Filtrate clearness during the operation can be evaluated through a glass window at the filter outlet.

- A simple tubular heat exchanger in the oil line can be used to cool the oil from the filtration temperature (22–24 °C) to the storage temperature. Common tap water (15–16 °C) can be used for this.

A range of pressure values in the system arrangement of Figure 14.2 can be as follows:

- at pressure gauge 1 (PG_1), 0.1 atm relative pressure (RP)
- at PG_2, 0.4 atm RP
- at PG_3, 0.1 atm RP
- at PG_4, 0.05 atm RP.

The assembly of a filtration system based on hydraulic pressure (no pump)

The arrangement of this system is presented in Figure 14.3 in which the milling and storage rooms are on different floors.

In this case, the hydrostatic pressure and the nitrogen pressure provide the filtration driving force and the pump can be eliminated, thus reducing the mechanical shearing action upon the oil.

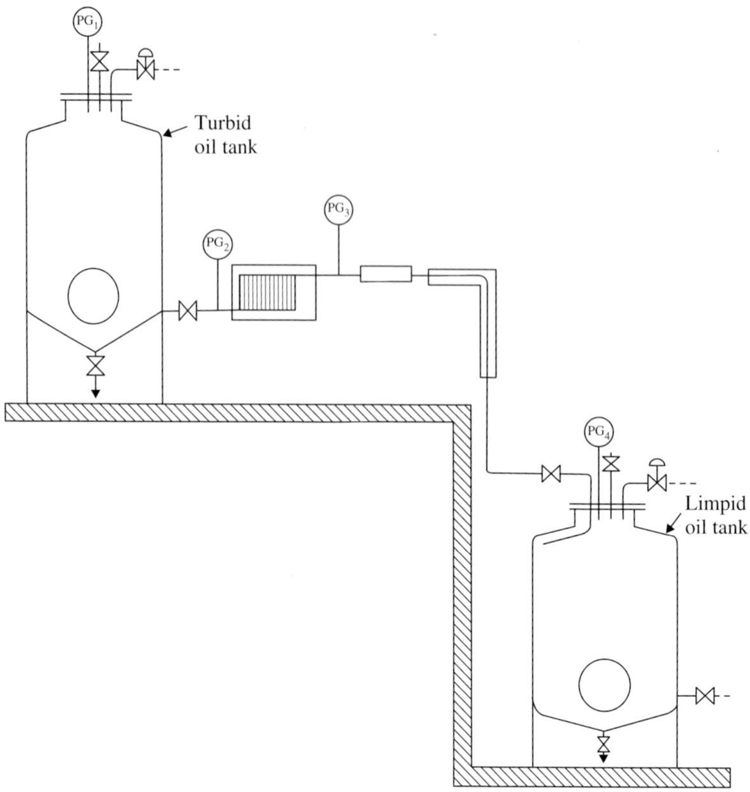

Figure 14.3 The assembly of a filtration system based on hydrostatic and nitrogen pressure (no pump).

The range of pressure values in the system arrangement of Figure 14.3 can be as follows:

- at pressure gauge 1 (PG_1), 0.1 atm relative pressure (RP)
- at PG_2, 0.5–0.2 atm RP
- at PG_3, 0.1 atm RP
- at PG_4, 0.01 atm RP

Hydrostatic pressure at the bottom of a tank in which a liquid of density ρ is at height h, is:

$$P = h\,\rho\,g$$

in which g is the acceleration due to gravity = 9.81 m/s^2. If we assume, for instance, that h is 3 m and the oil density ρ is 951 kg/m^3, the hydrostatic pressure is:

$$P = 3 \times 951 \times 9.81 = 27\,988 \text{ Pascal} = 0.28 \text{ atm}$$

(1 atmosphere is 101325 Pascal).

If we assume that the buffer tank is under a nitrogen pressure of 0.2 atm, the total pressure at the filter inlet varies from $0.28 + 0.2 = 0.48$ atm when the buffer tank is full, to 0.2 atm when the buffer tank is empty. The reduction of pressure during a filtration cycle may not be a disadvantage, because the accumulation of retained solids makes the filter-solids a highly compressible medium with enhanced clogging effects if the pressure is too high.

The alla barese *filter*

The *alla barese* filter is an old gravity filter that may be used for very small-scale, artisanal production of olive oil. Cotton mats are used as filtration media. The surface of the filtering area is small and filtration times are consequently long. Exposure to the air increases the risk of oxidation. However, with this filter there is no pump and the shear stress on the oil is negligible. If temperatures are kept low (20–22 °C) and filtration times are in the range of very few hours, it may still be used in small-scale production with considerable improvement in the appearance and stability of the oil.

A typical arrangement of the *alla barese* filter is shown in Figure 14.4.

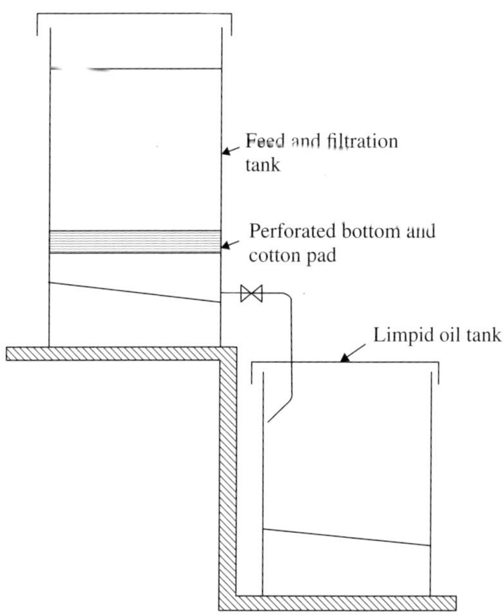

Figure 14.4 The *alla barese* filter.

14.6 Conclusion

The following list reports ten rules for the best practice for extra-virgin olive oil filtration:

1. Reduce as much as possible the suspended material in the oil by using effective finishing centrifuges.
2. Carry out the filtration as soon as possible after the finishing centrifugation.
3. Optimal temperature for extra-virgin olive oil filtration is 22–24 °C (72–75 °F).
4. Use appropriate filter sheets for deep-bed filtration made of pure cellulose or cellulose-diatomaceous earth, asymmetric profile, 10–30 μm porosity.
5. Preferably filter the oil under inert-gas blanketing in both the feeding and the receiving tanks.
6. Use only positive displacement pumps at low rotating speed and low flow rate so that the pressure at the filter inlet never exceeds 0.5 atm relative pressure; adjust the pump speed variator to meet this condition.
7. If possible, due to the available hydrostatic pressure head and nitrogen pressure, eliminate the pump.
8. Based on product characteristics and operator experience, decide on the best combination of assembled filtration area and the volume of the batch to be filtered so that the filter retention capacity is fully exploited and filtration time does not exceed 4 h. Never reuse a filter assembly that has been previously used and has remained inactive for hours (for example overnight). The oil and deposit in the filter medium can spoil rapidly and the spoiled oil can contaminate the following filtration batch.
9. Store the filtering sheets and material in a clean, dry, safe place. Filtering media are a favourite refuge for insects and other pests. At the end of the filtering period each day, thoroughly disassemble, wash and clean the filter. Keep it in a dry, clean environment.
10. Take great care in assembling the filter pack: coarser surface-inlet side/finer surface-outlet side.

Further reading

Earle, R.L., Earle, M.D. (n.d.) *Filtration, in Unit Operations in Food Processing*, www.nzifst.org.nz/unitoperations (accessed 29 September 2013).

Peri, C. (1983) *La filtrazione nelle industrie alimentari*, Edizioni AEB, Brescia.

Peri, C. and Zanoni, B. (1994) *Manuale di Tecnologie Alimentari*. CUSL, Milan, Parts I and III.

15

Extra-virgin olive oil storage and handling

Claudio Peri
University of Milan, Milan, Italy

Abstract

Storage is critical for maintaining the quality of extra-virgin olive oil. Spoilage of oil should be minimized by carefully avoiding temperature abuse, exposure to air (oxygen), exposure to light, the presence of water and organic residues in the oil (cloudiness and deposit), lack of hygiene in the oil environment, exposure to a contaminated atmosphere, and mechanical stress during transfer, pumping or transportation. Time-temperature relationships and inert-gas blanketing are thoroughly discussed. Annex 15.1 is devoted to pumps, tanks and piping assembly for extra-virgin olive oil transfer circuits.

15.1 Introduction

Storage is critical for maintaining the quality of extra-virgin olive oil. The main reason for the frequent spoilage of olive oil during storage, transportation and distribution, as well as during its storage and use at home, is the underestimation of the perishability of extra-virgin olive oil on the part of consumers and experts alike. The fact that microbial degradation is rare in olive oil has contributed to spreading the erroneous belief that olive oil is a stable product and that it can be submitted to mechanical or thermal abuse without penalty. In restaurants, olive oils are normally served in bottles where the oil remains for days or weeks at the wrong temperature and in contact with the air. In these conditions, finding sensory defects happens frequently and the most common fate of an extra-virgin olive oil in a restaurant is the loss of its most interesting sensory notes and health-promoting properties.

The Extra-Virgin Olive Oil Handbook, First Edition. Edited by Claudio Peri.

If an excellent quality oil obtained at the producer's factory is to reach the consumer's table, a revolution in the mentality of olive oil traders and users must take place. They have to start considering that the most interesting characteristics and properties of an excellent olive oil are highly vulnerable. Losing them results in changing an excellent extra-virgin oil to an ordinary extra-virgin oil or, sometimes, to a virgin olive oil with detectable sensory defects. This point is thoroughly discussed in Annex 18.1 of Chapter 18.

The second reason for degradation is the underestimation of the role of time. While the production process from olives to oil is a matter of hours or of a very few days, storage of the oil goes on for months or even years. Time is one of the variables determining the extent of spoilage, while temperature determines the rate of the spoiling reactions. A small increase in temperature has negligible effects if it occurs over a short period of time (hours or days), but it can have catastrophic effects on quality if it continues for weeks or months.

The third reason for oil degradation is the underestimation of some of the spoiling conditions, such as the presence of excess water and solid residues in unfiltered oils, or contamination and poor hygiene of the storage environment.

All things considered, the level of quality of an extra-virgin oil reaching the consumer's table depends most of all on the storage and handling conditions during distribution, transportation, and selling, storing and using at home or at the restaurant.

The consequences of improper storage conditions are summarized in Table 15.1. The main risks factors can be listed as follows:

- temperature abuse;
- exposure to air (oxygen);
- exposure to light;
- presence of water and organic residues in the oil (cloudiness and deposit);
- lack of hygiene in the oil environment and exposure to contaminated atmosphere;
- mechanical stress during transfer, pumping, transportation.

It must be noted that the above risk factors have synergistic effects in the sense that spoiling reactions are greatly accelerated by the simultaneous exposure to several risk factors.

15.2 Prevention of temperature abuse

Temperature abuse is the single most frequent factor of spoilage and therefore it is suggested that compliance with the correct temperature conditions be documented.

Temperature should be maintained within a suitable range without interruption from production to consumption. Low and high temperatures should be avoided. However, the effects of low or high temperature are very different. At temperatures

Table 15.1 Spoilage phenomena and risk factors in extra-virgin olive oil storage and handling.

Spoilage mechanisms in extra-virgin olive oil storage and handling	Risk factors of oil spoilage
Oxidative degradation: loss or reduction of sensory quality and of antioxidant potential; sensory defects	Oxygen concentration, temperature, daylight (UV rays), mechanical stress
Enzymatic degradation: sensory defects	Water and solid residues, temperature, mechanical stress
Contamination by volatiles: toxic substances, sensory defects	Atmospheric contamination by volatile compounds (solvents, lipophilic components of smoke, PAHs)
Chemical contamination: toxic substances, sensory defects	Grease from pump leaks, plasticizers from plastic materials, lipophilic components of smoke, pesticides from aerosols or in-house treatments
Particulate contamination: nonconformity to hygienic standards; negative impact on consumer perception	From the environment, people, insects and rodents, fine smoke particles, dust

lower than 12 °C (53.6 °F), crystallization of triglycerides causes the oil to solidify with consequent difficulties in handling and pouring, but other spoiling reactions are inhibited in these conditions. On the contrary, temperatures higher than the optimal range of 14–16 °C (57–61 °F) accelerate degrading reactions with more serious consequences to the oil quality.

In general, an exponential relationship exists between the time and temperature of any chemical transformation. Hence, a very common way to represent time/temperature relationships of technological operations is in a semilog plot, with temperature on the abscissa in a linear scale and time on the ordinate in logarithmic scale.

The time–temperature relationship that allows the best conditions for oil storage is shown in Figure 15.1 (Peri 2013). Temperatures on the abscissa are between 10 and 25 °C (50 to 77 °F), while time is on the ordinate, in a log scale, between 1 and 1000 days. The two oblique lines divide the diagram into two areas: the area below the oblique lines is compatible with good storage of excellent oil and the area above is incompatible. The space between the two oblique lines is an area of variability depending on the chemical stability of the oil and its antioxidant content.

Temperatures higher than 25 °C (77 °F) are not considered in the graph because they are incompatible with the storage of excellent olive oil. The shaded area below 12 °C (53.6 °F) should also be avoided because crystallization and solidification of triglycerides start at this temperature.

This graph should be used as a reference for planning and controlling the storage temperature. For instance, an excellent oil should not be allowed to remain at 24–25 °C (76 to 77 °F) for more than 2–3 days. It can also be seen that if the oil is stored for 2 months before consumption, the temperature should not be higher than 18 °C (64 °F). It is also evident that at 14–15 °C (57 to 59 °F) an excellent oil can be safely stored for 1–2 years.

The temperatures indicated in Figure 15.1 refer to the oil. Often, the temperature of the storage room is measured and controlled by an automatic air-conditioning

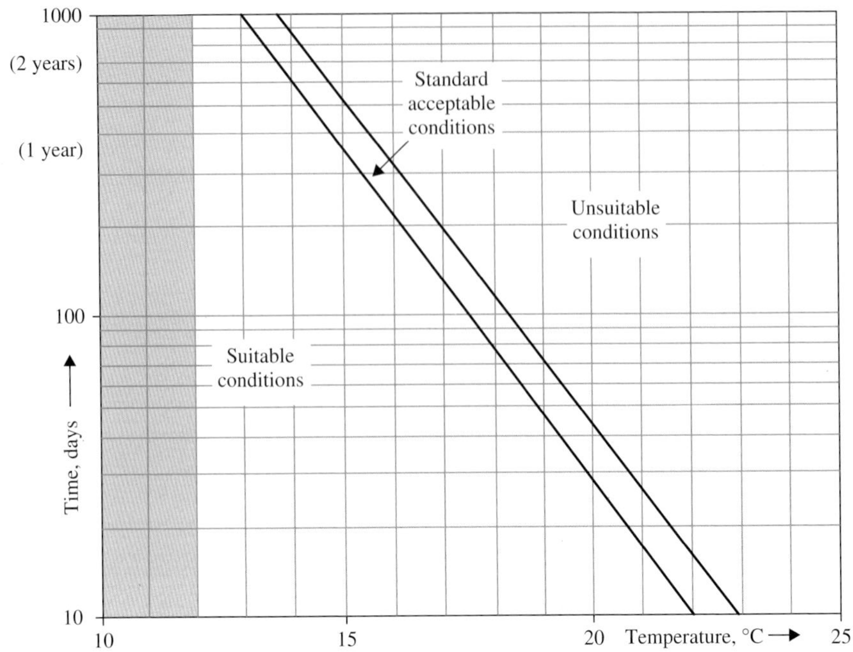

Figure 15.1 The time-temperature relationship of oil storage (*Source*: Peri, C. (2013). Reproduced with permission from Wiley).

system. The right temperature of the storage room generally results in the right temperature of the oil, but it may not be so if the oil is transferred or decanted or filtered or mixed. In these cases, there may be a difference between the temperature of the oil and the temperature of the storage room, which must be taken into consideration for the best control and management of the storage conditions. Continuous recording of the temperature of the storage rooms and periodical measurement of the temperature of the oil is the most appropriate control procedure in handling excellent extra-virgin olive oil.

Control of temperature is easier if the oil is stored in bulk than in bottles. In fact, bottles have a much higher surface-to-mass ratio than tanks and contain much lower quantities of oil. They therefore have a much lower thermal inertia. Consequently, changes in temperature are much faster in bottles than in large containers. This point is discussed further in Chapter 16.

15.3 Prevention of exposure to air (oxygen)

There are two systems for protecting an excellent extra-virgin olive oil against the loss of quality due to oxidation: the first is inert-gas blanketing of containers and the second is using floating roof containers.

Inert-gas blanketing consists in substituting the air present at the top of the storage containers with an inert atmosphere. An 'inert atmosphere' refers to a nonreactive gas that contains little or no oxygen. In extra-virgin olive oil storage, nitrogen is the most common inert gas, but argon is sometimes used. In Table 15.2 the characteristics of nitrogen, oxygen and argon, the three most abundant components of air, are compared. It is worth emphasizing that nitrogen is 78% of normal air, whereas argon is less than 1%. The most important parameter to be considered is their specific gravity (gravity compared to the gravity of air at 21 °C, conventionally taken as equal to 1). The specific gravity of nitrogen is slightly lower than that of air, whereas the specific gravity of argon is higher. This means that in the presence of a small quantity of air in the head space of an oil container, if the inert gas is nitrogen, the small amount of residual air will probably be in contact with the oil surface, while if the inert gas is argon, the small amount of residual air will be pushed away from the oil surface. In the case of nitrogen, flushing of residual air with inert gas (vent open) before seal closing of the tank can minimize this effect. Nitrogen is more widely used due to its availability and relatively low cost.

Table 15.2 Some characteristics of oxygen and inert gases.

Gas	Specific gravity[a]	Concentration in the air (% by volume)
Nitrogen	0.972	78.08
Oxygen	1.105	20.95
Argon	1.377	0.93

Note: [a]specific gravity relative to air at 21 °C (70 °F).

All inert-gas blanketing systems need three controlling devices:

- A pressure reducing valve. Since gas sources provide gas at a much higher pressure than desired, a pressure-reducing valve is needed to decrease the inlet pressure to the tank.
- A blanketing regulating valve. The inert gas pressure inside the container should be slightly higher than atmospheric pressure (10–20 cm of a water column – which is 10–20 millibar – is enough). This is a measure to prevent contamination. If leaks should occur, some gas will leak out rather than contaminants infiltrating the container. This condition is automatically assured by a blanketing regulating valve. When the pressure inside the container drops below a set point, the valve opens and blanketing gas enters. Once the pressure reaches the set point, the valve closes.
- A pressure vent. This consists of a vent that opens when the pressure inside the container exceeds a maximum set pressure. This helps prevent the container from rupturing due to high pressure.

A second system for preventing the risk of oxidation is by using containers with a floating roof that rises and falls with the level of the liquid, thereby eliminating air in contact with the oil surface. A sliding seal between the roof and the tank shell assures a hermetic separation between the full and the empty part of the tank.

Nitrogen sources

Nitrogen for inert-gas blanketing can be supplied in traditional gas cylinders or may be produced at the oil factory. In this case, two options are available:

- *PSA (pressure swing adsorption) technology* uses two towers filled with carbon molecular sieve (CMS). Compressed air enters the bottom of the 'online' tower and flows up through the CMS. Oxygen and other trace gases, which are small molecules, penetrate the pores of CMS, while the larger nitrogen molecules by-pass the CMS and emerge from the top of the column as nitrogen gas. After a preset time, the online tower automatically switches to the regenerative mode, venting contaminants from the CMS. Pressure swing adsorption nitrogen generators typically produce very high purity nitrogen (higher than 99.5% purity).
- *Membrane technology* is based on filtration of dry compressed air through hundreds of thousands of hollow fibre membranes, the diameter of a human hair. Oxygen and the other trace gases permeate through the membranes, while nitrogen is retained and collected at the other end of the fibres. Membrane nitrogen generators are typically used in applications where the purity requirement is below 99.5%

For oil storage, both systems have an acceptable performance.

The final choice between cylinder supplied or nitrogen production with one or the other of the two methods depends on cost, and this in turn depends on the location of the factory (proximity to a nitrogen supply service) and the volume of nitrogen needed.

15.4 Prevention of exposure to light

Exposure of extra-virgin olive oil to light favours photo-oxidation of the oil. The presence of photosensitizing pigments like chlorophyll and carotenoids can speed up oxidative phenomena. Short-wavelength radiation, particularly the UV part of the spectrum, energizes the oxygen molecules into a more reactive form called triplet oxygen, which greatly accelerates lipid oxidation. This is usually not a problem at the oil factory where oil is stored in stainless steel containers.

For bottles, different coloured glass is used: from brown to dark brown, from dark green to violet. The effectiveness of the various glass types is the object of further insight and discussion in Chapter 16. In any case, in the absence of precise information, the best recommendation is to store the bottles in a dark place.

15.5 Prevention of water and organic residues in the oil

Water and suspended solids may spur microbial and enzymatic spoilage of the oil. It is therefore advisable that, after the final centrifugal finish, the oil is filtered with pure cellulose pads or other suitable material. This should not be considered as an option, but as an essential condition for good oil storage (Chapter 14).

15.6 Prevention of exposure to contaminated atmosphere and poor hygienic standards

Extra-virgin olive oil must be stored in hermetically sealed containers in order to prevent any particulate or volatile contamination from the environment or the atmosphere. Covers simply resting on top of containers, a common system in small olive oil factories, do not safely protect the oil from atmospheric contaminants or intrusion of pests. At the same time, the overall hygiene of the oil factory must be implemented according to the requirements and practices recommended in Chapter 21.

15.7 Prevention of mechanical stress

Turbulence and shearing actions on oil should be avoided. They may negatively influence the perceived thickness of the oil and favour dispersion of air. In the presence of water and solid residues, turbulence and shaking favour dispersion of the solids in the oil.

Negative effects of turbulence and shearing stress are linked to the high viscosity of oil and to the use of unsuitable pumps. This problem is discussed in detail in Annex 15.1.

Annex 15.1: Pumps, tanks and piping

Introduction

During the storage and commercial life of an oil, several operations requiring mechanical action on the oil are carried out, for example pumping, blending, pouring, filtering and bottling. The physical property that is more directly related to the effects of mechanical stress upon the oil quality is viscosity.

Viscosity (symbol: μ) may be defined as the resistance of a liquid to flow. It may be considered as a sort of internal friction and is expressed in a unit called 'poise' (actually dyne-second per cm^2). The most commonly used unit is the centipoise (symbol: cP), which is one hundredth of a poise. This practical unit is used because the viscosity of water (something that is familiar to everybody) is 1 cP at 20 °C.

At 20 °C, extra-virgin olive oil has a viscosity of 84 cP and is therefore much more viscous than water. This means that, under the same driving force as, for example, gravity, oil flows much more slowly than water. The high internal friction is also the reason why oil flows from a dispenser as a thread instead of breaking into drops.

When a viscous product like extra-virgin olive oil (or any other vegetable oil) is transferred through pumps, pipes and valves, great care should be taken to avoid any turbulence and shaking. In fact, these actions not only cause large energy dissipation but may also cause air absorption and emulsion formation, especially in the presence of suspended solids and residual water in nonfiltered oils. These effects may impair oil stability and determine undesirable changes in sensory perception.

In Table A4 in the Appendix, the viscosity of extra-virgin olive oil is reported as a function of temperature. It is evident that at the optimal storage temperature (15 °C, 59 °F) the oil viscosity is high and handling should be particularly careful. Extra-virgin olive oil is an inherently 'slow' product and forcing it to flow faster can only result in a stability and quality risk.

Pumps

The choice of suitable pumps is a critical detail for avoiding mechanical stress to the oil.

Centrifugal pumps (never to be used for extra-virgin olive oil transfer)

In an olive mill there are many centrifugal pumps. They are used for all water transfer duties: in the plant of olive washing, in heating and cooling circuits, in wastewater discharge, and so on. Centrifugal pumps are the best for pumping water. Their main advantages are: simplicity, low cost, uniform (nonpulsating) flow, small floor space, low maintenance cost, quiet operation. Figure 15.2 presents the functional scheme of a centrifugal pump. It consists of a case containing a rotating blade impeller. Power from an outside motor is applied to the impeller. The rotating blades produce

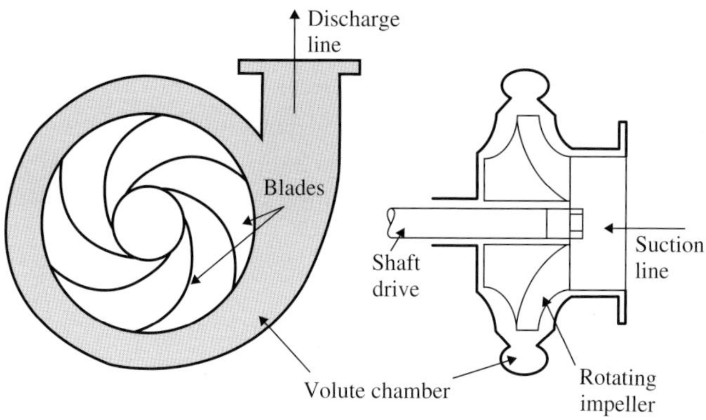

Figure 15.2 Orthogonal sections of a centrifugal pump.

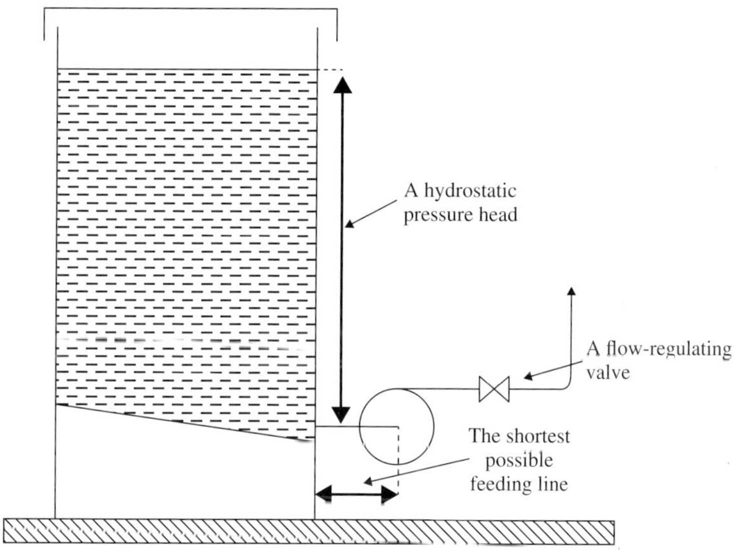

Figure 15.3 The three essential features of a centrifugal pump circuit.

reduced pressure at the entrance of the impeller, thus causing the inlet liquid to be sucked from the feeding pipe. The liquid follows a spiral pattern with increasing tangential velocity. The velocity (kinetic) head is transformed into a pressure head as the liquid passes into the volute chamber and is pushed out to the discharge pipe.

Figure 15.3 is a schematic representation of the correct assembly of a centrifugal pump. There are three main precautions that must be followed when using a centrifugal pump: (i) the pump must be installed as closely and directly as possible to the feed tank; (ii) the pump should be at a level lower than the feed liquid level in order to facilitate pump priming; and (iii) a regulating valve should always be present at the outlet of the pump in order to create suitable pressure head losses and to adjust the discharge of the liquid at the desired flow rate.

The performance data of a centrifugal pump consist of flow rate and pressure head: flow rate decreases with increasing pressure head losses (resistance) in the discharge circuit.

Centrifugal pumps cannot be used with viscous and air-sensitive fluids because they generate very high turbulence with mechanical stress, air mixing and emulsification effects. This is why centrifugal pumps should never be used for pumping extra-virgin olive oil.

Positive displacement rotary pumps

These are the most suitable pumps for extra-virgin olive oil. They assure a constant (positive displacement) and continuous (rotary) flow, while minimizing shearing effects.

Figures 15.4 and 15.5 represent gear and lobe pumps, respectively.

Two impellers with the shape of toothed-gear wheels (Figure 15.4) or of lobed cams (Figure 15.5) rotate in the case with very low clearance between them and the casing. The pumped liquid flows into the spaces between the impeller teeth or lobes as these cavities pass the suction opening. The liquid is then carried to the discharge port and forced into the discharge circuit.

Figure 15.6 represents a single rotor screw pump ('mono' pump) that can be used for viscous fluids like extra-virgin olive oil and also for semi-liquid slurries as, for example, olive paste or pomace. In the functional scheme represented in Figure 15.6, a wormlike spiral (screw) impeller rotates in an elastomeric stationary sleeve. The liquid is transferred by means of the helical progress of concrete cavities as the rotor is turned. In the representation of Figures 15.6 a and b, the screw position is represented 180 ° (half a revolution) apart. In Figure 15.6b, the liquid which was in

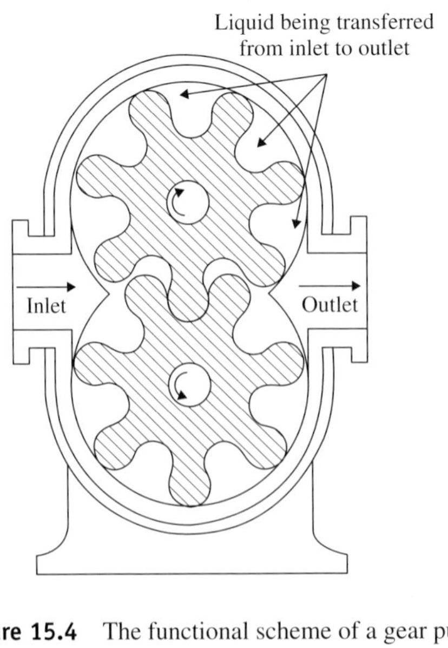

Figure 15.4 The functional scheme of a gear pump.

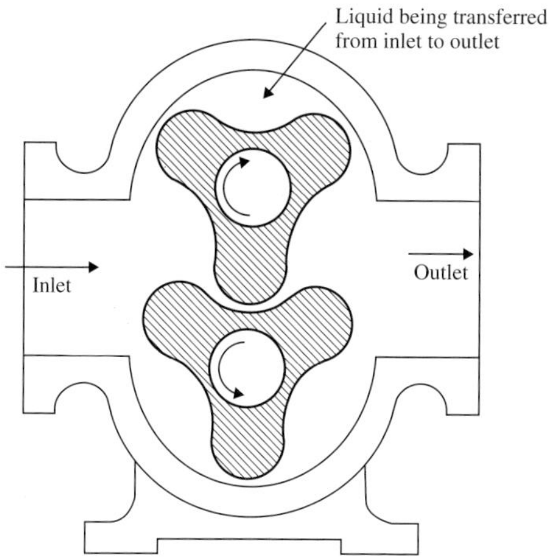

Figure 15.5 The functional scheme of a lobe pump.

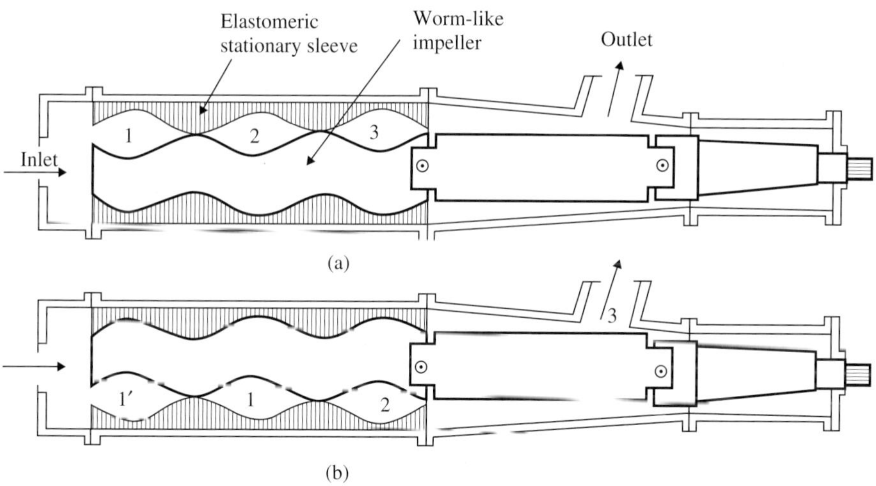

Figure 15.6 The functional scheme of a mono pump in two subsequent positions 180° (half a revolution) apart.

position 3 in figure 15.6a has been discharged. The product that was in position 2 and 1 has progressed one position toward the outlet; and a new liquid portion is entering in the new 1′ position.

This pump is extremely versatile. In the food industry it is often used for gentle handling of large solids suspended in a liquid and in all cases of shear-sensitive materials.

Figure 15.7 presents the functional scheme of a peristaltic pump. In this pump, the liquid is contained in a flexible tube fitted inside a circular casing. A lobe rotor

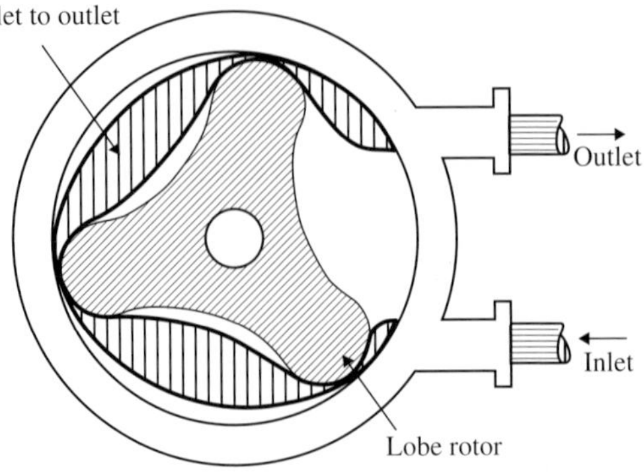

Figure 15.7 The functional scheme of a peristaltic pump.

compresses the flexible tube, thus trapping and transferring discrete portions of the liquid towards the outlet.

An important difference between centrifugal and positive displacement pumps

Unlike centrifugal pumps, in positive displacement pumps the flow rate is constant independent of the head losses in the discharge circuit. Flow rates cannot be adjusted by using a valve at the pump outlet. Such a valve will have practically no effect on the flow rate and closing it completely will involve generation of very high pressures. To prevent this risk, positive displacement pumps are often fitted with cut-off pressure switches or a bypass pipe that allows a variable amount of fluid to return to the inlet. Flow-rate variation in pumps for extra-virgin olive oil requires that the pump has a speed variator of turning impellers, so that the rotating speed is adjusted according to the desired flow rate.

Tanks and piping

Extra-virgin olive oil tanks are often vertical and cylindrical in shape, with fixed or floating roofs. They should have hermetically closed tops to prevent contamination of the oil from the environment. Tanks may have a flat bottom, a cone-shaped bottom or a sloped bottom. They must have rounded corners going from vertical walls to the bottom for easier cleaning and to withstand the hydraulic pressure of the liquid. The transfer of oil should be carried out at low flow-rate and low flow velocity, through smooth pipes, avoiding as much as possible sudden changes in pipe diameter or in the direction of flow, minimizing the distances and the number of obstacles to flow, including measurement devices and valves. When discharging the oil into a tank, the incoming flow should be directed close to the tank wall so that the oil slides down the sides without splashing.

Figure 15.8 shows a typical piping arrangement for the transfer of extra-virgin olive oil. In particular, notice the indication of a circuit-emptying valve, which must always be present at the lowest point of the circuit so that all the oil present in the circuit (including pipes and pump) can be discharged by gravity at the end of the operation. The oil should never remain in the circuit or part of it for a long time because the circuit conditions do not usually guarantee the proper temperature. All piping and valves should be part of the cleaning-in-place system.

Floating roof containers have some advantages when transporting oil. In fact, during transportation, the presence of a head space either filled with air or with an inert gas favours shaking the oil, which may accelerate spoiling effects especially in the presence of water and solid residues. With floating-roof arrangements, in the absence of a head space, the shaking effect is minimized.

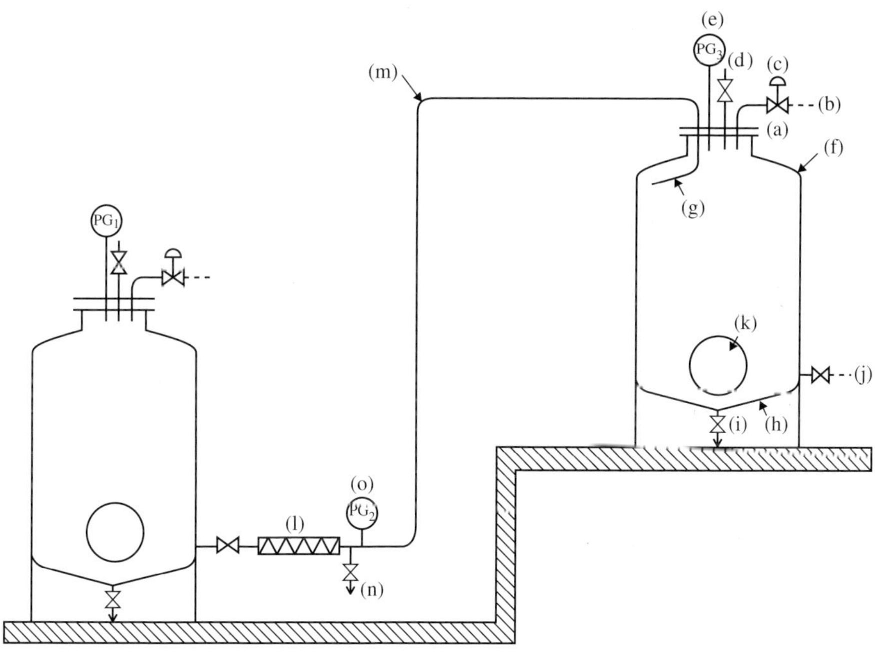

Figure 15.8 The essential features of a circuit for transferring extra-virgin olive oil from one tank to another. (a) an hermetic seal; (b) the inert gas inlet; (c) the automatic pressure regulating valve of inert gas; (d) vent; (e) pressure gauges; (f) rounded corners; (g) oil discharge close to the wall so that oil slides down the sides without splashing; (h) cone shaped bottom; (i) valve for total emptying of the tank; (j) tank discharge valve; (k) opening for access, repair, maintenance; (l) a positive, rotating, low-shear, pump; (m) smooth pipes with large radius curves; (n) the valve in the lowest position for total emptying of the oil circuit; (o) an automatic regulation of maximum pressure.

Finally, principles of hygienic design discussed in the annex of Chapter 21 must be accurately applied.

Reference

Peri, C. (2013) Quality excellence in extra-virgin olive oil, in *Olive Oil Sensory Science* (eds E. Monteleone and S. Langstaff), John Wiley & Sons, Ltd, Chichester.

Further reading

Earle, R.L. and Earle, M.D. (n.d.) Fluid-flow application, in *Unit Operations in Food Processing*, www.nzifst.org.nz/unitoperations (accessed 29 September 2013).

Peri, C. and Zanoni, B. (1994) *Manuale di Tecnologie Alimentari*, CUSL, Milan, Parts 1 and 3.

16

Extra-virgin olive oil packaging

Sara Limbo,[1] Claudio Peri[2] and Luciano Piergiovanni[1]
[1] *Department of Food, Environmental and Nutritional Sciences, University of Milan, Milan, Italy*
[2] *University of Milan, Milan, Italy*

Abstract

The packaging process is described as a sequence of operations from the purchase order to shipment of the product to the customer. The characteristics and performance of glass, metal and plastic containers in the packaging of extra-virgin olive oil are presented and compared in terms of cost, protection from light and oxygen, possibility of recycling and reuse, mechanical resistance and inertness. Various filling and closing techniques are described. Finally, the limitations and possibilities of labelling are discussed.

16.1 Introduction

Packaging operations can be classified in three levels as:

- primary packaging, which is in direct contact with the product
- secondary packaging, containing several primary packages – for example, a corrugated case;
- tertiary packaging, containing a number of secondary packages – for example, a stretch-wrapped pallet of corrugated cases.

Only primary packaging will be dealt with in this handbook. Primary packaging must satisfy five principal and interconnected functions: containment, protection, convenience, communication, and sustainability.

The Extra-Virgin Olive Oil Handbook, First Edition. Edited by Claudio Peri.

Containment

This is the basic function of packaging and it mainly concerns the size, weight and shape of the package. In the case of extra-virgin olive oil, this point deserves careful consideration. In fact, for a product that is consumed at a rate of 10–20 ml per person per day, the sizes and volumes presently used in distribution and retail outlets, as well as in restaurants, are totally disproportionate. In many restaurants, a 750 ml bottle is used to serve customers by putting it on the restaurant's tables for days or weeks at an improper temperature and in large contact with air. These are unsuitable conditions for handling good or excellent extra-virgin olive oil.

The appropriate solutions for restaurants and families are twofold: (i) use smaller containers so that the contents are used before being spoiled, and (ii) store extra-virgin olive oil in suitable dispensers kept at the proper storage conditions, from which single or multiple portions can be taken to be served at the table according to the need. The role of packaging in this scenario is essential.

Protection

Edible oils in general, and extra-virgin olive oil in particular, need containers with specific protection properties at every step of the chain from production through storage, transportation, distribution, and final selling and use:

- First of all, the oil should be protected from contaminating substances from the environment and especially moisture, oxygen, odours, smoke, dust, micro-organisms.
- Secondly, the oil should be protected from nonpolar, hydrophobic contaminants such as solvents or additives, monomers and degradation products deriving from contact materials.
- Thirdly, the oil should be protected from light, which accelerates photo-oxidation.
- Finally, like any other food whose excellence is based on origin and authenticity, containers must be fraud and tamper proof.

Convenience

Characteristics that make packages easy to carry or hold or open and reclose or that make it easier to pour the oil without dripping are so important and influential in consumer choice that this function is referred to as the 'smart function' of packaging. Small size, single-portion containers and the dispensers discussed under point 16.1.1 (containment) are also convenience requisites.

Communication

A frequently cited sentence is: 'A package must protect what it sells and sell what it protects'. Distinctive branding and labelling on containers should enable supermarkets to function on a self-service basis. At the time of purchase, consumers are greatly influenced by easy-to-catch and easy-to-read information on the package. Nutritional information, country or region of origin, date for best consumption, harvesting year, certification marks and so forth are all crucial points of information for consumer decision. Communication also includes the United Product Code (UPC), which can be rapidly read by modern scanning equipment at the retail checkout.

Sustainability

Sustainability has become a critical issue in packaging research and development, but also an essential factor in consumer choice. The possible adverse impact of packaging on the environment is a hot technical, regulatory and political issue. Not surprisingly, food-packaging materials are the focus of many texts on environmental protection. Reducing the amount of waste due to packaging materials and recycling of packages is key to future development.

Economic sustainability is another aspect to be considered. The cost of the package should be in a reasonable relationship with the cost of the product.

16.2 The packaging process

The packaging operation, consisting in the sequence of filling-closing-labelling is part of a more complex packaging process whose optimization criteria are much different from those of olive production and milling. It requires relatively complex materials and equipment, significant investments and close connections with marketing.

At the same time, packaging is so critical for marketing success that many small-sized companies producing relatively low amounts of excellent olive oil, have their own bottling and packaging facilities. Large and small packaging plants with operating systems from full automation to manual coexist in the extra-virgin olive oil world.

Co-packaging is a new approach to packaging in areas where the size of manufacturing companies is too small to afford the cost and operation of a modern packaging plant. A contract packer, or co-packer, is a company that specializes in packaging products as an outsourcing service for other manufacturing companies.

A co-packer works under contract with a hiring company and is fully responsible for the packaging process. The main benefits of co-packaging are:

- sharing investments to reach a high technological level, which may be needed for packaging containers of special size and form, inert-gas blanketing, refilling-proof closures, and so on
- bringing together know-how and experience in a unique place, developing new practices and solutions
- raising the quality and safety level of the packaging process
- improving the efficiency of the packaging process due to large volumes and economies of scale.

A co-packaging centre can hire staff and workers with specific knowledge and skills in packaging; it can more easily obtain packaging material at good economic prices from selected suppliers.

Figure 16.1 shows the flowchart of the olive oil packaging process. It shows that a packaging process starts from a purchase order from a customer, defining the quality and quantity requested (as well as time and cost). The end of the process is the packaging company forwarding the extra-virgin olive oil in packages to the customer, complying with the legal and technical terms according to agreed upon requirements.

The first critical operation of the process consists in defining and forming a 'packaging lot', which is a homogeneous batch of oil conforming to customer requirements. There are two possibilities: the requirements are met by an existing storage batch or a new 'packaging lot' must be obtained by suitably blending oils from various storage batches.

Blending is a complex operation requiring high-level skill. It can be carried out according to two very different approaches or any suitable combination of the two:

- The most sophisticated approach consists in applying a computational optimization procedure. A set of optimization algorithms elaborate analytical data and the cost of oils from different storage batches. On this basis, the percentage weight of different batches that must be blended are defined in order to obtain a packaging lot conforming to the customer requirements at the lowest cost.
- The simplest approach is based on highly skilled expert tasting of different blends until the required sensory profile is achieved. This procedure is often applied to small-scale production of excellent olive oils.

After the new packaging lot is made, it undergoes a final and complete evaluation to characterize the analytical and sensory profile of the oil. If the result of this evaluation is not completely satisfactory, some feedback and refining of the blending may be necessary. If results conform to the requirements, the oil is sent to the packaging operation.

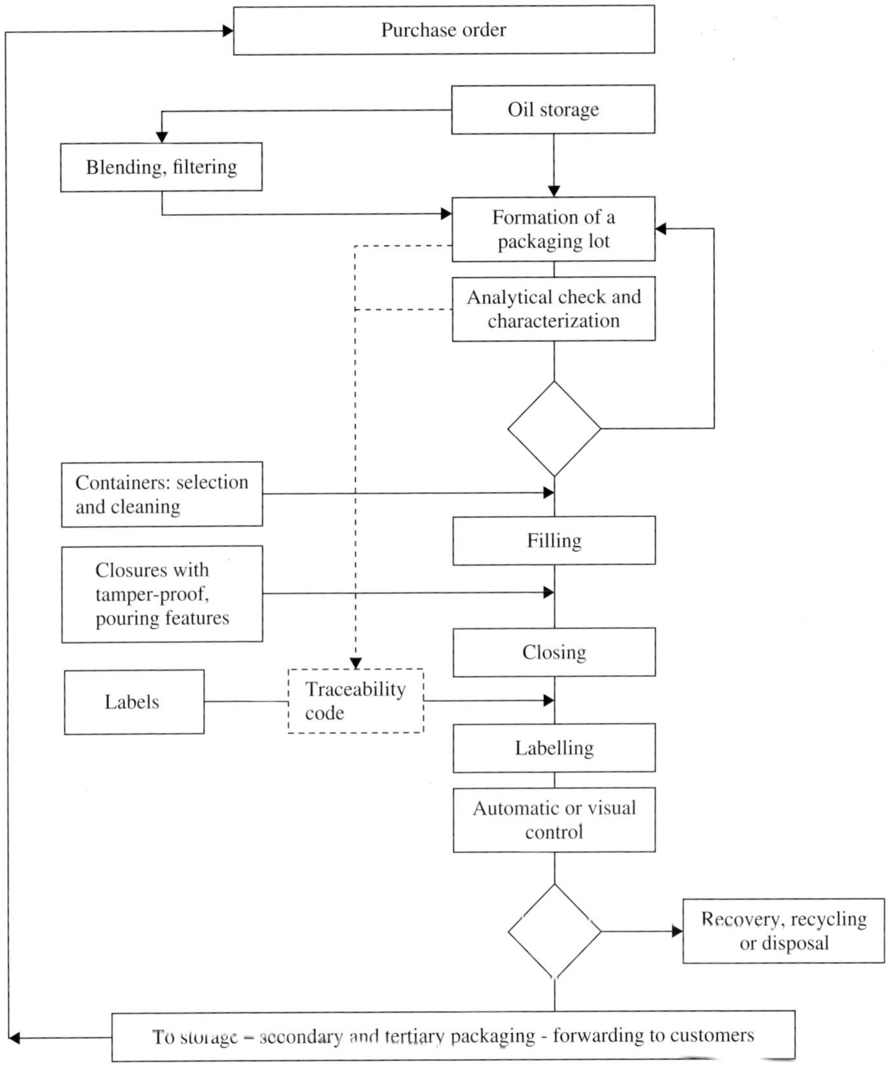

Figure 16.1 The flowchart of the packaging process.

The formation and characterization of packaging lots is the most critical step of product traceability (Chapter 19). This is the step in which the identity and authenticity of the product are verified and guaranteed to the final consumer. An accurate and reliable record of this step represents the point of connection of the marketing and the production segments of the olive oil chain.

After the filling-closing-labelling operation, a final automatic or visual control has to be carried out to verify conformity of the packages, integrity of the containers, correct positioning of labels, and so on. Packages that do not conform to standards are discarded, recycled and recovered or disposed of.

Packaging As Soon As Possible or As Late As Possible: ASAP or ALAP?

From the quality point of view, is it preferable to store the oil in tanks and to bottle it just before shipping it to the customers, or is it preferable to bottle the oil as soon as possible and store it in the bottles before shipping them to the customers?

If critical storage conditions are under control in both cases, the two options can be considered as equivalent but the risk of non-conformity is not the same.

Prevention of light exposure and photo-oxidation

In a stainless steel tank, protection from light is assured 100%; in a glass or plastic bottle this depends on the light shielding characteristics of the packaging material. Unfortunately, this risk cannot be avoided during the marketing life of the bottled product and during exposure on the retail shelves. From this point of view, the closer the oil is bottled before selling to the final consumer, the safer it is from light exposure and abuse.

Prevention of oxygen contact

In a stainless-steel tank, which is duly maintained under an inert atmosphere at a pressure slightly higher than atmospheric pressure, this risk is negligible. The risk of oxygen permeating plastic containers is high. Glass and metal containers guarantee suitable protection only if air is flushed from the headspace of the bottles and replaced with inert gas. Even in this case, however, the plastic components of the closure (including the pouring device) can be a risk for oxygen permeation.

Prevention of temperature abuse

A storage tank and a bottle inside an air-conditioned storage room that is maintained at the right storage temperature ($15 \pm 2\,^\circ$C) have the same level of protection from temperature abuse. The vulnerability to temperature changes is, however, very different and much higher in a bottle than in a tank for two reasons:

- A bottle has a much higher surface-to-volume ratio than a tank and this means a much higher heat transfer rate. It can be easily calculated that the surface-to-volume ratio in a cylindrical 0.5 litre bottle is about 30 times higher than in a storage tank with a 5 m^3 capacity.
- The total heat capacity of the oil contained in a tank is much higher than the heat capacity of the oil contained in a bottle. Being that the heat capacity is proportional to the mass, the comparison of the two situations presented above shows that the heat capacity of the oil in a 5 m^3 tank is 10 000 times greater

than the heat capacity of a 0.5 litre bottle. This means that the same amount of heat has a negligible effect on the oil temperature in a tank, but a relevant effect on the temperature of the oil in a bottle.

It is evident that temperature changes in the environment may be much more detrimental to oil quality if the oil is in bottles than in tanks.

Both the tank and the bottle can be considered a safe protection from moisture or contaminants or micro-organisms from the environment.

In conclusion, oil should be stored in optimal conditions in bulk and bottled as late as possible (ALAP) before the shipment date to customers. When in bottles, great care should be given to protect the bottles from natural or artificial light and from temperature change.

16.3 The packaging materials

Packaging materials, including closing and pouring features that are in contact with the oil, may influence its quality and shelf-life. Careful consideration should be given to the choice of the most suitable material considering the environmental conditions and the shelf-life from packaging to consumption. A first crucial point is about consumer preference and the perception that consumers have about what is the most suitable container for extra-virgin olive oil. An investigation among regular consumers in Italy gave results that are summarized in Table 16.1 (elaborated from Esposito 2012). A conventional value of 100 is given to the most preferred material (dark glass).

Despite scarce information and available knowledge, Italian consumers seem to give a sharply differentiated rating to their preference of packaging material. Preferences may be different in different countries, but in any case this point has a decisive influence on the choice of packaging material for extra-virgin olive oil.

For general works on packaging materials see Piergiovanni and Limbo (2010a, b), Pistouri *et al.* (2010) and Robertson (2010).

16.3.1 Glass bottles

Glass provides a total barrier to moisture and gases, including oxygen. However, glass bottles do have some weak points (Gawel 2010). Transparent glass and exposure to light (both in the UV and visible wavelength range) lead to photo-oxidation.

Ultraviolet (UV) light is electromagnetic radiation with a wavelength shorter than that of visible light – that is, lower than 400 nm. The ISO standard on determining solar irradiances defines Ultraviolet A (UVA) radiation between 400 and 315 nm, Ultraviolet B (UVB) radiation between 315 and 280 nm, and Ultraviolet C (UVC) radiation between 280 and 100. Radiation in the range of 400–300 is defined as ‘near ultraviolet’.

Table 16.1 The preference scale of Italian consumers for extra-virgin olive oil containers (*Source*: Reproduced with permission from Alessandro Esposito).

Container material	Preference scale
Dark glass	100
Clear glass	40
Metal	30
Plastic	0.7

Light in the UV region and in the visible region can contribute to oil oxidation through different pathways. The most harmful wavelengths are usually in the ultraviolet part of the spectrum, just below the visible range, due to the high-energy content, which is capable of splitting certain chemical bonds. The shorter the wavelength, the higher is the energy of light and the higher its oxidation-promoting effect.

The energy associated with the UV region can trigger the formation of free radicals, which play a fundamental role in the oxidative reactions. The visible region of the spectrum is characterized by wavelengths with low energy but it is also harmful, especially for oils and coloured beverages (Limbo *et al.* 2007). In fact, their sensitivity to light in this region is due to their pigment content, which is stimulated by quantas of the visible spectral region, acting as photosensitisers (Thron *et al.* 2001). The oxygen that is dissolved in the oil is 'energized' when hit by light and transformed into an active state, which is called singlet oxygen. In this form, oxygen triggers oxidative chain reactions leading to discoloration, off-flavour formation, rancidity and loss of antioxidants. For this reason, photosensitizing pigments like chlorophyll accelerate these phenomena. In general, the reaction rate of singlet oxygen with some food components is much greater than that of triplet oxygen. Singlet oxygen can attack double bonds directly; its reactivity with linoleic acid is about 1450 times faster than that of triplet state oxygen (Min and Boff 2002).

Many producers think that the darker the glass, the more effective it is in preventing photo-oxidation. This is not true and the distinction between the light-transmission properties of packaging materials in the UV and visible regions are fundamental in determining the right protection for oils. Glasses with a low trasmittance (high absorption) of light in the UV range are the best for protecting oil from radical oxidation and this they can be made by adding special additives in producing the glass. However, exposure to UV light during storage is not so frequent, especially at the retail level, where the lamps used to display products are poor in UV emissions. On the other hand, while displayed on sales shelves these packages are exposed to visible light that may impair oil quality.

Figure 16.2 shows the spectra of light transmittance of coloured glass (see Gawel, 2010 and Piergiovanni and Limbo 2010a, Pistouri *et al.* 2010). The glass labelled 'flint' is clear glass. By observing the spectra variation as a function of wavelength, it can be easily established that blue and emerald green glasses are very ineffective in filtering the most dangerous UV light, in the range of UVA and near-UV. Dark

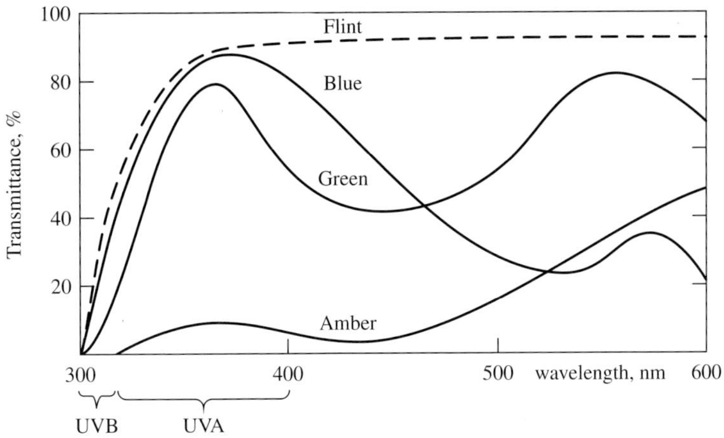

Figure 16.2 Light transmittance of various collared glass (*Source*: Robertson, G.L., (1993). Reproduced with Kind permission from Taylor & Francis).

blue bottles are effective filters of visible light but this has a low protective effect on the oil. The green colour and the dark green, similar to the colour of olives, have a similar but slightly better performance. The brown amber bottles have the best light-protecting effect. The thickness of the glass wall has little influence on protection from light.

Other weak points of glass are fragility and high specific weight, which present problems in handling and carrying. Glass fragility may lead to contamination of the oil by glass fragments. Automatic or visual inspection followed by air blowing just before filling should be considered as a critical control point.

Glass is a perfect barrier to gas and oxygen, but the polymeric closing and sealing materials can be considered as a minor risk to oxygen permeation.

Finally, temperature abuse can be a problem for glass bottles, especially if the control of storing and transportation conditions in the commercial network is not satisfactory.

From the environmental point of view, glass containers represent an almost ideal recyclable product; glass can be fully recycled when properly disposed of.

16.3.2 Metal containers

Metal containers are light in weight, have good mechanical resistance and are a total barrier to light and oxygen. They are made of tin plate or tin-free steel (TFS) based on chromium instead of tin plate or also aluminium or aluminium alloys. The containers are protected from corrosion by an internal coating with food grade enamels (lacquers).

Metal containers are supplied by the manufacturer top-bottom closed with only one orifice to be used for filling and capping or just bottom closed. The latter allows inspection and air blowing before filling, which is a useful precaution. Bottom-closed containers are closed with top installation after filling.

Metal containers reach the oil company ready for use and must be protected in the original package for storage in a clean warehouse.

Metal containers are normally used in sizes from 1 to 20 litre. However, special brands have recently adopted metal containers as small as 0.1 litre.

Metal containers can be recycled, but corrosion of the internal coating is a limitation to reuse.

16.3.3 Plastic containers

Plastics are relatively new materials for the packaging of edible oils, but seldom used for extra-virgin olive oil. The polymers most frequently used are polyethylene terephthalate (PET), high-density polyethylene (HDPE) and polyvinylchloride (PVC). They have some advantages over metal and glass containers, namely a lower price, a lower weight and a higher mechanical resistance.

Polyethylene terephthalate, in particular, is receiving some attention for its brilliance and transparency, enhancing the aesthetic appeal of the oil (Gawel 2010; Rizzo *et al.* 2013). However, plastic containers do not provide as long a shelf life as metal or glass. Their major drawbacks are scalping, migration of contaminants, oxygen permeability and light transmission.

Scalping consists in the absorption of volatile and nonvolatile components of the oil at the plastic surface with loss or reduction of desirable flavour compounds.

Migration of plastic additives, monomers or oligomers not chemically bound to the polymer matrix may contaminate the oil. This is a serious safety threat and a violation of olive oil legislation. Migration of acetaldehyde from PET bottles can seriously damage the sensory profile of the oil. Hydrophobic affinity of the oil and the plastics greatly favours scalping and migration phenomena.

Oxygen permeability is a major limitation in the use of plastic containers for extra-virgin olive oil. Recent advances in the development of active packaging have led to plastics, especially PET bottles, with ‘oxygen scavengers’ that not only avoid oxygen permeation from the outside, but also remove the dissolved oxygen in the oil. The action of the oxygen scavengers can be further potentiated by coating with high-barrier resin. Results seem to be encouraging, but consumers are still very reluctant to accept plastics as extra-virgin olive oil containers.

Light transparency is also a weak point of plastics. For instance, PET, which has a very good filtering capacity in the wavelength range of UVB, is highly transparent to light in the UVA region. Built-in UV blockers could possibly assure satisfactory light-filtering performance.

Plastic containers are the unique category of containers that can be manufactured directly at the oil factory, just before filling, using preformed pieces that are heated and blown on site. This feature provides an extra guarantee against contamination of solid powder and particulates. Of course, preformed containers are also available and they have the same safety/cleaning problem as glass and metal containers. Plastic containers can be easily recycled and incinerated, if properly collected and managed. They cannot be reused.

16.3.4 Bag-in-box (BiB) containers and dispensers

The bag-in-box is a container consisting of a strong bladder (a plastic bag) made of several layers of metalized film or other plastic seated inside a corrugated paperboard box. The bag is supplied to the filler as an empty pre-made bag. The filler removes the tap, fills the bag and replaces the tap. The bags are available for semi-automatic machines or for automated filling systems. There is also a version in which the bags are manufactured online from reels of film and then the tap is inserted and finally filled in a rotary head filler. This technology is widely used with wine. The absence of light is assured by the paperboard box.

Due to its collapsible characteristics, the plastic bag shrinks when the oil is poured out so that the presence of air in the headspace is avoided. It must be noted, however, that oxygen permeates through the plastic and therefore, extra-virgin olive oil should not be kept in a BiB for longer than 3 months.

In 3–5 litre containers, a built-in spigot allows the oil to be poured according to the needs of use or consumption in families and restaurants. The use of BiB as an oil dispenser allows the oil to be served at the table in portions, while keeping the rest of the oil in the absence of air and at the optimal temperature. The ease of printing on the paperboard box allows attractive figures to be drawn and to give the consumer suitable information. These features are fostering increasing success of the BiB system as an oil dispenser in restaurants and families (Monini SpA 2013).

16.3.5 The choice of the most suitable package

Packages of various sizes and material are generally used by extra-virgin olive oil companies, depending on customer needs, the use of oil, the marketing turnover, the conditions of storage and transportation. The decision about packages has very significant consequences on oil cost, as is clearly shown in Chapter 23.

Such a decision should take into account the most suitable characteristics in the different cases. Table 16.2 presents a synthesis of some relevant aspects to be considered in package selection and choice. The evaluation of suitability is very simply expressed as high, medium, or low.

16.4 The packaging operation

For general works on packaging operations see Piergiovanni and Limbo (2010a) and Robertson (2010). The packaging operation consists in the following series of steps: depalletization, decasing, cleaning, filling, closing, coding, fill checking,

Table 16.2 A qualitative comparison of performance of various materials for extra-virgin olive oil packaging.

Package material performance	Rating of suitability		
	Glass containers	*Metal containers*	*Plastic containers*
Reputation and consumer preference	High	Medium	Low
Economy (minimum cost per unit quantity of oil)	Low	Medium	High
Protection from light	Medium	High	Low
Protection from oxygen	High	High	Low
Recycling/reuse possibility	High	Medium	Medium
Mechanical resistance	Low	Medium	High
Lightness in weight	Low	High	High
Inertness (minimal release of metal or plastic components)	High	Medium	Low
Prevention of scalping	High	High	Low
Hermetic sealing	High	High	High
Accessibility for cleaning of empty containers	High	Low	Medium
Length of turnover periods (average time from packaging to oil use)	High	High	Low

labelling, final checking, casing and palletization. In highly mechanized and automated systems, a well-planned operating rate of the various machines should assure smooth functioning of the entire packaging line. In particular, as filling is the most important and the slowest step in the line, the preceding operations of depalletizing, decasing and cleaning should operate at a rate about 10% faster than filling so that stable feeding of containers is assured to the filler. Similarly, the following steps, particularly closing and labelling, should also operate at a rate about 10% faster than the filler so that filled containers do not accumulate at the exit of the filling station. Much less stringent rules are applied in semi-automatic or manual operations of the packaging line.

Systematic cleaning of the various machines, especially the filler, should be carried out in order to avoid oil contamination or oil spoilage in dead spots of the circuit during intervals from one filling operation to the following one. This is a critical point in small factories, where long periods of time separate one packaging operation from another. Complete cleaning procedures with appropriate emptying, detergency, sanitizing, rinsing and drying should precede and follow each packaging operation.

The most important requirements in extra-virgin olive oil packaging are: (i) flexibility, which is the ability to use the same filling-closing-labelling machines for containers of various forms and sizes; and (ii) precision in filling of predetermined weights or volumes because underfilling raises legal problems while overfilling causes economic losses.

A good rule is to set the average fill weight or volume at least three standard deviations above the declared value on the label. Setting the average fill three standard deviation above the declared value on the label gives a probability less than 0.13% of amounts less than the declared quantity on the label. In large-scale production, statistical process control must be adopted with frequent checking of proper filling performance.

Packaging lines can be organized as straight-line or rotary configurations. The latter are generally used in high capacity lines. A wider diameter of the rotating turret, with a higher number of filling valves, operates at higher filling speed.

16.4.1 Bottles filling

The filling of extra-virgin olive oil in bottles is generally based on a predetermined level in the container. A critical condition in oil packaging is to avoid any shaking or turbulence and aeration during filling because air mixing spurs oil oxidation. Temperature variations should be avoided because they determine viscosity variations with fluctuation of the filler performance. Also, too low a temperature of the oil should be avoided during filling because a subsequent increase in temperature can cause oil expansion and improper pressure on the closure. The best would be to have a temperature of 20 °C (±1 °C) during the packaging operation, before sending the bottled oil to a storage room at 15 °C (±2 °C).

The two systems most commonly used in filling extra-virgin olive oil in glass bottles are pure gravity filling and vacuum filling to a constant predetermined level. Figure 16.3 shows the functional scheme of pure gravity filling.

A spring-loaded sleeve valve consists of two concentric tubes, the outer one for the downward flow of the oil from the oil supply tank into the bottle and the inner tube for venting air from the bottle during filling. The valve opens when the bottle is pushed up into the sealed position against spring pressure. The oil flows down without turbulence along the bottle wall. When the air vent tube is reached by the oil level, the flow stops because the residual air cannot escape from the container and thus no further liquid can flow into the container. At that point the container is lowered and disengaged from the valve that automatically closes the oil outlet. Pure gravity fillers are easy to clean and operate both in automatic and manual machines. They are also very accurate.

In other plants, level sensing fillers trigger a control system that shuts down the oil flow at the predetermined level. In this case there is no need to seal the container by pressing it against the filling valve. This system can therefore also be used for plastic containers. Figure 16.4 shows the functional scheme of pure vacuum filling.

A filling valve sealing against the bottle neck connects the bottle with two lines: a feeding line coming from the supply tank, which is at atmospheric pressure, and a vacuum line connected to a vacuum pump through a vacuum chamber. The vacuum in the bottle sucks the oil from the supply tank until the oil level reaches the vacuum port in the filling valve. This is the constant level of oil in the bottle. The overflow of oil goes to the vacuum chamber and is recycled to the supply tank. Vacuum filling is

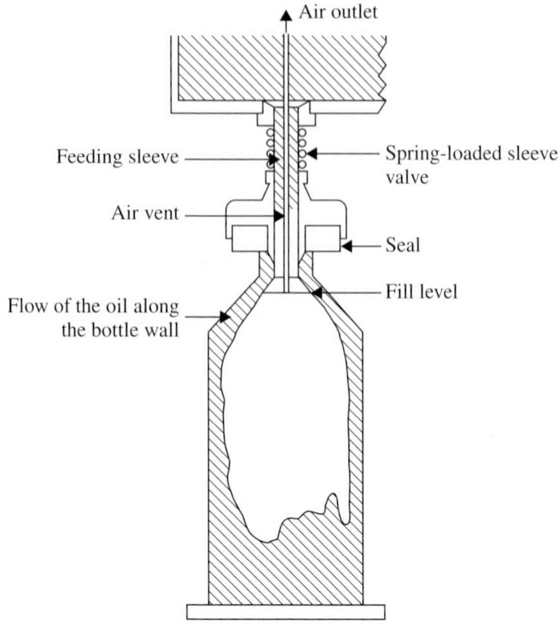

Figure 16.3 The functional scheme of pure gravity filling (*Source*: Lee, D.S. *et al.* (2008). Reproduced with Kind permission from Taylor & Francis).

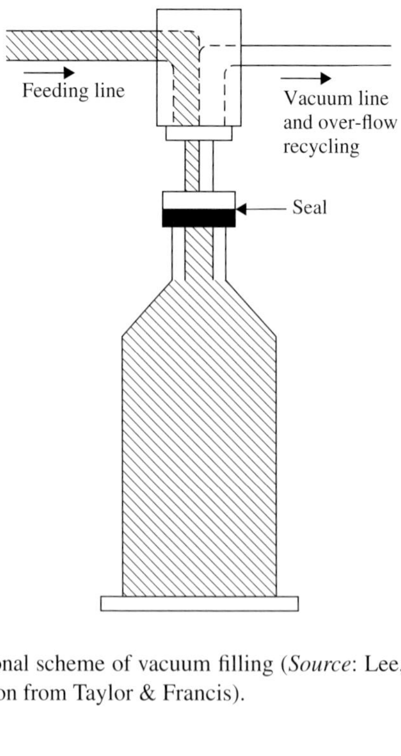

Figure 16.4 The functional scheme of vacuum filling (*Source*: Lee, D.S. *et al.* (2008). Reproduced with Kind permission from Taylor & Francis).

faster than gravity filling and prevents cracked or defective bottles from being filled because of the impossibility of obtaining a vacuum in the bottle.

This system cannot be used for plastic containers because they are not rigid enough to withstand the vacuum.

16.4.2 Bottles closing

Closure should provide hermetic sealing of oil bottles. Closure of extra-virgin olive oil bottles should also include pouring and tamper-evident devices.

Tamper-evident is a device that makes unauthorized opening of the bottle or packaging easily detected. Tamper-evident design can be vital for the consumer to know that the product has not been altered since it left the manufacturer. Consumers should be educated to watch for signs of tampering.

In order to ensure the authenticity of extra-virgin olive oil, the EU is considering the possibility of requiring the mandatory use of nonrefillable containers for extra-virgin olive oil in hotels, restaurants and cafes. These are containers with closures that cannot be removed without breaking.

In bottling extra-virgin olive oil, threaded closures are generally used. They are put on and taken off by screwing on the mouth of the bottle. Both application and removal torque are important.

Torque is the tendency of a force to rotate an object about an axis. Mathematically, torque is the cross product of the lever-arm distance and the force which tends to produce rotation. Torque is a measure of the turning force needed to apply a cap to the bottle (application torque) and the force needed to twist off the cap (removal torque). A removal torque that is too low presents a risk of accidental loosening of the closure, whereas a torque that is too high causes difficulty in opening the container.

An effective protection from oxidation is nitrogen flushing and air removal from the head space in the bottle neck just before cap installation. This is particularly useful if the oil has been effectively degassed as, for example, in vacuum filling.

The most common closure for glass bottles of extra-virgin olive oil are screw caps, also called continuous thread cap, and roll-on cap.

Screw or continuous thread (CT) caps

Screw cap closures attain a seal with the bottle by engagement of its thread with the corresponding threads of the bottle neck. Screw caps are tinplate or tin-free steel or aluminium. Hermetic sealing is due to close contact between the closure and the bottle mouth. A resilient liner compressed between the closure and the bottle mouth provides a tight and secure fastening (Figure 16.5).

Liners consist of backing and facing materials. Backing materials are soft and elastic to provide cushioning under compression. Facing materials should have an

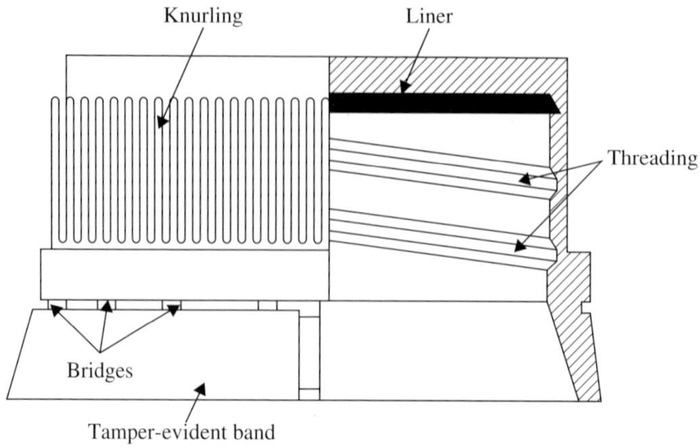

Figure 16.5 A section view of a screw cap (*Source*: Lee, D.S. *et al.* (2008). Reproduced with Kind permission from Taylor & Francis).

adequate barrier property and should combine closely and tightly with the pouring device that is mounted on the bottle mouth.

An oxygen-absorbing liner may be inserted into the cap to protect the oil from oxygen diffusion and permeation from the outside atmosphere.

The lower part of the cap skirt is linked through bridges to the cap body and separates when the cap is twisted for opening. Tamper-evidence is given by breaking of the bridges and the drop-down band may remain on the bottle neck as evidence of tampering.

Roll-on caps

A roll-on closure is formed into final shape by the pressurized rolling of the preformed aluminium shell onto the bottle threads. A top pressure assures the seal between the container finish and lining compounds of the cap. A breakable skirt for tamper-evidence is similar to continuous thread closure. The closing operation is commonly achieved by a chuck that grips the closure and turns pushing it into the desired position. The capping head tightens the closure until the desired torque is reached.

16.4.3 Labelling

Labelling is one of the most controversial issues of extra-virgin olive oil marketing.

This presentation is based on European legislation (Commission Regulation (EC) No 182/2009 amending Regulation (EC) 1019/2002 on marketing standards for olive oil, and Commission Implementing Regulation (EU) No 29/2012 on marketing standards for olive oil), which is a reference for legislation on olive oil worldwide.

The discussion about labelling is divided in two parts: the first one concerns the mandatory information on the label and the second concerns the voluntary information that can be put on an olive oil label.

Mandatory information on the label

The labelling of olive oil must bear 'in clear and indelible lettering' the name of the category of the oil, based on technology and analytical/sensory standards. One of the following denominations should appear on the label:

- 'Extra-virgin olive oil' together with the following information: 'superior category olive oil obtained directly from olives and solely by mechanical means'. This oil has a maximum free acidity, in terms of oleic acid of 0.8%, and the other characteristics that comply with those laid down for this category (Chapter 2).
- 'Virgin olive oil' together with the following information: 'olive oil obtained directly from olives and solely by mechanical means'. This oil has a maximum free acidity, in terms of oleic acid of 2.0%, and the other characteristics which comply with those laid down for this category (Chapter 2).
- 'Olive oil composed of refined olive oil and virgin olive oils' together with the following information: 'oil comprising exclusively olive oils that have undergone refining and oils obtained directly from olives'. This oil has a maximum free acidity, in terms of oleic acid of 1.0%, and the other characteristics which comply with those laid down for this category (Chapter 2).
- 'Olive pomace oil' together with the following information: 'oil comprising exclusively oils obtained by treating the product after the extraction of olive oil and oils obtained directly from olives'. This oil has a maximum free acidity, in terms of oleic acid of 0.3%, and the other characteristics that comply with those laid down for this category (Chapter 2).

The other mandatory information on the labels is:

- the name under which the oil is sold;
- the net quantity;
- the date of minimum durability;
- the name or business name and address of the producer;
- the number or code of the identity and traceability of the oil.

The name under which the oil is sold, the net content and the traceability code should be in the same visual field.

In Europe and in many other countries, the size allowed for containers destined to the final consumer is established by law.

Table 16.3 A series of examples of solutions and difficulties in the labelling of extra-virgin olive oil.

Claims	Comments about their trustworthiness
Claim about the origin	The origin of olives and oil is important information from the consumer's point of view. Origin means tradition, biodiversity, oil-culinary connections, and a specific sensory style. Origin cannot be identified with certainty by analysing the oil. It can only be identified through a combined system of documented balance of material lots and the analytical finger-printing of each lot. European Union legislation provides a very detailed definition of the different cases of designation of origin, from the Protected Designation of Origin (PDO) of oils from areas or regions to blends of oils from the same EU country or from various EU countries and also blends of olive oils of EU and non-EU origin. No claims of territorial origin are allowed in Europe outside this legal frame.
Claims about the cultivar(s)	In principle, these claims are not allowed in the EU legislation, unless they refer to the cultivars formally defined in the PDO regulations. This is a serious gap of information for consumers because cultivars are the most effective and appealing image of biodiversity. A cultivar cannot be identified with certainty by analysing the oils. It can only be identified through a combined system of documented balance of material lots and the analytical fingerprinting of each oil lot.
Claims about sensory characteristics	The Commission Implementing Regulation No 29/2012 allows indications concerning the sensory properties referring to taste and/or smell of the oil to be written on the label of extra-virgin olive oils. This must be done by using the terms set up by the law and assessed by applying the method provided by the law. Extra-virgin olive oil is an unusual case in which sensory perceptions must be described according to a predefined and mandatory glossary of sensory notes.
Claims about chemical standards	The indication of maximum acidity may appear on the label only if accompanied by an indication, in lettering of the same size and in the same visual field, of the peroxide value, the wax content and the ultraviolet absorption determined according to the official methods of analysis (Commission Regulation (EC) No 182/2009 amending Regulation (EC) 1019/2002 on marketing standards for olive oil).
Claims about the harvest date	Product freshness is an important feature of oil quality and excellence, but claims of harvest date are not allowed. This is a classic case in which the concern to avoid deceiving the consumer deprives him/her of useful information. The harvest date cannot be identified with certainty by analysing the oils. It can only be identified through a combined system of documented balance of material lots and the analytical finger-printing of each oil lot.

(*continued overleaf*)

Table 16.3 *(continued)*

Claims	Comments about their trustworthiness
Claim of 'cold extraction'	According to the EU legislation, the claim 'cold extraction' may appear only for extra-virgin olive oils obtained at temperatures below 27 °C (80.6 °F). This is a good example of how bad the result can be of an over-regulating system. In the first place, there is no evidence that 27 °C is better than 30 °C or 32 °C. It is just another limitation of the producers to experimenting for optimization. In the second place, there is no way to convey to consumers the perception that 27 °C is 'cold' while 30 °C is 'too warm'. In the third place, saying that an oil has been obtained by 'cold extraction' leads the consumer to think that there must be some kind of a 'hot extraction', which, in fact does not exist. Finally, controlling and certifying the process conformity to this requisite is very difficult or impossible if normal temperature control devices are used.
Claims of 'nonfiltered oil'	Oddly enough, the legislation has no objections to this claim, which is, in fact, misleading to the consumer because it leads to the understanding that filtration is detrimental to oil quality, while the opposite is true. On the other hand, what is the sense of writing something when it is self-evident from the turbidity of the oil?
Suggestions about storage conditions	This critical information for oil quality should hopefully become mandatory and be well evidenced on the label.
Good readability	Overloading a label with excessive information and analytical data written in microscopic characters could be considered another way of misleading the consumer. Too much information dangerously borders on not enough.
Nutritional information	The information about the nutritional content and profile should be given according to precise legal indications. It may concern the energetic and major nutrient content as well as figures of mono and polyunsaturated fatty acids, vitamins (such as, for instance, α-tocopherol). Minimum content is often prescribed or included in label information. (See Regulation (EU) No 1169/2011).
Other information	Whereas No 12 of the Commission Implementing Regulation No 29/2012 says that 'all other indications appearing on the label should be corroborated by objective elements in order to ensure that consumers are not misled and that competition on the markets in the oils concerned is not distorted'. Phenolic content can probably be considered as complying with this requirement.

Voluntary information on the label

It is evident that the mandatory information listed earlier gives only limited information about the product characteristics and properties. It is therefore understandable that producers and retailers would like to give more detailed information about the process and the product and this is the point that causes most of the confusion and debate. The problem is that many claims creating important consumer appeal and marketing effectiveness are difficult (and sometimes impossible) to demonstrate through reliable analytical means. As a consequence fraud is not detectable. This has caused the creation of compulsory rules, either too rigid or too restrictive, with a hindering effect on simple and clear communication to the consumer.

The rationale for the European legislation is clearly explained in the 'whereas' section, point 10 of the Commission Regulation 29/2012:

indications shown on the labelling should not mislead the purchaser, particularly as to the characteristics of the olive oil concerned, or by attributing to it properties which it does not possess, or by suggesting that it possesses special characteristics when in fact most oils possess such characteristics.

This point, which is important and can be shared by everyone, leads the legislator to express the following concept:

Certain commonly used, optional indications that are specific to olive oil also require harmonized rules to precisely define such claims and ensure that their accuracy can be verified ...

This wise proposal contains a serious risk: that an agreement on 'harmonized rules' could never be attained and that important claims in the consumer's interest and wishes would never be allowed.

Some of the controversial issues about labelling of extra-virgin olive oil are summarized in Table 16.3.

References

Esposito, A. (2012) La nuova legislazione del settore dell'olio extra vergine di oliva: analisi empirica degli effetti sulle preferenze dei consumatori italiani, PhD Thesis, University of Udine.

Gawel, R. (2010) Are All Dark Glass Bottles Used for Storing Extra Virgin Olive Oil the Same? http://www.aromadictionary.com/EVOO_blog/?p=357 (accessed 11 October 2013).

Limbo, S., Torri, L. and Piergiovanni, L. (2007) Light-induced changes in an aqueous beta-carotene system stored under halogen and fluorescent lamps, affected by two oxygen partial pressures. *Journal of Agricultural and Food Chemistry* **55**, 5238–5245.

Min, D.B. and Boff, J.N. (2002) Chemistry and reaction of singlet oxygen in foods. *Comprehensive Reviews in Food Science and Food Safety* **1**, 58–72.

Monini SpA (2013) *La nuova Bag in Box di Monini – Dieci litri d'olio, niente sprechi*, www.italiaatavola.net/articolo.aspx?id=24757 (accessed 11 October 2013).

Piergiovanni, L. and Limbo, S. (2010a) *Food Packaging – materiali, tecnologie e qualità degli alimenti*, Springer Verlag, Milan.

Piergiovanni, L. and Limbo, S. (2010b) Packaging and shelf life of vegetable oils, in *Food Packaging and Shelf Life – A Practical Guide* (ed. G.L. Robertson), CRC Press, Boca Raton, FL.

Pistouri, G., Badeka, A. and Coutominas, M.G. (2010) Effect of packaging material headspace, oxygen and light transmission, temperature and storage time on quality characteristics of extra virgin olive oil. *Food Control* **21**, 412–418.

Rizzo, V., Torri, L., Licciardello, F. *et al.* (2013) Quality changes in extra virgin olive oil packaged in coloured PET bottles stored under different lighting conditions. *Packaging Technology and Science* http://onlinelibrary.wiley.com/journal/10.1002/%28ISSN%291099-1522/earlyview (accessed 11 October 2013).

Robertson, G.L. (ed.) (2010) *Food Packaging and Shelf Life – A Practical Guide*, CRC Press, Boca Raton, FL.

Thron, M., Eichner, K. and Ziegleder, G. (2001) The influence of light of different wavelengths on chlorophyll-containing foods. *Lebensmittel – Wissenschaft und Technologie* **34**, 542–546.

Yam, K.L. (ed.) (2009) *Wiley Encyclopedia of Packaging Technology* (3rd edn), John Wiley & Sons, Inc., New York.

Further reading

Lee, D.S., Yan, K.L. and Piergiovanni, L. (2008) *Food Packaging Science and Technology*, CRC Press, London.

Mendez, A.I. and Falqué, E. (2007) Effect of storage time and container type on the quality of extra-virgin olive oil. *Food Control* **18**(5), 521–529.

17

The olive oil refining process

Claudio Peri
University of Milan, Milan, Italy

Abstract

Olive pomace extraction and olive oil refining processes are described through flow-charts and discussion of the main process operations. Chemical and physical refining are compared and a final note is devoted to the quality and virtues of refined olive oil.

17.1 Introduction

According to European legislation (Council Regulation (EC) No 1234/2007 of 22 October 2007 (Single CMO Regulation), consolidated version 2013-01-26), the products of the refining process are:

- *Refined olive oil.* Refined olive oil is the olive oil obtained from virgin lampante oil by refining methods that do not lead to alterations in the initial glyceridic structure. It has a free acidity, expressed as oleic acid, of no more than 0.3%. A high proportion of the oil produced today in the Mediterranean area belongs to this category.
- *Refined olive-pomace oil.* Refined olive-pomace oil is obtained from oil extracted with solvents from pomace and then refined. It has a free acidity, expressed as oleic acid, of not more than 0.3%. It is refined with the same methods used for refining lampante oils.

Both refined olive oil and refined olive-pomace oil are tasteless, so they are seldom used as such; more frequently they are blended with flavourful extra-virgin or virgin olive oil, thus obtaining:

- *Olive oil composed of refined olive oils and virgin olive oils.* This category of olive oil consists of a blend of refined olive oil and extra-virgin or virgin

The Extra-Virgin Olive Oil Handbook, First Edition. Edited by Claudio Peri.

olive oil. It has a free acidity of no more than 1.0%. Most of the olive oil sold in the world falls into this category. Different blends are made, with more or less extra-virgin or virgin oil, to achieve different tastes at different prices. Oils described as 'Light' or 'Extra Light' in the United States fall into this category and are most likely made with a large proportion of refined oil.

- *Olive-pomace oil.* Olive pomace oil is a blend of refined olive-pomace oil and virgin olive oil. It has a free acidity of no more than 1%.

Confusing terminology

It is not easy to explain to consumers the difference between a 'refined olive oil', a 'virgin olive oil' and an 'extra-virgin olive oil'. Most consumers do not know that there are various olive oils with quite different composition, sensory and health-promoting properties. In many cases, a mixture of refined and virgin olive oil has a more attractive appearance (limpid and pale yellow) than an extra-virgin olive oil and this may be misleading for the consumer's perception of quality. For some consumers, the term 'refined' has a positive meaning (something refined is understood as being better than nonrefined) and the term 'virgin' is incomprehensible in this context. Even more deceptive is the attribution of the 'light' or 'extra light' name to these oils based on their very mild taste. In general, consumers know that a 'light' food has a reduced caloric content compared to the 'regular' food. This is a totally inappropriate interpretation, because an olive oil consists in triglycerides and has the same caloric content of any edible oil or fat, which is 9 kcal per gram.

17.2 The process of extraction of crude pomace oil

The production of pomace oil requires a solvent extraction process before refining. The refining processes applied to lampante oil and to the crude oil extracted from pomace are very similar.

Figure 17.1 presents the process of oil extraction applied to pomace deriving from two-phase decanter processing. In this case, the extraction is carried out in two steps: the first one, based on malaxation and three-phase decanter separation, allows the recovery of about 50% of the extractable oil; the second step, based on solvent extraction, allows recovery of the residual 50% of the extractable oil. In the first step there is a significant reduction in water content and, consequently, a reduction in the cost of drying, which must always precede the solvent extraction step. In the case of pomace deriving from three-phase processing, the extraction process starts from the drying and solvent extraction operations.

The main operations in Figure 17.1 are described in the following sections.

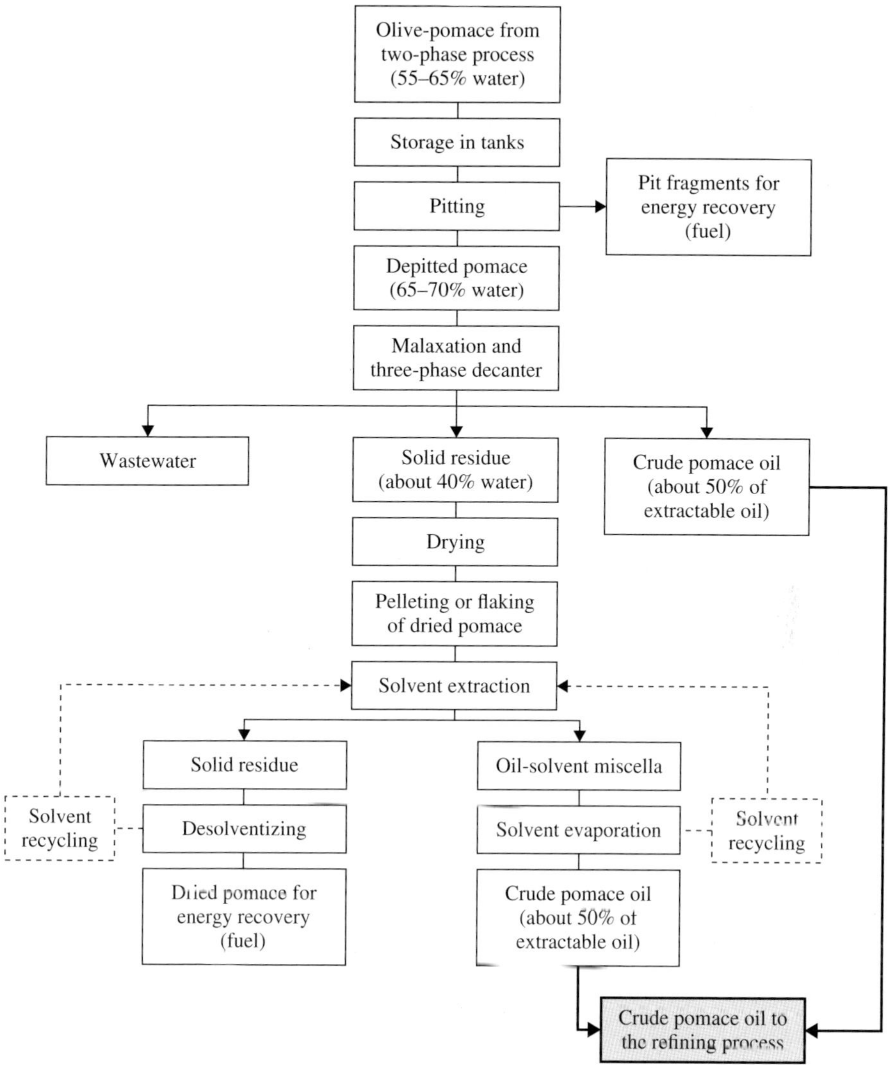

Figure 17.1 The pomace oil extraction process.

Pomace storage

The two-phase olive pomace is a semi-liquid slurry containing about 55–65% water and 30% solids, composed of olive skin, pulp, stone, kernel, and a small percentage of soluble components such as salts, sugars, and phenolic compounds. Due to the presence of sugar, pomace can ferment, however, with difficulty because the polyphenols have an inhibiting effect on microbial growth. Pomace is stored for 7 to 9 months in tanks or ponds because very large amounts of pomace are produced in a short period of time, while extraction must be carried out according to the capacity of the solvent-extraction plant.

Pomace pitting

The first operation of the process is pitting of the pomace. Pitting is carried out as a milling operation with suitable rotating speed and suitable grid openings of the mill so that fine soft particles pass through the grid, while coarse hard particles of olive stones are retained and separately discharged. Pitting causes a 13–18% reduction in weight of pomace and a 5–10% proportional increase in water because pit fragments are an inert woody material with a very low water content. The recovered olive pit fragments are a good combustion material due to their high calorific and low ash content; in suitable burning conditions they can produce a flame up to 800 °C.

First oil extraction step by two-phase decanter separation

Pitted pomace undergoes malaxation to favour oil coalescence, followed by decanter centrifugal separation. In this case, three-phase decanters are used, with separation of vegetation water, oil and a solid concentrated slurry. This operation has several advantages: (i) wastewater from two-phase pomace has a fairly high content of phenolic compounds and could therefore be used for the extraction of useful antioxidants; (ii) about 50% of the oil in the pomace can be separated by this operation, which is carried out at a relatively low temperature (40 °C) with minimal damage to the oil quality; (iii) the three-phase system allows a considerable reduction of the water content with a significant reduction in the energy consumption in the following drying operation.

Pomace drying

Starting from this operation, the process is the same for two- and three-phase pomace. Drying of pomace is necessary to use the solvents efficiently. Pomace is air-dried in continuous drum-drying plants with a reduction in the water content to 6–8%, which is considered as optimal for solvent extraction. Drying is an energy-intensive operation and different energy-saving systems can be used. Most frequently, hot drying gases can be produced by mixing air with the gases obtained by burning olive stones and dried exhausted olive pomace.

Drying temperatures are high (up to 200–300 °C in normal conditions) and therefore a significant degradation of the oil takes place with an increase in free acidity, peroxide value and spectrophotometric values and the formation of brown compounds to be removed in the refining process. If temperatures go beyond 400 °C, formation of polycyclic aromatic hydrocarbons (PAHs) may become significant.

Dry pomace pelleting or flaking

The dried pomace is a powdery material consisting of different sized dried particles, which are unsuitable as such for solvent extraction. This powdery material tends to lump and pack together, thus hindering close contact between the solids and the solvent. Therefore, dried pomace is steam treated and formed into pellets or flakes by pressing in extrusion or flaking operations. At the same time, pellets should be compact enough to maintain their form during extraction and porous enough to allow the solvent to penetrate and impregnate them fully.

Solvent extraction

Solvent extraction is a mixing-and-separation operation of a solid (pomace) and a liquid (solvent) phase with an extraction phase in between. The following steps take place during the operation: (i) mixing of pomace pellets and the solvent; (ii) the solvent penetrates and impregnates the solid phase; (iii) dissolution and diffusion phenomena allow the oil to pass from the solid into the solvent phase; (iv) separation by mechanical means or by gravity of the oil-solvent solution (the 'miscella') from the solid phase, which consists of pomace impregnated by a portion of solvent.

Extraction is carried out in a semi-continuous, counter-current mode, with a series of tanks where the fresh solvent is pumped into the last tank where the solids are already partially depleted of oil, while the miscella is drained off the first tank that receives the fresh pomace.

The solvent generally used is n-hexane which has a good selectivity for oil, a low boiling temperature (60 °C) and latent heat of evaporation, and no effects on oil quality.

Desolventizing of the exhausted pomace and the oil-solvent miscella

The exhausted pomace pellets are desolventized in a heat exchanger where the solvent evaporates at the boiling point. Heat exchangers are jacketed cylinders with the heating steam condensing in the jacket, while solids are stirred inside the cylinder and pushed toward the outlet by a screw conveyor. The solvent vapours are condensed and the solvent is recycled. The dried exhausted pomace is used as fuel for energy or for compost production

The oil-solvent miscella is desolventized in a multiple-effect evaporator, under vacuum. The solvent vapours are condensed and the solvent is recycled. The crude pomace oil is sent to the refining process.

17.3 The refining process

Lampante olive oil and crude olive-pomace oil must be refined in order to become edible. The principle of the process is the same for both products, even if the operating conditions may be quite different. In any case, the refining processes for the two oils must be carried out separately because the final products must be separated according to the international legislation. Table 17.1 presents a list of undesirable substances in lampante oil and in crude olive-pomace oil and the operations for removing them.

Table 17.1 Undesirable substances in lampante oil and crude olive-pomace oil and the corresponding operations for their removal.

Undesirable substances	Refining operation
Suspended materials	Gravity settling or (rarely) filtering
Free fatty acids, monoglycerides and diglycerides, gums and phosphatides	Demucilagination and chemical neutralization in the chemical refining process or steam distillation in the physical refining process
Natural pigments and brown compounds formed at high temperatures especially during the drying and desolventizing operations in the extraction of crude olive-pomace oil. Traces of metals. Polycyclic aromatic hydrocarbons (PAHs) formed during pomace drying	Bleaching with bleaching earths with or without addition of activated carbon
Off-odours and, in the physical refining process, also free fatty acids	Vacuum distillation with steam stripping

Figure 17.2 presents the flow-chart of the refining process. The main operations can be described as follows.

Gravity settling

The crude olive-pomace oil is generally clear due to the filtering effect of the solvent extraction operation where the miscella is drained out through the porous bed of pomace pellets. Lampante oil, instead, needs some gravity settling (or filtering) in order to avoid cloudiness and sediment in the final product.

Chemical neutralization

The aim of neutralization is to eliminate the free fatty acids and the monoglycerides and diglycerides that are formed in the olives and the oil by endogenous or exogenous lipases. The operation is carried out by adding a calculated amount of sodium (most frequent) or potassium or calcium hydroxide to the oil at a temperature in the range of 65–90 °C. In the reaction of the lye with the free fatty acids, soaps are formed that are soluble in water and can therefore be separated by centrifugation. The water-soluble fraction that is separated by centrifugation is called soapstock and contains not only the soaps but all the other water-soluble components such as monoglycerides and diglycerides, phospholipids, and sterols. Soapstocks are used for the production of free fatty acids, which are an important raw material for chemical syntheses. From an economic point of view, it is important that the losses of neutral oil and free fatty acids are minimized.

If the addition of lye is preceded by an acidification step with phosphoric or citric acid, gums and phosphatidic acids can be previously removed by centrifugation.

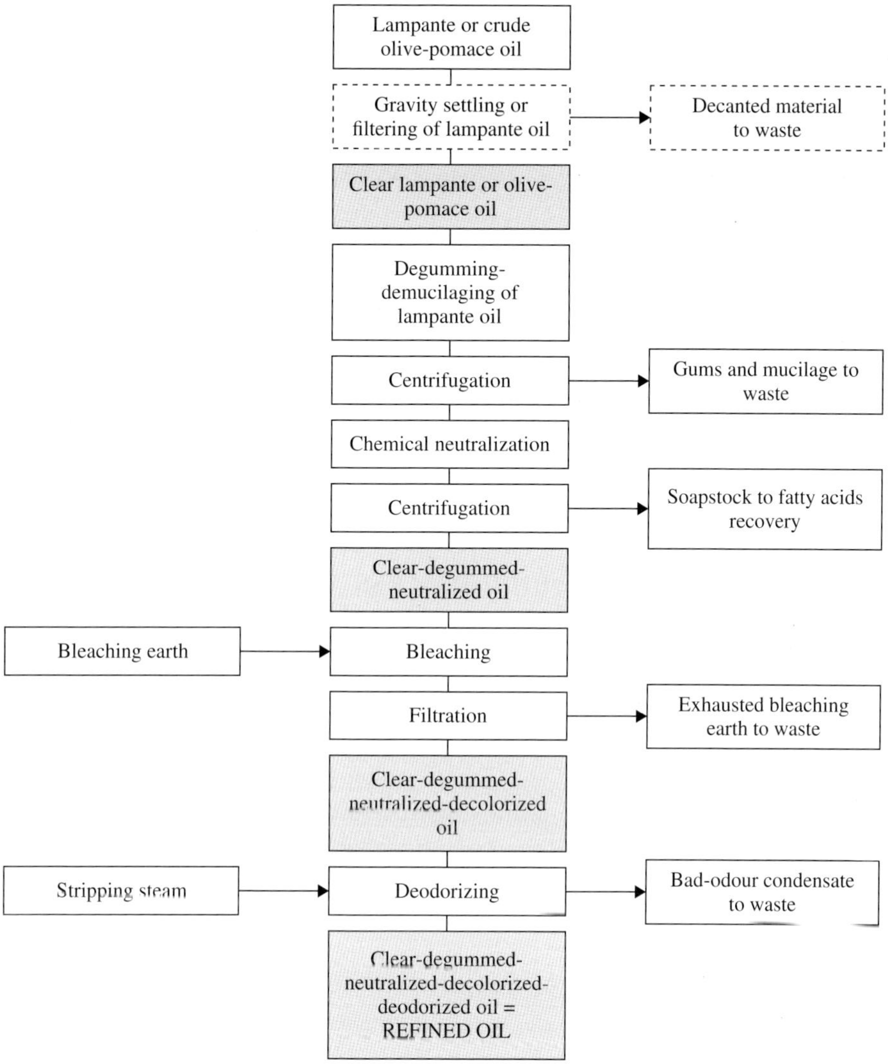

Figure 17.2 The olive oil refining process.

Bleaching

The aim of bleaching is to eliminate all the coloured compounds so that the refined olive oil becomes colourless. This operation is carried out by adding to the oil 0.5 to 1.5% of bleaching earth, a mineral clay (bentonite or montmorillonite) with a high absorption selectivity and capacity. Natural coloured compounds such as carotene, xanthophyll and chlorophyll are removed as well as newly formed brown compounds due to the high-temperature processing during the drying and desolventizing operations. Removal of the coloured compounds is due to surface absorption of the coloured material by the earth particles. If the oil contains polycyclic aromatic hydrocarbons (PAHs), formed during pomace drying, 5 to 10% activated carbon is added to the bleaching earth, which greatly increases the absorption capacity of the mixture. Activated carbon, in fact, is extremely porous and has an enormous surface area per gram with great absorption capacity.

Bleaching is carried out in discontinuous operations, under vacuum, at a temperature of 90–110 °C for 20–30 min, with low speed agitation in order to maintain the bleaching earth in suspension. The suspension is then cooled to 70 °C and the bleaching material is separated from the decolourized oil by filtration with filter-plate or leaf filters.

Deodorizing

The aim of the deodorizing operation is the removal of off-odours that were formed in the previous operations of the refining process. Deodorizing is the last operation of the process. It is a continuous vacuum distillation with steam stripping of volatiles.

17.4 The physical refining process

The physical refining process differs from the chemical refining process by the fact that the deodorizing operation is carried out in conditions (higher temperature and lower pressure) that allow not only the volatile, bad-smelling compounds, but also the free fatty acids, to be removed. With physical refining, the chemical neutralization operation is avoided. The neutralizing effectiveness of the physical process is very good. The free acidity of the oil, after the physical refining process, is in the range of 0.03–0.05%. The conditions of the deodorizing operation for the two cases are compared in Table 17.2.

Table 17.2 Comparison of the conditions of the deodorizing operation in the chemical and physical refining processes.

Operating conditions	Chemical refining	Physical refining
Temperature	220–230 °C	240–250 °C
Pressure	3 mbar	2 mbar
Steam	10 kg per ton of oil	15 kg per ton of oil
Time of the operation	60 min	60 min

17.5 The quality and uses of refined olive oil

An in-depth discussion of the effects of solvent extraction and refining on the various components of olive oil is beyond the purpose of this handbook. Only a few points will be given to indicate the differences between refined and extra-virgin olive oils.

The refining process eliminates most of the antioxidant compounds. Polyphenols are eliminated by both the neutralization and the bleaching operations. The tocopherols are also stripped off during the deodorization process. Research is being carried out to improve the refining process and to reduce the loss of useful compounds like sterols, polyphenols and tocopherols.

Concerning unwanted changes, formation of *trans*-fatty acids and especially *trans* isomers of linoleic and linolenic acids may exceed the legal limits when temperatures go beyond the smoke point (Scott and Porter 2012). Formation of PAHs during drying of the pomace must be followed very carefully and activated carbon in the bleaching operation must be used to effectively remove them.

Despite these limitations, refined olive oils may be considered as a very interesting product for a series of reasons:

1. Refined olive oils are the only possibility for the olive grower to gain income in cases in which extra-virgin olive oils cannot be obtained, such as:
    - In case of serious fly infestation.
    - In case of olive spoilage due to unfavourable climatic events (including hail and frost).
    - In less-developed, poorly mechanized, areas, where harvesting extends well beyond the optimal maturity of the olives.
    - In more-developed areas where the availability of the workforce is scarce or too costly to satisfy the needs of a short optimal period of olive maturity. In these circumstances, olive growers often decide to harvest only part of their production during the optimal maturity period in order to obtain a high quality extra-virgin olive oil to be sold at a high price. The rest of the production is harvested after the optimal period and used for the production of refined olive oil to be sold at a lower price. This combination of extra-virgin and refined olive oil is the basis of the economical balance for many olive growers.
2. The olive oil refining process is similar to the processes that are applied for the production of all the other edible vegetable oils on the market. All of them are obtained by solvent extraction and refining. The process is considered as perfectly safe and has been well tested during decades of research, development and field experience.
3. Refined olive oil is a very good oil in nutritional terms. It is not comparable to extra-virgin because it lacks most of the minor antioxidant compounds that make extra-virgin olive oil a unique product. However, the high content of oleic acid, the good balance of mono- versus polyunsaturated fatty acids, the presence of an effective, lipid-soluble, antioxidant like squalene, make refined olive oil an interesting source of food lipids.

4. Finally, refined olive oils have a flat sensory profile. Nothing to be compared with the richness, variety and specificity of the sensory profiles of excellent extra-virgin olive oils. However, as thoroughly discussed in Chapter 24, concerning the culinary uses of olive oils, the absence of a strong sensory characterization may be an advantage when oil is used as a preservative or as a cooking medium.

In conclusion, extra-virgin olive oils are the top in the range of olive oils. But, second place, even before the 'virgin' olive oils, can probably be assigned to good 'refined olive oils' or to 'olive oil – composed of refined olive oils and extra-virgin olive oils'.

Reference

Scott, J.H. and Porter, S.E.G. (2012) Heat induced cis/trans isomerisation in vegetable oils and oleic acide. *Journal of Undergraduate Scholarshiup* http://blogs.longwood.edu/incite/2012/01/30/heat-induced-cistrans-isomerization-in-vegetable-oils-and-oleic-acid/ (accessed 11 October 2013).

Further reading

Antonopoulos, K., Valet, N., Spiratos, D. and Siragakis, G. (2006) Olive oil and pomace oil processing. *Grasas y Aceites* **57**(1), 56–67.

Sánchez Moral, P. and Ruiz Méndez , M.V. (2006) Production of pomace oil. *Grasas y Aceites* **57**(1), 47–55.

Part III

The process control system

18

Process management system (PMS)

Claudio Peri

University of Milan, Milan, Italy

Abstract

The structure of a process management system (PMS) is presented, with applications to the extra-virgin olive oil process. In a unique management scheme, the system integrates various goals and conditions. Risk analysis is suggested as an effective tool for identifying critical points and for achieving effectiveness and efficiency in pursuing the company's goals with minimal formality and essential documentation. Annex 18.1 presents a standard model of excellence in olive oil. A quality-proximity matrix is suggested for chain control and for conveying excellence to the consumer's table. Annex 18.2 presents an example of integrated risk analysis of extra-virgin olive oil processes including the control of: quality, safety and hygiene, traceability, worker's health and safety, and environmental impact.

18.1 Introduction

This chapter has a central role in the structure of the handbook. Process management and control determine the conditions for producing good or excellent extra virgin olive oil, winning the consumer's trust and achieving effective competitiveness.

At the beginning of the 1990s, the standards of quality management systems (QMS), which were already common practice in the most advanced industrial sectors such as aeronautics and the nuclear industry, started being applied to all business activities, including the food industry.

Within a few years, thanks to the unifying efforts of the International Standards Organization (ISO), QMSs were applied in hundreds of thousands of manufacturing and food industries around the world. The ISO standards have become the model for many quality management systems set up by public and private organizations in all sectors of human activities.

The Extra-Virgin Olive Oil Handbook, First Edition. Edited by Claudio Peri.

The reason for such an impressive spreading of the new culture of quality management systems was globalization of the market and the consequent need to use the same approaches and tools for giving assurance of product quality in business negotiations across countries and continents.

At the same time, implementation of a QMS forced companies to adopt systematic, documented procedures of process management with great benefits in terms of:

- planning of product and process requirements;
- communication among trade partners;
- optimization of resources
- less post-process verification and losses;
- focusing on end results;
- control based on critical points;
- more efficient and dynamic control of hazards.

Since the early 2000s, the concept of food quality has changed considerably. Consumers no longer just consider product quality and safety, but also convenience, communication, product origin (traceability), the ethical behaviour of producers and retailers towards the environment and the workers, biodiversity, and the value of tradition. Therefore, planning a management system nowadays entails a number of objectives and constraints, control points and procedures. That is why the title of this chapter is not 'quality management system' but a more general and comprehensive title 'process management system'.

A process management system (PMS) is defined as a system of resources and activities that organizations apply to meet product and process requirements. The expression 'management system' is considered as a synonym of 'control system'. Management consists, in fact, in keeping a process under control.

Based on Deming's PDCA wheel, which is the foundation of all management systems, the principles and methods of a process management system can be applied to any process or company, big or small (Peri *et al.* 2006).

18.2 The structure of a PMS

The structure of a PMS (Figure 18.1) consists of four interconnected parts: the company's policy about product and process; the product requirements and standards; the process requirements and standards and the management system which, in turn, includes:

- control of product conformity to standards;
- control of process conformity to standards at critical points;

- general control procedures; and
- audit and system review.

Two feedback circuits operate in the system: one concerning the audit and system review compares the process results with the company's policy and strategic objectives; the other continuously reporting the results of control of product and process to the technical and operating level.

18.2.1 The company's policy (point 1 of Figure 18.1)

In the first part concerning the company's policy, the following points are defined:

- long-term production and marketing objectives;
- product requirements and marketing focus;
- participation in product and/or process certification schemes;
- investments and innovation of resources, including plant and people;
- services in outsourcing.

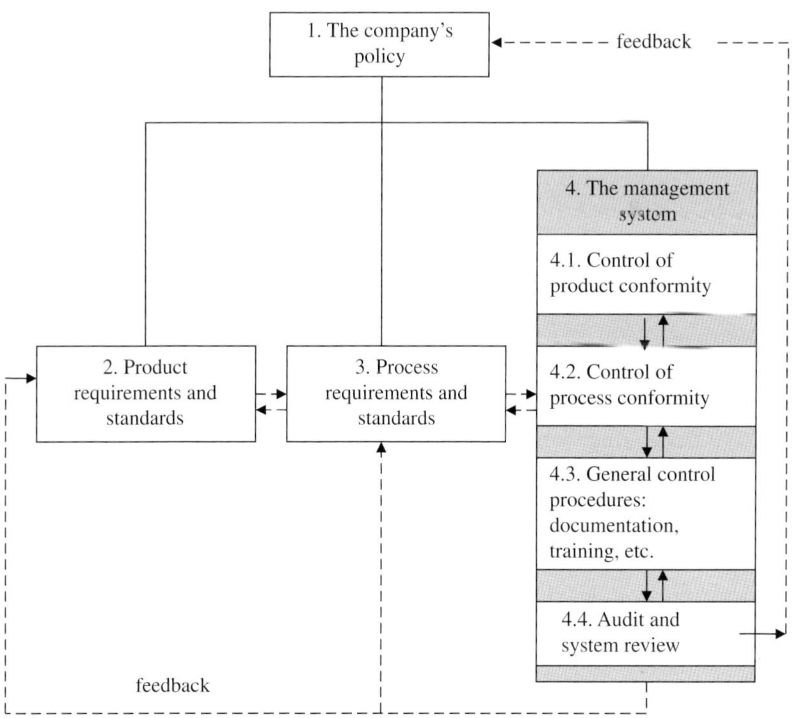

Figure 18.1 The structure of a process management system.

18.2.2 Product requirements and control (points 2 and 4.1 of Figure 18.1)

Product requirements should be defined in terms of composition, quality and safety standards. Packaging and labelling requirements are also included because a 'product' in marketing terms is the combination of the product and the container.

Product requirements can be considered as the process objectives and product conformity to the predefined standards is the most important measure of process effectiveness.

Nonconformity to product standards should lead to a critical appraisal of the causes of nonconformity: if nonconformity is due to excessively demanding standards, product standards must be modified. If nonconformity is due to mistakes or deviation from the standard operating conditions, operating behaviour must be corrected.

Reacting to nonconformities with corrective actions or with redefining/reinforcing preventive actions is the most important duty of a process manager. On the contrary, accepting a product that does not conform to the predefined standards, without an appropriate reaction, deprives a management system of its meaning and effectiveness. It also allows operators and workers to think that the system management rules can be disregarded without consequences.

18.2.3 Process requirements and control (points 3 and 4.2 of Figure 18.1)

Process requirements consist of:

- operating procedures and rules that the technical staff has defined in order to achieve the predefined results. In order to achieve the maximum control effectiveness at the minimum cost, control procedures should be applied to *critical points* selected through a systematic risk analysis as discussed in detail in Section 18.3;
- procedures and conditions established by the laws or by certification bodies (Chapter 20);
- product traceability (Chapter 19);
- process yield;
- processing cost (Chapter 23).

Process nonconformities should lead to a critical appraisal of the causes. If nonconformities are the consequence of procedures or operating conditions that are too demanding, the operating standards must be modified. If nonconformities are due to mistakes or deviation from the predefined operating procedures and conditions, operating behaviour must be corrected. If nonconformities are due to structural deficiencies, more appropriate procedures should be defined and structural changes should be planned with appropriate investment and timing.

18.2.4 General control procedures (point 4.3 of Figure 18.1)

Procedures that should be applied regardless of the sector of the company's activity or its size must be defined. Two of them (documentation and training) are key to a satisfactory management system implementation.

Documentation

Documentation must concern:

- Prescriptive documents, for example a management system manual and standard operating and control procedures. Prescriptive documents are, at the same time, a guide for operators and proof of the company's commitment.
- Records of operating conditions.
- Analytical data of products.
- Records of nonconformities and of the corrective actions undertaken.

Effective management of the documentation requires simple but strict rules should be established on:

- who has the responsibility of managing and updating the documentation;
- where and when data and facts should be recorded;
- how data should be recorded, identified, classified and placed in suitable files;
- to whom they should be submitted for evaluation and decision.

Training of staff and workers

The objective of training people is to improve their capability and performance. Training may include transfer of information from an instructor or a book or a document. In a workplace, however, the most important way of training is through experiential learning focused on how to master skills which can only be acquired by doing. With such training the operators' skill is continuously upgraded and updated.

Training is a very essential part of a management system, as good training makes it superfluous to write complex procedures about activities because well-trained people know how to carry them out almost automatically. A good training programme can greatly contribute to reducing the documentation of a Process Management System.

A good training programme should define:

- what are the subject matters and activities requiring training;
- who is responsible for the training programme;

- to whom should the training programme be addressed;
- how, where and when should training be carried out;
- how should the effectiveness of the training programme be evaluated.

A most effective way of learning is by correcting mistakes. Mistakes can be seen as unintended deviations from planned procedures and goals. They can also be incorrect actions as a result of insufficient knowledge. Two kinds of 'mistake cultures' exist in companies: a mistake management culture and a mistake aversion culture. In organizations that practise the mistake management culture, people communicate about mistakes, share their mistake knowledge, help others in mistake situations and can quickly detect and handle mistakes. Though this is a positive approach, most organizations follow the other kind of culture, which is the mistake aversion culture. In these organizations, people are anxious when mistakes happen. Superiors tend to punish mistakes and blame people for them. Mistakes are seen as being due to unwanted personality traits, insufficient knowledge and skills or low intelligence. This approach is the worst enemy of management systems, which can easily become a system for hiding instead of displaying and correcting nonconformities.

In conclusion, a good training programme should activate and manage three complementary learning methods:

- learning from information
- learning by doing
- learning by correcting mistakes

18.2.5 Audit and system review (point 4.4 of Figure 18.1)

An audit is an 'objective' evaluation – based on data and facts – of the effectiveness of a process management system. 'Objective' should be considered here as equivalent to 'independent' of personal interest or advantage or preference. Audits are essential:

- for verifying systematic conformity of product and process to planned requirements
- for comparing process results and process costs
- for providing evidence concerning reduction or elimination of problem areas, non-conformities and losses; and the effectiveness of corrective actions.

Audits can be carried out by internal or external auditors. External auditors are usually preferable. They are not influenced by the routine practices in the company and hence they are in a better position to identify wrong habits and to avoid indifference to routine mistakes as well as to potential and positive opportunities. An external auditor can bring in fresh expertise and precious critical thinking about the management system rules, procedures and standards.

Audits, together with learning from mistakes, are the most important tools for improving a company's performance. Audits must be carefully planned by defining:

- who is responsible for audit planning and who should be charged for carrying them out
- when they are to be carried out
- how the audit report should be written also suggesting an order of priority of mistakes to be avoided and points to be improved
- to whom the audit report should be sent for analysis and decisions.

In general, in extra-virgin olive oil companies, audits must be carried out during the harvesting and milling period, when operating conditions and critical points can be verified. A second audit should be carried out during the period of oil storage and handling for marketing and distribution, in order to verify two critical points:

- the storage and handling conditions of the oil in bulk or in bottles
- the product's constant traceability until it reaches the final consumer.

The results of the audits should be considered as the main feedback contribution to system improvement and control and they should therefore be addressed to the company's leaders and owners for possible system review and improvement. By comparing the management system prescriptions and the results of the audits, changes and improvement of the management system should be suggested and made effective the following year.

18.2.6 The need for simplification

The experience of many companies with management systems has often resulted in a formalistic application of procedures, poor understanding and a passive attitude of staff and workers alike.

Product and process certification is often considered as a burden imposed by fussy customers or legal requirements, instead of an opportunity for the company's improvement and competitiveness. This is probably the consequence of excessive detail, excessive formality and documentation. How can a management system be effective and simple at the same time?

The answer is to base control on risk evaluation with selection and control of critical points. Risk control introduces the principle of doing only what is needed to avoid mistakes and nonconformities. Assuming that in a complex process 'trivial are many, but critical are few', focusing on critical points with rigor is the effective and efficient way of achieving results.

18.3 Control of critical points

Critical points are resources or activities or conditions where the risk of a negative impact upon a company's objectives is considerable. A negative impact means the loss of something valuable to the company, such as product quality or safety, traceability, workers' health, environmental protection and also the company's reputation.

Critical points are identified by applying a risk analysis procedure and control of relevant process conditions is carried out as a risk treatment procedure. This is an important methodological step entailing that a process manager be an expert in risk analysis and risk management.

Critical point identification and the definition of risk prevention and treatment require a three-step procedure according to Figure 18.2.

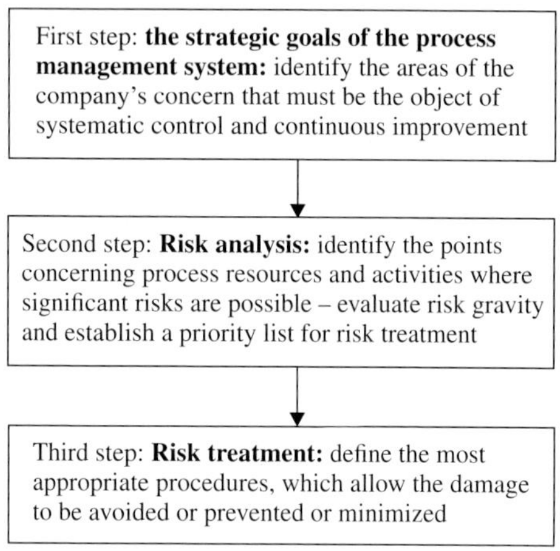

Figure 18.2 Three steps of a risk management system.

18.3.1 Identify the areas of company's concern

This step is closely related to the company's policy and vision. The planning of a management system comes after defining the company's policy as well as the product and process requirements (points 1, 2 and 3 of Figure 18.1). Deciding if product quality or hygiene, or workers' well-being or environmental protection or participation in a product certification scheme should be included in the process management system is in the discretionary power of the company's ownership.

The company's marketing approach, ethics and reputation, strong and weak points and the position of major competitors should be taken into consideration.

18.3.2 Identify the points at risk, evaluate risk gravity, identify critical points

For each area of concern, a detailed analysis should be carried out on resources and activities in order to identify points where there is a risk of possible negative consequences on process results. This step is based very much on the analyst's experience and judgment. One point to consider is that process management systems have a self-improvement ability based on detecting and correcting mistakes: if critical points have been neglected at the planning stage, they will become a source of nonconformities and will therefore be the object of a later corrective action and subsequent inclusion in the process system management.

Evaluation of risk gravity consists in determining, through the systematic use of available information, how often negative effects may occur and the magnitude of their consequences. A very simple mathematical relationship is generally applied:

$$RG = LC \times GC$$

in which the risk gravity (RG) is the product of the likelihood of consequences (LC) times the gravity of consequences (GC).

Likelihood is a qualitative description of probability or frequency. Probability is defined as the ratio of negative outcomes to the total number of outcomes. It is expressed as a number between 0 and 1, with 0 indicating an impossible outcome and 1 indicating that an outcome is certain. Frequency is a measure of the number of negative occurrences in a given time.

Calculation of probability or frequency requires precise knowledge about negative occurrences as a function of the total number of occurrences or as a function of time. This information is seldom available in risk analysis of food processes and therefore the risk analyst should give a qualitative or semi-quantitative estimate of likelihood.

The same can be said for evaluation of the gravity of consequences. The scale from negligible loss or damage to extreme gravity as, for instance, the death of people or permanent contamination of the environment, cannot be defined in precise mathematical terms. In this case, too, qualitative or semi-quantitative estimates should be given.

Qualitative estimates use words to describe the likelihood and the consequences of negative outcomes. Semi-quantitative estimates consist in giving numbers to the evaluations expressed with words by the qualitative estimate. Tables 18.1 and 18.2 present the scale of values that are suggested for likelihood and gravity of consequences in evaluating risk.

The nine possibilities (nine levels of risk gravity) resulting from the product of the three levels of likelihood times the three levels of gravity of consequences are presented in Table 18.3 and further classified in four levels as follows:

a. Risk gravity is low (the value of (LC $\times$ GC) is ≤ 2) when it can be managed by routine procedures based on people's knowledge and normal operating skill;

b. Risk gravity is moderate (the value of (LC × GC) is > 2 and ≤ 4) when management responsibilities must be specified and formal procedures must be defined.

c. Risk gravity is high (the value of (LC × GC) is > 4 and ≤ 6) when senior management attention is needed and formal procedures must be defined.

d. Risk gravity is extreme (the value of (LC × GC) is > 6) when immediate action is required: the process must be stopped immediately, emergency procedures must be implemented.

Table 18.1 Three levels of likelihood of negative consequences.

Definition (qualitative analysis)	Description	Estimate of frequency	Numerical value (semi-quantitative analysis)
Extremely unlikely	May occur only in exceptional circumstances	Once in decades	**1**
Very unlikely	Could occur at some time	Once in years	**2**
Unlikely	Might occur at some time	Once in months	**3**

Table 18.2 Three levels of gravity of negative consequences.

Definition (qualitative analysis)	Description	Numerical value (semi-quantitative analysis)
Minor	Consequences are minor losses and damage, low financial loss, first-aid treatment . . .	**1**
Moderate	Medium-high financial losses, medical treatment required . . .	**2**
Major	Extensive injuries, loss of production capability, major financial loss . . .	**3**

Table 18.3 The gravity of risk as the product of the level of likelihood times the level of gravity of negative consequences.

Likelihood	Consequences		
	Minor *Level 1*	*Moderate* *Level 2*	*Major* *Level 3*
Unlikely Level 3	Risk gravity: **Moderate** (LC · GC) is 3 × 1 = **3**	Risk gravity: **High** (LC · GC) is 3 × 2 = **6**	Risk gravity: **Extreme** (LC · GC) is 3 × 3 = **9**
Very unlikely Level 2	Risk gravity: **Low** (LC · GC) is 2 × 1 = **2**	Risk gravity: **Moderate** (LC · GC) is 2 × 2 = **4**	Risk gravity: **High** (LC · GC) is 2 × 3 = **6**
Extremely unlikely Level 1	Risk gravity: **Low** (LC · GC) is 1 × 1 = **1**	Risk gravity: **Low** (LC · GC) is 1 × 2 = **2**	Risk gravity: **Moderate** (LC · GC) is 1 × 3 = **3**

On the basis of risk gravity, the most appropriate decision about risk treatment is made in the third step.

18.3.3 Define appropriate risk treatments

Risk treatment involves choosing among four possibilities:

1. The first possibility is to avoid the risk (risk elimination) by deciding not to proceed with the activity or the material, which generates the risk.

Examples of risk elimination

- The risk to the safety of consumers and workers and the risk of environmental pollution associated with pesticide treatments can be avoided by switching to organic agriculture. In this case, however, reduction of quality and yield due to pest attack must be taken into consideration and possibly controlled by appropriate management decisions about harvesting and milling conditions.
- Outsourcing (which is a risk transfer to other specialized subjects) can be usefully applied in many cases requiring risk prevention as, for example: pest control, plant maintenance, oil bottling, waste disposal.
- The risk to workers' safety can be prevented by using plants with built-in, automatic safety devices.

2. The second possibility is to reduce the likelihood of negative occurrences (risk prevention). This category includes all the control measures, which can be of varying complexity depending on the complexity of the operation and the gravity of possible consequences. The risk is not avoided, but activities are carried out with special attention so that the likelihood of negative occurrences is substantially reduced.

Examples of risk prevention

- Regular and careful cleaning reduces the likelihood of product contamination from the environment or from the plant or the building.
- Blowing filtered, sterile air into bottles before filling them reduces the likelihood of oil being contaminated by dust, soil and glass fragments.
- Monitoring pest attack, suitable and timely use of pesticides, reduces the likelihood of product losses and, at the same time, product contamination, poisoning of workers and pesticide pollution.

It is impossible to give simple suggestions about the format for risk prevention procedures. They may consist of very simple notices (for example: 'do not enter' or 'pets not allowed' or 'do not activate this pump unless . . .') or a warning signal, or they can be very complex entailing a written text with:

- the name of the person responsible for the procedure
- the activities that must be carried out for the proper implementation of the procedure
- the data to be monitored and recorded
- the procedure to be applied in case of negative occurrences, losses or damages.

The technical literature has numerous examples of procedures for the prevention of quality or hygiene risk as well as for workers' health and safety risk or environmental pollution risk. Choosing the right format and complexity, detail and language is the responsibility of the system manager, according to the company's needs and the operators' experience and skill.

3. The third possibility is to reduce the gravity of the consequences. This case applies when risk cannot be prevented at a reasonable cost, for instance in the case of product loss due to atmospheric phenomena, especially storms and hail or to external factors and agents that are difficult to control, such as theft or fire or fraud. In this case, risk treatment may consist of contract conditions, design features, insurance coverage, emergency plans.
4. The fourth possibility is to accept the risk and its consequences because they are considered less expensive than setting up preventive measures. This possibility should be applied with great caution as all the statistics about risk demonstrate that the most frequent reason for serious damage is underestimation of the risk. In case it is decided to use this option, a good rule is to apply it only to risks whose consequences are of minor or moderate gravity.

18.4 Risk analysis: a blanket rule for management decisions

In summary, the process management system presented in this handbook is based on two essential methodological features:

- a structure in four points according to the scheme of Figure 18.1: policy, product, process and control. This last point comprises, in turn, four sections: control of product, control of process, general control procedures, and audit with system review.

- A method of risk analysis and risk treatment for identifying and keeping under control the process points that have a critical impact on process results.

The risk analysis approach should be considered as a general method of process management. It should be applied not only to product safety but also to all the other company's commitments concerning product quality, workers' safety, environmental sustainability, and so on.

Furthermore, the risk analysis criterion can be applied not only to process-control decisions, but to any decision implying choice, effectiveness, and cost. For instance, in deciding which analyses should be carried out on products, the probability of product nonconformity should be taken into consideration. Analysis of parameters that commonly conform to standards should seldom be the object of analytical evaluation. On the contrary, analyses of parameters at risk of nonconformity must be frequent and systematic. This criterion also applies to management procedures. If, for instance, the risk of confusion in recording and documentation is high, the management procedure of these activities should receive much care and attention.

Annexes 18.1 and 18.2 give further examples of process management and risk analysis applied to extra-virgin olive oil production.

Annex 18.1: Excellence in extra-virgin olive oil

Introduction

Excellence may be defined as a level of high quality within the category of extra-virgin olive oil. Good and excellent oils as well as common and anonymous oils are all included in this category. It is therefore of no surprise that so many proposals and attempts are being made to foster niches of excellence in local or global olive oil markets.

This annex summarizes the results of an experiment carried out in the years 2004–2011 by extra-virgin olive oil producers from Italy, Spain, Greece, California, France and Australia.

The experiment, planned and implemented by the nonprofit association 3E (Ethics – Excellence – Economics) received support and approval from several organizations including the Georgofili Academy of Florence, the Olive Center of the University of California Davis, the Culinary Institute of America at Greystone (California), and the Spanish Interprofesional del Aceite de Oliva. The theme of extra-virgin olive oil excellence was presented and debated in five annual meetings of the International Conference 'Beyond Extra-Virgin' (Association 3-E 2008).

The 3E model

The 3E model of olive oil excellence can be described as a three-step procedure concerning product and process requirements (Peri *et al.* 2010).

First step: basic product requirements

Table 18.4 presents the five basic requirements for excellence in extra-virgin olive oils according to the 3E model.

The five basic requirements have different meanings:

- documented chain traceability from the field to the consumer's table is the most important proof of the authenticity of the oil and its correspondence with claims
- the three analytical values of free acidity, peroxide number and K_{232} absorption are proof of the excellent quality of the olives and good processing conditions
- the absence of sensory defects is proof of the hygienic design and proper operating conditions of the mill.

The above requirements should be considered as common to all excellent extra-virgin olive oils, independent of origin or cultivar or processing technology.

Second step: sensory style

In order to add characterizing requirements to the excellence of oils, the sensory profile and style should be defined. A characterizing sensory profile is considered as the

Table 18.4 The five basic requirements for excellence in extra-virgin olive oil according to the 3E model.

1. Chain traceability	Documented material balances from the field to the consumer's table
2. Free acidity	Less than or equal to 0.3 (± 0.02)
3. Peroxide number	Less than or equal to 7.5 (± 0.2)
4. K_{232} absorption value	Less than or equal to 1.85 (± 0.02)
5. Sensory defects	Absent according to the IOC panel test

essential characterizing requirement for culinary uses and for communication and dialogue with the consumer. Under 3E standard requirements, a well-defined sensory style that can be reproduced year after year is necessary as a brand-qualifying asset.

Third step: an effective process management system

Excellence cannot be the result of chance or of a lucky combination of casual events. In the 3E model, excellence must be the result of a well-planned management system, with careful control of a series of critical points, including:

- availability of suitable human and technical resources
- traceability of batches through documented material balance
- control of the harvesting operation and careful planning of the harvesting-milling link
- control of pests and diseases of the trees and the fruit
- control of olive handling and storage
- control of time-temperature relationships and residence time in critical processing steps
- control of oil handling and storage
- standards of workers' safety
- standards of environmental sustainability.

In the experiment of Association 3E, evaluation of these points was carried out by applying a semi-quantitative scale of rating agreed upon both by producers and customers. Continuous improvement was required with commitment to progressive elimination of nonconformities. Failing to meet the standard of traceability determined the automatic exclusion from selection for excellence.

Updating the 3E model: the quality-proximity matrix

In 2012 the 3E system was further developed by introducing the quality-proximity matrix concept (Peri 2012). The inspiring principle of the quality-proximity matrix

PROXIMITY LEVEL QUALITY LEVEL	**FIRST LEVEL:** Control of storage conditions from field to bottle	**SECOND LEVEL:** Control of storage conditions from field to retail shelves	**THIRD LEVEL:** Control of storage conditions from field to table
EXCELLENT (see annex on excellence)		**AREA OF EXCELLENCE**	
GOOD (e.g. protected designation of origin)			
COMMON (e.g. legal standards for extra virgin olive oil)			

Figure 18.3 The quality-proximity matrix.

is that the quality of an extra-virgin olive oil should be judged from the point of view of the consumer at the moment of its use. It should therefore be acknowledged that quality levels of extra-virgin olive oil, controlled, defined, verified and certified at the producer's gate may not be sufficient to guarantee the highest level of quality at the consumer's table. In fact, the oil can undergo changes and degradation in the long and uncertain distribution chain. This is especially true for excellent extra-virgin olive oil, whose sensory and nutritional characteristics are particularly fragile and perishable.

The quality-proximity matrix (Figure 18.3) suggests that the quality at the consumer's table is based on a combination of two steps of the control system:

- the quality control, evaluation and guarantee at the production stage, and
- the control of the commercial chain ending as close as possible to the moment of consumption (proximity). The closer an uninterrupted control of storage and handling conditions is to the consumer's table, the higher is the probability that the excellence achieved by the producer may become available to (and enjoyed by) the final consumer.

As can be seen in Figure 18.3, different quality levels can be achieved in the production of extra-virgin olive oil. For example:

- a level of common quality conforming to the legal definition of extra-virgin olive oil

- a higher level as, for example, in product certification according to the standards of local and regional designations of origin
- a further and higher level of quality according to the standards of excellence of the 3E model.

For each of these levels, the control of handling and storage conditions can be carried out.

- From the field to the producer's facility, when bottles are shipped. This is the most frequent case in practice but also the least reliable in terms of quality at the consumer's table.
- Controlling and monitoring of the handling and storage conditions during the commercial life of the product from production to the retail shelf. This would be a great step forward in terms of guarantee.
- Finally, control of handling and storage in the last segment of the chain – restaurants, foodservices and families. This would be the most effective way of providing the final consumer with the best quality. The area of excellence is a combination of the highest level of quality at the production, followed by the control of the conditions of commercialization as close as possible to the moment of consumption.

The application of the quality-proximity matrix principles and practice requires a substantial improvement in oil marketing conditions and use.

Annex 18.2: An exercise of integrated risk analysis applied to the process of extra-virgin olive oil

Introduction

The method presented in this paragraph is a training exercise. Experts may find it too theoretical or obvious. They may prefer a less formal approach with simple discussions and brain storming with interested people. A list of critical points emerging from a discussion with people with extensive experience can perfectly satisfy the needs of process management system planning. However, reading and thinking about this annex can be a good memorandum of what should be taken into consideration and, maybe, a stimulus to further attention and innovative approaches.

Risk analysis and risk treatment

According to the three-step procedure described in Figure 18.2, the risk analysis of an extra-virgin olive oil process can be carried out as follows:

First step: identify the areas of concern

Suppose the owner of an olive milling company has decided to minimize risk in the areas of concern reported in Table 18.5. A few comments may help explain the rationale of the method. The first five points represent traditional areas of concern; putting them under the control of an effective management system seems a very reasonable decision. Point 6, instead, seems out of the usual management schemes. On the contrary, it is as critical as anything else given its effects on the company's image and reputation. An owner may very well consider it a crucial point and decide that it must be closely controlled by the management system.

Second step: risk analysis and risk treatment

This part consists of a detailed analysis of the structural resources (buildings, plants, human resources) and the activities that may become the cause of a possible risk.

Table 18.5 The six areas of concern of a hypothetical olive miller to be integrated into the company's process management system.

No.	Area of concern
1	Product quality
2	Product safety and process hygiene
3	Product traceability
4	Waste disposal
5	Workers' health and safety
6	Reception and welcoming of visitors and customers in an oil selling facility

The points to be considered may be presented in the form of a checklist or with more sophisticated representations as layouts and flowcharts. It is important that any point that can be a source of risk be identified. Resources include all the structural components of a process – in particular, buildings, the external environment, logistics, equipment and people, including workers and staff. Activities include the operating procedures and conditions as well as the monitoring and control activities.

The checklists of points concerning resources and activities are presented in Tables 18.6 and 18.7, respectively. As is evident from Table 18.6, the analysis of resources may include material structures such as buildings and plants, design features such as hygienic design or ergonomics and even psychological resources like people's motivation.

Some critical points, which are indicated only once in a checklist, may concern, in fact, various points in the process. This is, for instance, the case of ergonomics, which is only cited at point 11 of Table 18.6 but it may concern many points and steps of the process from harvesting and handling of the olives to handling of the oil.

In other cases, many points refer to the same requirement. For instance, traceability – an essential requirement of product authenticity and guarantee – may be at risk at several critical points. Similarly, what has been indicated as 'overall hygiene of the oil factory' involves many critical resources and activities as described in detail in Chapter 21 on the hygiene of the oil factory.

Presentation and discussion of the various points in Tables 18.6 and 18.7 can be considered as a good training programme for the staff and the workers of an extra-virgin olive oil company.

Once the lists have been compiled, the analyst should go through it point-by-point and ask: 'In the situation of the factory under examination, should this point be considered as critical or not for the achievement of one of the planned product or process requirements?' The experience of the analyst and of the company can suggest a 'yes' or 'no' answer.

A second question should follow: 'In the situation of the factory under examination, will the gravity of risk be high enough to require a risk treatment?' And again, the experience of the analyst and of the company can suggest a 'yes' or 'no' answer. This may be a fast way of carrying out the analysis. However, this type of judgment should be pondered carefully in order to avoid charging the management system with excessive and useless control duties, but, also, on the other hand, the really critical points should not be neglected. Taking enough time for analysis and discussion with informed people is always useful.

An exercise of risk analysis

For the sake of training, in the following tables risk gravity evaluation is presented referring to the six areas of concern mentioned in Table 18.5 and to the case of a factory pursuing quality excellence. In the same tables, suggestions are given for possible risk treatment.

Table 18.6 Resources of an extra-virgin olive oil company and sources of potential risk.

Resources	Potential crucial points are where mistakes or nonconformities are frequent, with consequent losses, or complaints, or costs ...[a]
Environment (outside of the factory)	1. Potential sources of biological or chemical or particulate contamination
	2. Unsuitability of logistics
Buildings	3. Unsuitability of the layout (availability of rooms and space, separation of dirty from clean areas, state of maintenance of buildings and services, protection from outside contamination, etc.)
	4. Temperature control in the storage area
	5. Work environment: temperature, humidity, light, airflow, noise
	6. Suitable space and facilities for waste discharge and treatment
Equipment	7. Equipment available and operating characteristics
	8. Suitable control of operating parameters
	9. Structural precautions for workers' safety
	10. Protection from environmental contamination
	11. Ergonomics
	12. Hygienic design
	13. Cleaning-in-place features
	The analysis must be carried out on each component and step of the process, such as: olive harvesting, handling and transportation, cleaning, milling, paste malaxation, decanter fractionation, centrifugal finishing, oil filtration, oil handling and storage, oil bottling and packaging, measuring and control devices, pumping and transferring equipment, storage tanks, etc.
People	14. Professional and personal skills
	15. Duties and responsibilities
	16. Motivation
	17. Training
Energy and water supply	18. Available energy and water resources
	19. Energy and water saving/recovery systems

Note: [a]A nonconformity is failure to meet predefined requirements of resources or activities due to accidental causes, human error or equipment failure.

Table 18.7 Operations and activities of an extra-virgin olive oil company to be considered as sources of potential risk.

Operations	Potential critical points
Maintenance and hygiene	1. Outside space and facilities
	2. Buildings
	3. Plants and equipment during operation and standstill
	4. Hygiene and the HACCP system
	5. Waste management and treatment
Harvesting	6. Ripeness evaluation and monitoring
	7. Inconsistencies in the harvesting decision
	8. Workers' safety
Olive handling	9. Time-temperature control of harvested olives
	10. Storage and transportation conditions of harvested olives
	11. 'Milling lots' (olives) identification and weighing
Olive cleaning and washing	12. Rinsing effectiveness and water replacement rate
	13. Waste disposal
Olive milling	14. Flexibility (options available, including pitting) and control of the milling operation
Olive paste malaxation	15. Time-temperature control
	16. Residence time distribution
	17. Workers' safety
Decanter separation of the oil	18. Control of the water content of the olive paste
	19. Effective oil recovery
	20. Residence time distribution of olive paste and oil
	21. Waste disposal
	22. Noise control
Centrifugal finishing	23. Equipment adjustment and tuning
Oil filtration	24. Control of time, temperature, shear stress and environmental contamination
	25. Disposal of exhausted filter pads
Olive evaluation and weighing	26. Oil evaluation by rapid, in-line analyses
	27. 'Milling lots' identification and weighing
	28. Oil assignment to 'storage lots'

(*continued overleaf*)

Table 18.7 *(continued)*

Operations	Potential critical points
Oil storage	29. Environmental hygiene conditions
	30. Inert atmosphere supply and control
	31. Time-temperature control
Blending	32. Final off-line evaluation and identification of 'bottling lots'. Analytical and sensory criteria.
Waste treatment plant	33. Effectiveness at peaks of waste discharge
Outsourcing[a]	34. Outsourcing services may include: chemical and sensory analyses, building and equipment maintenance, pest control, bottling, selling and transportation, workers' training, etc.

Note: [a]Outsourcing is when a particular activity or process is purchased as a service. Outsourcing is increasingly popular among producers. Companies pay only for the service they need, when they need it. It reduces the need to hire and train specialized staff, brings in fresh expertise and reduces capital and operating expenses.

A preliminary remark

The purpose of this exercise is to stimulate the interest for risk analysis. By no means the evaluations presented in this exercise should be considered as generally valid. Applying this procedure to the reader's experience is a useful educational excercise and may lead to very different results, depending on the factory size and location, and on the manager's culture and experience.

Critical points of product quality

If a company aims at producing an extra-virgin olive oil of excellent quality (Annex 18.1), it should be presumed that all the basic conditions for quality are met, such as an appropriate system for olive harvesting and handling, including a short interval of time between harvesting and milling, a modern milling plant and good control of the milling conditions. Excellence, however, is a very demanding objective in which the company pursues standards that are well beyond those required for extra-virgin olive oil. For instance, excellence requires values of free acidity, peroxide number and spectrophotometric values much lower than those required for common extra-virgin olive oil. Not only should sensory defects be absent, but the sensory profile should correspond to the sensory style of the brand. Finally, the concept of excellence should apply not only to production of the oil, but also and especially to consumption of the oil. This entails special attention to the handling conditions during the commercial life of the oil and the application of the principle of the quality-proximity matrix.

The results of a semi-quantitative evaluation of risk gravity and some indication of risk treatment options are reported in Table 18.8. The gravity of consequences is

Table 18.8 Critical points and gravity of risk for oil quality.

Critical point	Gravity of risk for quality (likelihood × gravity)	Risk treatment
Pest attack	2 × 3 = 6 (high)	Monitoring pest attack and timely decision for pesticide treatments
Timing of harvesting	2 × 2 = 4 (moderate)	Monitoring of olive ripeness and conditions for the harvesting decision
Handling of harvested olives and time between harvesting and milling	1 × 2 = 2 (low)	Not a critical point
Residence time in malaxation	2 × 2 = 4 (moderate)	Hygienic design of the plant. Residence time control in the malaxing operation. Frequent emptying, washing and cleaning of the plant
Time-temperature relationship in malaxation	1 × 2 = 2 (low)	Not a critical point
Oil handling and storage in the commercial chain from production to consumption by the final consumer	3 × 2 = 6 (high)	Control of time-temperature relationship, inert atmosphere, hermetic seal, protection from light absence of suspended material

given a value of 1 when quality fails to meet high quality standards. Failure to meet the common extra-virgin standards which implies selling the oil as 'virgin' is given a gravity value of 2, while classification of the oil as 'lampante' and the consequent need to refine it is given a gravity value of 3.

Data presented in Table 18.8 show that:

- Pest attack of the olives is the first risk to be minimized as a prerequisite of quality and even more so for achieving a level of excellence. The problem may be particularly difficult to deal with in the case of organic agriculture.
- The choice of harvesting time is a critical point because excellence requires obtaining a characterizing sensory profile. A wrong or uncertain definition of the maturity stage of the olives is still a rather common occurrence.
- The residence time and the time-temperature relationship of the olive paste in the steps from milling of the olives to the centrifugal finishing of the oil are critical conditions for excellence. During this phase, the oil is in contact with the watery phase and in the presence of lipolytic and oxidizing enzymes. There may be spoiled oil in some dead spot(s) of the plant and sensory defects may be transferred to the mass of good oil being processed.

- Storage and handling conditions during the commercial distribution, including storage and transportation operations, are often far from ideal for the best preservation of oil quality. In terms of excellence, this is very often the weakest point of the olive oil chain.

In conclusion, in the case considered in this example, excellence of the oil depends very much on two major points: control of pest attack in the olive grove and handling and storage conditions of the oil during its commercial life. Contrary to what is usually said about malaxation, the critical problem is the control of residence time in the malaxers and not the time-temperature relationship, which is automatically controlled in modern milling plants.

Critical points of hygiene and consumer safety

Judgment based on knowledge and experience suggests that points that may be critical for the achievement of good hygienic standards and consumer safety objectives are:

- Control of pesticide residues. According to official records, pesticide contamination from treatments in the olive grove is a rare occurrence. A more dangerous and seldom-discovered condition of contamination is when pesticides are used by unskilled people to fight pests at the milling and oil storage facility.
- Particulate contamination during the bottling operation due to the presence of particulate contaminants in the bottles. The problem may be of some gravity in the case of glass fragments.
- Oil contamination can take place on several occasions during processing, when the oil is exposed to contamination from the environment, pests, people, or the atmosphere.
- Toxic contaminants from the environment (for example, PCBs or PAHs) are extremely unlikely to contaminate, except the case of factory location in densely populated and industrialized areas.

It is assumed that contamination with nontoxic materials represents a gravity of consequences of 2, while the presence of toxic contaminants represents a gravity of 3. The level of gravity of 1 has been eliminated from the analysis of safety risk as a precautionary measure. In fact, a safety problem is never considered as minor by consumers. Results of a semiquantitative evaluation of risk gravity and some indications of risk treatment options are reported in Table 18.9.

It is interesting to observe that most of the problems concerning hygiene and oil safety may be avoided by an effective overall hygiene programme at the oil factory. It is also worth considering that a significant risk may be due to the presence of particulate contaminants in the bottles; this problem is often underestimated.

Table 18.9 Critical points and gravity of risk for consumer safety.

Critical point	Gravity of risk to consumer health (likelihood × gravity)	Risk treatment
Pesticide residues	2 × 2 = 4 (moderate)	Strictly follow directions for pesticide treatment in the field. Leave the responsibility for pesticide and rodenticide treatments at the milling and oil storage factory to experts
Bottling	2 × 2 = 4 (moderate)	Prevent particulate contamination in bottles, especially glass fragments
Overall hygienic conditions at the milling, oil storage and bottling factory	3 × 2 = 6 (high)	Prevent contamination from the environment, people and the atmosphere (particulate, volatiles, smoke, dust). Apply the HMS or HACCP system and procedures
Toxic environmental contaminants such as PCBs, PAHs, etc.	0.5 × 3 = 1.5	Not a critical point

Table 18.10 Critical points and gravity of risk for product traceability.

Critical point	Gravity of risk of loss of traceability (likelihood × gravity)	Risk treatment
Recording of quantitative and compositional data at critical chain points	2 × 3 = 6 (high)	A unique record and control system 'from the field to the consumer's table', including: harvesting and milling lots, storage, bottling and selling lots.

Critical points of traceability

The application of risk analysis to oil traceability should take into consideration a distinctive feature of this requirement: traceability is based on recording amounts at some precise points of the oil chain 'from the field to the consumer's table'. Missing even one single point invalidates the whole procedure and causes the loss of traceability even if the traceability procedure has been duly applied at the other points. In terms of risk prevention traceability is an all-or-nothing alternative because a partial or incomplete traceability is meaningless and does not correspond to consumer requirements. For this reason, the gravity of consequences can only be of level 3 (loss of traceability).

The suggested risk treatment makes true traceability a possible objective only for short chain organizations, as pointed out in the first chapter of this handbook. In the context of industrial production and global marketing, true traceability is a very difficult and sometimes impossible task.

Critical points for workers' health and safety

Judgment based on the company's experience and official statistics suggest that critical points for workers' health and safety are:

- Pesticide treatments. Skin contact or inhalation of pesticides may have serious consequences on the workers' health. Very strict rules must be followed in handling, using and spraying pesticides. The same applies for the use of pesticides and rodenticides in milling and oil storage factories.
- Accidents during harvesting can have serious consequences when using ladders and harvesting machines. Improper ergonomics, especially with hand-held harvesting machines, can also have health consequences.
- Many operations in the milling process may result in accidents and chronic diseases. Major risk factors are: excessive weight lifting; intense and continuous noise; improper ergonomics and other organizational inconsistencies with consequences of stress, especially at peak times in milling operations.
- Maintenance of the plant and equipment by inexpert operators. Mechanical accidents or electrocution can result from improvised maintenance or repair operations.

Gravity of consequences is 1 in the case of a minor injury and a brief pause (hours) from work and first-aid treatment. Gravity is 2 in the case of a serious accident or illness requiring medical treatment and / or hospitalization and absence (days) from work. Gravity is 3 in the case of very serious injury or illness causing permanent disability or death. The results of a semi-quantitative evaluation of risk gravity and indication of risk treatment options are presented in Table 18.11.

Serious risks to workers' health and safety may derive from pesticide treatment at the olive grove or at the oil factory. The harvesting operation can be associated with a significant risk. Yearly maintenance of the milling plant has been given a relatively low risk gravity because these operations are usually carried out by experts. Working conditions can be very different from one factory to another and a much more detailed analysis is suggested for this particular concern. In fact, in terms of workers' health and the company's responsibility, very heavy consequences can derive from accidents, mistakes or non-conformities.

Table 18.11 Critical points and gravity of risk for workers' health and safety.

Critical point	Gravity of risk for workers' health (likelihood × gravity)	Risk treatment
Pesticide treatments	3 × 2 = 6 (high)	A standard operating procedure for handling, storage and use of pesticides
The harvesting operation	2 × 2 = 4 (moderate)	Safe use of machinery and safe access to olive trees
Poor protection from the equipment and poor ergonomics in the milling process and in handling of olives and oil	3 × 2 = 6 (high)	Control of the electrical system, machinery, slippery floors, storage tanks, excessive noise (decanter), breaking of bottles, fire
Yearly maintenance of the plant	1 × 2 = 2	Not a critical point

Critical points for environmental pollution

Judgment based on a company's experience and official records suggest that points that may be critical for reducing the impact of extra-virgin olive oil production on the environment are:

- Pesticide treatments may produce aerosol dispersion in the environment with negative consequences to wild fauna and people.
- Discharge of effluent and by-products of the milling process and poor hygiene of the milling site can cause water and soil pollution, infestation by insects and rodents, bad odours.

Gravity is given a value of 1 in the case of minor, short-term (days) environmental pollution. Gravity is given a value of 2 in the case of off-site release with detrimental effects no longer than one year. Gravity 3 is not considered as likely in the case of the extra-virgin olive oil process. The results of a semi-quantitative evaluation of risk gravity and indication of risk treatment options are reported in Table 18.12.

Although of moderate gravity in terms of overall impact to the environment, the problem of managing waste and by-products from extra-virgin olive oil factories should be seriously considered because waste discharge is concentrated in a short period of time corresponding to the harvesting and milling period. Improper handling of the waste and by-products not only produces an immediate and relevant increase in the polluting charge to the environment, but can also result in attracting pests and mites and odour pollution. Violation of the environmental protection laws and the negative image resulting from improperly handling waste can seriously damage the company's reputation.

Table 18.12 Critical points and gravity of risk of environmental pollution.

Critical point	Gravity of risk of environmental pollution (likelihood × gravity)	Risk treatment
Pesticide treatments	$2 \times 1 = 2$ (low)	Not a critical point
Discharge of waste from the milling process - poor hygiene outside the milling factory	$3 \times 2 = 6$ (high)	Control of disposal of pomace, vegetation waters, filtering residues, etc. Cleaning of the milling site, combating insect and mite infestation, control of unpleasant odours

On the other hand, waste and by-products of a milling factory can be used as raw materials for the production of energy or animal feed or fertilizing compost or for the extraction of biologically active compounds. These possibilities are discussed in detail in Chapter 22.

Critical points for visitors reception

This subject should be considered as increasingly important in small and medium-sized olive oil-factories, where direct selling at the factory is a good marketing opportunity and also an opportunity for direct information and education of customers and consumers.

The consumer reception area should be separated from the milling building. The milling, storage and bottling areas can be the object of guided visits, showing and explaining the extra-virgin olive oil process. Visitors must not be allowed to interfere with the working activities. On the other hand, the visitor reception area should be friendly and pleasant with possible preparation and tasting of simple dishes and oil. Good hygienic standards inside and outside of the oil factory, together with effective communication about the process and the oil, are appreciated by customers and consumers alike. On the contrary, poor hygiene and untrue or misleading communication are a serious threat to the company's reputation and consumer trust.

The gravity of consequences is 1 in the case of a minor non-conformity to hygienic standards, either inside or outside the olive oil factory. The gravity is 2 in the case of improper information about the process or the product and unplanned or unprotected visits to the working areas. The gravity is 3 in the case of misleading or false information about the process and the product and unplanned visits to the working areas with possible risk to visitors' safety (slippery floors, unprotected access to stone mills or malaxers or other dangerous equipment). The results of a semi-quantitative evaluation of risk gravity and indication of risk treatment options are reported in Table 18.13.

Table 18.13 Critical points and gravity of risk at visitors' and customers' reception area.

Critical point	Gravity of risk of environmental pollution (likelihood × gravity)	Risk treatment
Information for customers and consumers	$3 \times 2 = 6$ (high)	Carefully plan the mode and content of communication. Information about the product and the process must be clear and truthful
Overall hygienic conditions inside and outside the oil factory	$2 \times 3 = 6$ (high)	Clean the milling site, control bad doors. Prevent visitor access to the working areas and unprotected equipment
Risk to visitors' safety	$1 \times 2 = 2$	Not a critical point

The access of visitors (either experts or nonexperts) to an extra-virgin olive oil factory entails very demanding requirements in terms of management. They are:

- Information must be as simple, clear and truthful as possible. The risk of improper or misleading or false communication and of unjustified claims is high;
- The level of hygiene both inside and outside the olive oil factory has an immediate impact on visitors' impression and hence on the company's reputation.
- The risk to people visiting the olive oil factory, especially during the milling operations, entails further responsibility in terms of the visitors' safety. Glass windows can allow the visitors to view the process without having direct access to the working area.

Conclusion of the exercise

The results of the risk analysis according to Tables from 18.8 to 18.13 are summarized in Table 18.14.

The first conclusion is that calculation of risk gravity allows nine critical points to be selected, which are possibly associated to a high level of risk gravity. This is a small number compared to the six areas of the company's concern. The nine critical points are:

1. Pest attack
2. Oil handling and storage in the commercial chain from production to consumption

Table 18.14 Summary of risk gravity according to data reported in Tables 18.5–18.13.

Risk concerning	Level of risk gravity		
	High	*Moderate*	*Low*
1. Quality	2	2	2
2. Hygiene and consumer safety	1	2	1
3. Traceability	1	–	–
4. Workers' health and safety	2	1	1
5. Environmental pollution	1	–	1
6. Visitors' reception	2	–	1
Total	**9**	**5**	**6**

3. Overall hygiene conditions at the milling, oil storage and bottling factory
4. Recording of quantitative and compositional data at critical chain points
5. Pesticide treatments
6. Poor protection from the equipment and poor ergonomics in the milling process and in handling of olives and oil
7. Discharge of waste from the milling process
8. Information for customers and consumers
9. Poor hygiene outside the milling factory

The list can be further simplified by considering that points 1 and 5 refer, in fact, to the same problem and must be handled with a single procedure.

Points 3 and 9 also refer to the same problem and are to be handled with a suitable hygiene management system (Chapter 21).

The five critical points with an intermediate level of risk gravity are:

1. Correct timing of harvest
2. Excessive residence time in malaxation
3. Pesticide residues in the oil
4. Particulate contamination in bottles during the bottling operation
5. Safety during the harvesting operation

Points 1 and 5 concern the same operation and should therefore be part of the same procedure. Point 3 must be part of the pest control procedure, which has already been defined in the list of high-gravity risks.

It can be concluded that risk analysis can be a guide to focusing on the real critical points, with great simplification and precise targeting of preventive measures.

References

Association 3-E (2008), 'Beyond Extra-Virgin', Olive Oil Excellence and World Heritage Project – Fostering Innovation in Quality and Flavor Discovery at the Intersection of Agriculture, Science and Technology, and the Culinary Arts, a founding document presented by Academy of Georgofili (Florence), the UC Davis Olive Centre and the CIFAR of University of California Davis, The Culinary Institute of America and Association 3-E, Florence, June 7, 2008.

Peri, C. (2012) *La matrice qualità-prossimità: un nuovo modo di concepire l'eccellenza dell'olio di oliva*, Academy of Georgofili, Florence, www.georgofili.net/schedadigitale.asp?idv=3249 (accessed 11 October 2013).

Peri, C., Kicenic Devarenne A. and Pinton, S. (2010) 3-E Super-Premium Selection for Extra-Virgin Olive Oil. *Beyond Extra-Virgin, The Fourth International Conference on Olive Oil Excellence*, organized by Association 3E (Milan, Italy), the Academy of Georgofili (Florence, Italy), The Culinary Institute of America (St Helena, California) and the Olive Center of the University of California Davis, Verona, 22 September 2010.

Peri, C., Lavelli, V. and Marjani, A. (2006) *Qualità nelle aziende e nelle filiere agroalimentari*, Hoepli, Milano.

Further reading

ISO 9004:2009 (2009) *Managing for the Sustained Success of an Organization – A Quality Management Approach*, ISO, Geneva.

ISO 31000:2009 (2009) *Risk Management – Principles and Guidelines*, ISO, Geneva.

Peri, C. (2006) The universe of food quality. *Food Quality and Preference* **17**, 3–8.

19

Extra-virgin olive oil traceability

Bruno Zanoni
Department of Agricultural, Food and Forestry System Management, University of Florence, Florence, Italy

Abstract

Traceability consists of documented proof of the identity of a product and the responsibilities involved in the production chain 'from the field to the consumer's table'. Traceability is based on documenting material balances through discrete batch monitoring and management. Four steps are described as the basic frame of any possible traceability scheme of extra-virgin olive oil: harvesting batches, milling batches, storage batches and packaging batches. The role of analytical fingerprinting of batches is a complementary, not an alternative, tool for extra-virgin olive oil traceability.

19.1 Introduction

Traceability consists of documented proof of the identity of a product and the responsibilities involved in its production chain 'from the field to the consumer's table'.

The primary goal of traceability is of a juridical nature: blame in case of physical or economical or moral damage to the consumer due to wrongdoing or misleading information. The second goal is of a technical nature: identification of the causes of loss or spoilage in order to apply appropriate corrective action.

Since improper handling or fraud can take place at any step of the production chain, a true guarantee requires that traceability covers the whole chain from production of the raw material to consumption of the final product.

Traceability is based on documenting material balances through discrete batch monitoring and management. A batch is a portion of a given material having a specific composition and identity (Peri *et al.* 2010).

A good traceability system requires that a company is able to give documented information concerning all batches handled under its responsibility. These include

The Extra-Virgin Olive Oil Handbook, First Edition. Edited by Claudio Peri.

raw material, intermediate and final products. For each and every batch the following information should be available:

- when: the given time (a date);
- what: the name and identity of the material, including its analytical characteristics, if available;
- how much: amount in weight units, preferably kg;
- where: the location or the identification code of a batch;
- where it came from: a previous operation or outside supplier(s);
- its destination: a subsequent operation or outside customer(s).

The identity of a product changes when a batch is mixed with other batches. Therefore, in traceability systems, any mixing should be recorded with the relative amounts of the mixed components. New identification should be given to newly formed batches.

Traceability requires that quantitative changes be recorded from the formation of a batch until its depletion. Such information should allow the company to demonstrate a precise record of inputs and outputs, so that:

$$\text{Inputs} - \text{outputs} = \text{quantity available in the company's responsibility}$$

19.2 Four basic steps

The following four critical steps must be considered for extra-virgin olive oil traceability (Peri *et al.* 2010).

Step 1 – the harvesting batches

A harvesting batch is defined as the batch of olives harvested and delivered to the olive mill as a unit mass, under the responsibility of the olive grove owner or a person otherwise responsible for olive harvesting. This information allows the relationship between the quantity of harvested olives and the production potential of the olive grove to be documented. Data to be recorded at step 1 are summarized in Table 19.1

Step 2 – the milling batches

A milling batch is defined as the batch of oil obtained from the milling process and put in a single container. A milling batch of oil may correspond to a batch of olives

Table 19.1 Traceability data of harvesting batches.

Date	The (hour and) day of delivery to the olive mill
Responsibility	The name of the company or the person responsible for olive harvesting, transporting and delivering to the mill
Identification code	A progressive number or other suitable code identifying the harvesting batch at the olive mill reception
Quantity	The weight of olives in kg
Origin	The olive grove location and surface area or number of olive trees corresponding to the olive batch
Other identifying information	As, for example, the cultivar(s), the participation in a certification scheme, etc.

Table 19.2 Traceability data of milling batches.

Date	The (hour and) day of completing the milling batch
Responsibility	The name of the company or the person responsible for the milling process
Two quantities	The quantity of oil (kg) and the corresponding quantity (kg) of olives
Identification code	A progressive number or other suitable code. Some essential analytical data of the milling batch are recommended

or may derive from the milling of several batches of olives or from the milling of a fraction of a batch of olives.

The oil of each milling batch should undergo a simple analytical evaluation in order to avoid mixing of incompatible milling batches.

Data to be recorded at the second step are summarized in Table 19.2:

Step 3 – the storage batches

A storage batch is the batch of oil contained in a single storage tank, to be stored under suitable conditions before blending or packaging and selling. In general, storage batches are formed by mixing several milling batches.

The formation of storage batches is a pivotal step of the traceability chain. It is the ending point of the production process and the starting point of the marketing process.

Suitable process control requires that the oil of storage batches be identified by a standard analytical procedure and an identity card, including, at least: free acidity, peroxide value, spectrophotometric values and the oil category based on the sensory assessment of defects.

The storage company should comply with real-time recording of the identity of a storage batch through the complete formation-storage-depletion cycle. Data to be recorded at the moment of the storage batch formation and during batch depletion are summarized in Table 19.3.

Table 19.3 Traceability data of storage batch formation.

Date	The (hour and) day of completing the storage batch
Origin	The name of the company or the person responsible for the milling process
Responsibility	The name of the company or the person responsible for the storage process
Quantity	The quantity of oil (kg) contained in the storage tank as resulting from the sum of the quantities (kg) of the milling batches making up the storage batch
Identification code	A progressive number or other suitable code, a precise identification of the storage tank and its location
Analytical profile	The results of analyses demonstrating the analytical and sensory conformity of the product to standards
The storage tank depletion	Quantitative data of the oil taken for packaging or shipping should be recorded from the storage batch formation to its complete depletion. If the storage time is very long, further analysis at suitable intervals (for example, every 6 months) during the storage period is advisable.

Step 4 – the packaging batches

A packaging batch (or lot) is the batch of oil (usually a fraction of a storage batch or a blend of various storage batches) that is packaged into suitable containers to be sent to customer(s).

An analytical profile (or an official certificate, if necessary) is an essential element of this traceability step.

Data to be recorded at the fourth step are summarized in Table 19.4.

At this point, the link of the oil in a single bottle with the olive grove(s), the olive mill, the oil storage factory and the oil packaging factory, should be precisely and fully documented. In recent years, various information systems have been

Table 19.4 Traceability data of packaging batches.

Date	The day of the packaging operation
Origin	The name of the company or the person responsible for the storage process
Responsibility	The name of the company or the person responsible for the packaging process
Quantity	The quantity of packaged oil (kg)
Identification code	A suitable code should be defined and printed on the packaging label with precise identification of the packaging batch
Analytical profile	The results of analyses demonstrating the analytical and sensory conformity of the product to standards
Destination	The name of the customer and the destination of the packaged oil

implemented to allow the final consumer to use traceability codes on the bottle label to get information about the origin and history of the oil.

19.3 Comments and conclusion

The system in the four steps described earlier is the simplest possible case of extra-virgin olive oil traceability. Any mixing operation of olives and oils from different batches determines a change in the identity of the oil and introduces a risk of interrupting traceability. Oils can derive from olives of different producers or different producing areas. In industrial, large-scale, olive oil production, olive oils of different origin, obtained from different olive mills, are commonly blended in order to obtain a desirable composition and sensory characteristics. In these cases, the tracing of responsibilities may be extremely complex or even impossible. Similarly, the distribution and marketing systems may include the participation of several actors: importers, wholesalers, supermarkets and retailers.

Traceability is much easier in short chains where only two or three subjects are responsible for the whole chain 'from the field to the consumer's table' (Peri 2007).

The case of refined olive oils is obviously very different. In this case, the origin of the product is the refining factory and refined oils complying with the legal standards are essentially the same even if produced by different companies. Traceability of inputs and outputs of refining factories are the only meaningful traceability requirement of refined oils.

False claims have often been made in technical and scientific publications of reliable analytical fingerprinting of the origin and history of the oil. In fact, true traceability can only result from careful, precise, timely recording of material balances in the four steps described earlier. Analytical fingerprinting may be an effective supporting evidence of documented material balances. For this purpose, analytical fingerprinting does not need to be based on complex or sophisticated analytical methods and procedures. If sound records of material balances are available, the most common and routine analytical values may provide certain evidence of the identity of the batch.

References

Peri, C. (2007) Origin, method and prospects for short production chains. *Gastronomic Sciences* **1**(07), 36–45.

Peri, C., Kicenik Devarenne, A. and Pinton, S. (2010) '3E Super-Premium Selection for Extra-virgin Olive Oil', presented at Beyond Extra-virgin an International Conference on excellence in olive oil, fourth edition, BEV IV, 20–22 September, Verona, Italy.

Further reading

Consorzio Nazionale degli Olivicoltori (2013) *Traceability, Certification and Protection of the Quality of Olive Oil*, http://evoo.itchefs-gvci.com/index.php?option=com_content&view=article&id=608&Itemid=992 (accessed 11 October 2013).

ISO 22005:2007 (2007) Traceability in the Feed and Food Chain – General Principles and Basic Requirements for System Design and Implementation, ISO, Geneva.

Peri, C. (2010) *Strumenti gestionali e indicatori molecolari per la tracciabilità di filiera dei prodotti alimentari*, Academy of Georgofili, Florence.

20

Product and process certification

Ardian Marjani
Ardian Marjani & C. Sas, Milan, Italy

Abstract

The schemes of first-party, second-party and third-party certification are discussed. The structure of a third-party certification system is described with comments on the role of companies, certification and accreditation bodies, standardization organizations, and the Public Authority to control the fairness, reliability and transparency of the whole system. The ISO standards and other standards suitable for the certification of extra-virgin olive oil are listed. A flowchart describes the activities needed for implementing a third-party certification system.

20.1 Aims and approaches

Certification is a formal procedure by which an authorized person or organization assesses that the characteristics of goods or services, procedures or processes conform to established requirements or standards and attests this by issuing a certificate. Certification is a basic tool of guarantee in supplier-customer relationships.

More precisely, when a business enterprise or an organization is certified, it means that a management system has been successfully implemented within the organization, so that its guarantee of product and process requirements is reliable. In the language of certification, the supplier is called the first party, while the customer is the second party. Three different ways to certification can be followed:

First-party certification

In this case the supplier guarantees the customer that goods and processes comply with the agreed upon requirements and standards (Figure 20.1).

The Extra-Virgin Olive Oil Handbook, First Edition. Edited by Claudio Peri.

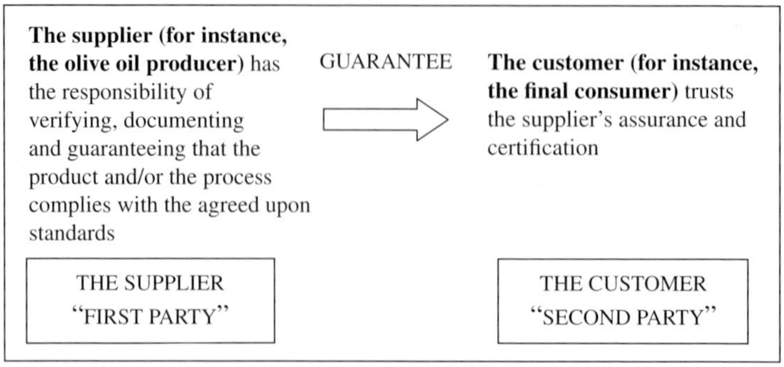

Figure 20.1 First-party certification.

First-party certification is a company's self-declaration of conformity. It often applies to small-scale extra-virgin olive oil businesses and very short chain arrangements, based on a direct connection between producer and consumer, backed by personal knowledge and reputation.

Second-party certification

In this case, the customer verifies, through product analyses and direct inspection carried out at the supplier's site, that the product and/or the process comply with the agreed upon requirements and standards. Such a guarantee is transferred to the final consumer (Figure 20.2).

This arrangement is being increasingly applied in the marketing of excellent extra-virgin olive oil. Traceability of the two marketing steps (from the producer to the retailer and from the retailer to the consumer) can be easily documented and proper handling conditions can also be assured 'from the field to the table' under the responsibility of the intermediate producer-consumer connector.

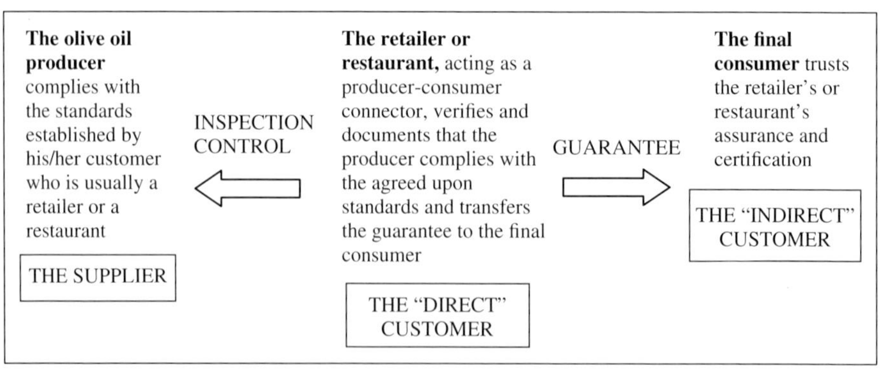

Figure 20.2 Second-party certification.

First- and second-party certifications are very effective systems of guarantee but require a high level of trust in the supplier-customer relationship. They can therefore only be applied in very short and direct chain arrangements.

In the most common situation, a single producer has many customers and a single retailer has many suppliers. Furthermore, direct connections are rare; more often a chain of several intermediate commercial connectors links the producer to the consumer. In this case, which is by far the most common in a global market, a third-party certification is required.

Third party certification

Third-party certification involves the independent evaluation of conformity by expert, unbiased organizations. The certification by independent organizations is an acceptable guarantee for all potential customers. In general, third-party certification is considered as the highest level of assurance and is particularly valuable when the business system is complex and involves international trade. Figure 20.3 shows a general scheme of third-party certification.

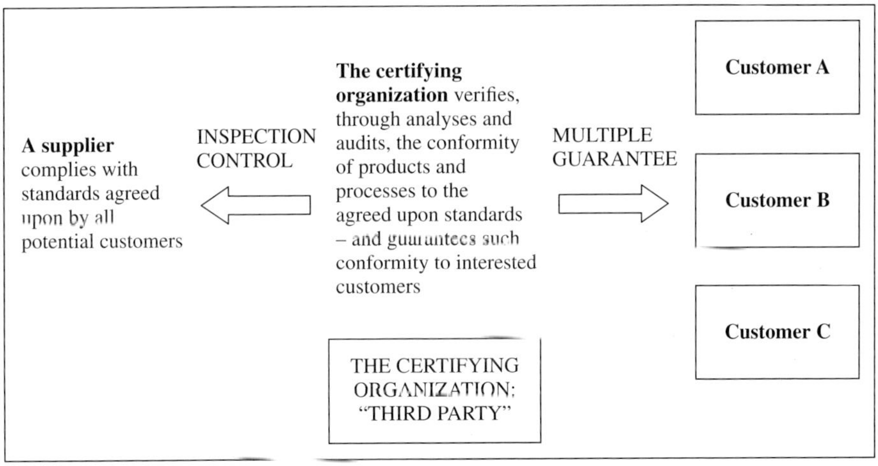

Figure 20.3 Third-party certification.

20.2 Product and process certification

Product certification is based on product conformity to voluntary or compulsory standards. It usually entails the analysis of samples of the product obtained from the factory and / or the open market. Process certification is based on conformity of the supplier's process and the supplier's management system with voluntary or compulsory standards.

However, it can be assumed that an effective guarantee of product quality can only derive from a system simultaneously certifying the quality of the product, the process and the management system. Product quality assurance is trustworthy only if the process is under control through an effective, documented management system.

In conclusion, the quality certification (or assurance or guarantee) of a food product or, in particular, an extra-virgin olive oil, entails the definition, implementation, testing, inspection and documentation of product standards, process standards and management system standards.

The structure of a third-party certification system

Since publication of *Certification – Principles and Practice* in 1980 (ISO 1980) the ISO committee concerned with certification has elaborated on the 'principles' and more specifically defined acceptable practices by issuing a number of guiding documents covering not only the major aspects of certification, but also the major aspects of a number of related activities by which verification of conformity of products and processes may appropriately be judged.

> The International Organization for Standardization (ISO) is a voluntary federation of standards setting bodies of some 130 countries. Founded in 1946–47 in Geneva as a UN agency, it promotes development of standardization and related activities to facilitate international trade in goods and services, and cooperation on economic, intellectual, scientific and technological aspects. It covers standardization in all fields including computers and data communications, but excluding electrical and electronic engineering and telecommunications.

Despite the great complexity and detail of the ISO standards, the structure of a third party certification system can be very simply represented as in Figure 20.4. With reference to the five points, the process of certification can be described as follows:

Companies pursuing certification operate according to standards agreed upon by their customers. Standards are relative to the product, process and management system. Certification bodies verify and certify that the company operates in conformity to the product, process and management system standards. An accreditation body verifies and certifies that the certification procedures are carried out according to agreed upon standards. A public authority at a suitable national or international level guarantees the fairness of the overall certification system with regard to the responsibility of the companies, of the certification and of the accreditation bodies and, most of all, with regard to the consumer's rights. A standardization organization provides the standards and guarantees their approval by all parties involved in the process. ISO is the most important and authoritative organization providing international standards but other organizations also define standards for special duties or marketing needs or supplier-customer relationships. The ISO

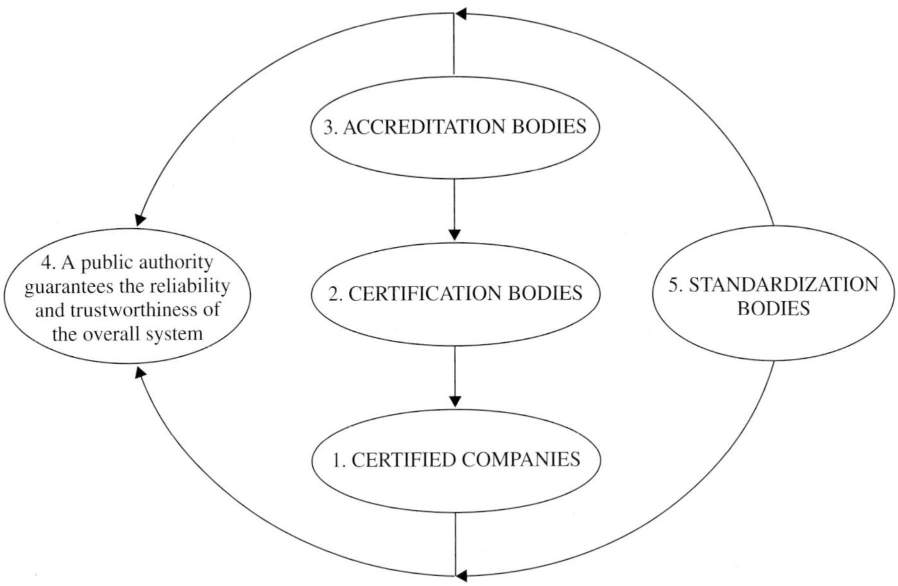

Figure 20.4 The structure of the third-party certification system (*Source*: Reproduced by permission of Hoepli, Milano).

organization and the ISO principles, however, remain the basic and inspiring source of standards.

An essential point of the effectiveness and reliability of the system is that all the actors in the system are required to implement an internal management system by which they control and constantly improve their ability to carry out their duties. Thus, a company must implement a management system to control the product and process. A certification body must apply a management system for the control of the certification procedure. An accreditation body must apply a management system for the control of the accreditation procedure. The public authority controlling the system must apply standard and transparent operating procedures. Finally, very strict procedures must be applied by ISO or other standardization bodies in the standardizing duty, because of the relevance and impact of standards on business and the importance of a general agreement of public and private organizations.

A point to be underlined is that ISO does not certify or carry out accreditation duties. Certification bodies are usually private organizations, while accreditation bodies are usually national and public organizations. However they both operate according to the certification and accreditation standards set up by ISO.

Figure 20.5 presents a five-level scale of the standards available for regulating the certification system according to ISO. A five-level scale of standards is as follows:

- At level 1, the supplier's self-declaration can be considered as the simplest certification system, which is suitable for first-party certification. In this case,

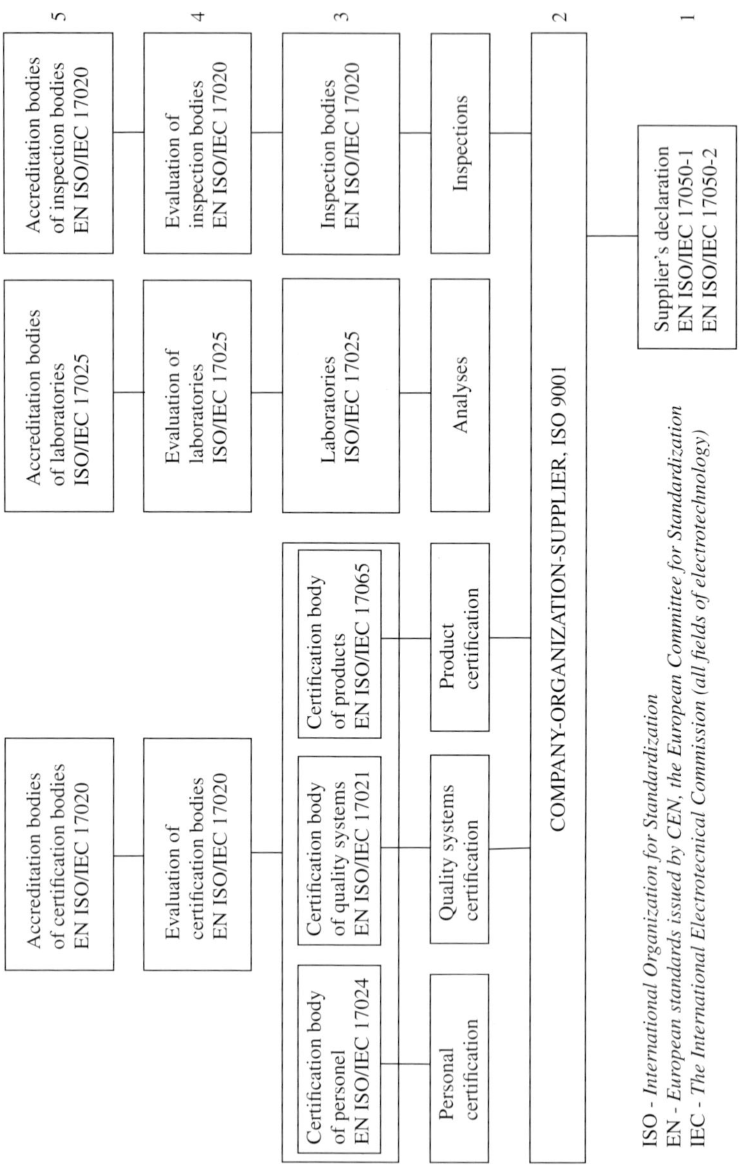

Figure 20.5 The ISO standards of the third-party certification system (*Source:* Reproduced by permission of Hoepli, Milano).

ISO suggests a standard format that gives an official character to the declaration and underlines the declarant's responsibility.

- At level 2, the standards of quality management systems are indicated. These standards should be applied by companies aiming at product and process certification, laboratories aiming at analyses certification and inspection bodies aiming at certified inspections and audits.
- At level 3, the standards to be applied to the certification process are reported. Certification can be applied to people or (quality system management) or product as well as to analyses and inspections or audits.
- At level 4, the standards for the evaluation of certification bodies, laboratories and inspection bodies are reported.
- At level 5, the standards regulating the accreditation processes are reported.

20.3 The selection of a certification system

A company should not consider certification as an objective but as a tool for meeting its objectives; nor should it consider it as a mere formality but rather as a challenge to its willingness to improve. A wide choice of standards and procedures is available to satisfy the different and complex supplier-customer relationships in the global market.

Table 20.1 presents some examples of standards available to companies in seeking product-process certification in the extra-virgin olive oil sector.

Point 1 presents the most common sources of product standards for extra-virgin olive oil. A company may decide to adopt the extra-virgin olive oil standard defined by the laws or it may choose to apply the standards of a Protected Designation of Origin (PDO) or those of excellent olive oil presented in Annex 18.1 of this handbook. Or, finally, a company can decide to apply its own standards for its private brand, within the limits of the legal definition of extra-virgin olive oil.

Point 2 presents some sources of process standards, in particular the PDO process standards or those applicable to organic agriculture. The company can obviously choose its own processing standards, provided that they comply with the legal definition of extra-virgin olive oil.

Point 3a presents the main sources of standards for the management system. A number of options are available for the quality and safety management system, which represents the basic management requirement of a food factory.

Point 3b suggests the standards in the case that the company wants or needs to activate a system for the management of environmental or social or workers' safety requirements. In this case, a good solution would be to adopt an integrated management system as suggested in Chapter 18 of this handbook.

After choosing the product, process and system management standards, the company should define its certification system. It can choose to adopt a first-party certification or it can agree with its customer on a second-party certification. In these cases, wide and discretionary space is available for establishing the most suitable formal

Table 20.1 Some sources of product, process and system management standards for the certification of extra-virgin olive oil companies.

Company's decision about	References	Options available
1. Product standards	The European Commission	1.1 Regulation (EC) no. 1019/2002 on marketing standards for olive oil 1.2 Regulation (EC) no 702/2007 amending Commission Regulation (EEC) No 2568/91 on the characteristics of olive oil and on the relevant methods of analysis 1.3 Regulation (EC) No 510/2006 on the protection of geographical indications and designations of origin for agricultural products and foodstuffs and Regulation (EC) No 1898/2006 laying down detailed rules of implementation of Council Regulation (EC) No 510/2006
	This handbook	1.4 The standards of excellence in olive oil given in Annex 18.1
	The company	1.5 A brand choice of analytical and sensory standards within the legal definition of extra-virgin olive oil
2. Process standards	The European Commission	2.1 Regulation (EC) No 510/2006 on the protection of geographical indications and designations of origin for agricultural products and foodstuffs and Regulation (EC) No 1898/2006 laying down detailed rules of implementation of Council Regulation (EC) No 510/2006 2.2 Regulation (EC) No 834/2007 on organic production and labelling of organic products
	The company	2.3 Choice of technological solutions, plants, operations and operating conditions
3a. Standards of quality and safety management	ISO	3.1 ISO 9001: 2008 on quality management system 3.2 ISO 22000:2005 on food safety management system 3.3 ISO 22005:2007 on feed and food traceability system

(continued overleaf)

Table 20.1 (*continued*)

Company's decision about	References	Options available
	FSSC 22000	3.4 Food Safety System Certification 22000. Certification scheme for food safety systems in compliance with ISO 22000:2005 and technical specifications for sector PRPs
	BRC food	3.5 British Retail Consortium (BRC) global standards for food safety, products, packaging, storage and distribution
	IFS food	3.6 International Featured Standard. Standard for auditing quality and food safety of food products
	This handbook	3.7 Chapter 18 on management systems
3b. Standards of integrated management system	ISO	3.8 ISO 14001:2004 on environmental management system 3.9 ISO is developing an ISO 26000 standard providing voluntary guidance on social responsibility (SR) 3.10 The ISO 31000 standard on risk management can be adapted for integrated system of risk management 3.11 The ISO 9001 standard on quality system management can be adapted for an integrated system of quality management
	Various standard and certification bodies	3.12 OHSAS 18001 on occupational health and safety 3.13 SA 8000, on social accountability
	This handbook	3.14 Chapter 18 of this handbook on process management system

requirements. In general, first- and second-party certification is based on reciprocal knowledge and trust; formalities are minimized. In case the company wants or needs a third-party certification, it does not have the right to establish the certification rules because they must comply with the ISO standards. However, there is space for free decision regarding the choice of the certification organization. In this choice, the company should consider not only the cost of certification but also, and foremost, the effectiveness and reliability of inspections and audits.

Finally, the accreditation step is out of the company's control: accreditation and control of the certification bodies must be carried out according to the ISO standards. However, it must be remembered that accreditation is not an obligation and if a certifying organization is not accredited it does not necessarily mean it is not reputable. It may be concluded that companies are fully responsible for the choice of a suitable and effective certification system.

20.4 The certification procedure

Figure 20.6 presents a flowchart of activities for implementing a third-party certification. The flowchart is self-explanatory. It should be pointed out that access to third-party certification is a very demanding task, often implying changes and adaptation of operating and documentation procedures. A two-year period from the decision to the routine implementation of certification may be considered as a reasonable period of time.

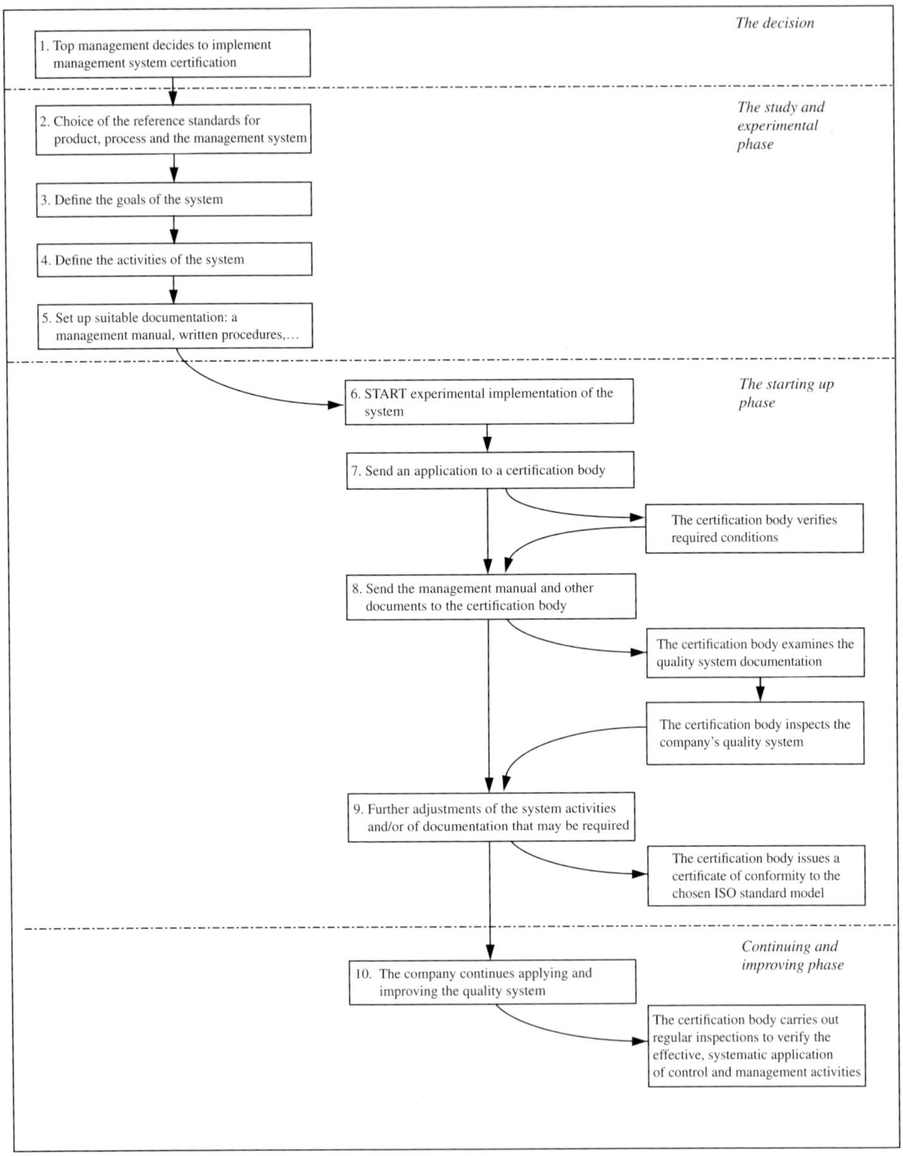

Figure 20.6 The flowchart of activities for implementing a third party certification system.

Reference

ISO (1980) *Certification: Principles and Practice*, ISO, Geneva.

Further reading

ISO 9000:2005 (2005) *Fundamentals and Vocabulary – Quality Management Systems*, ISO, Geneva.

ISO 22000:2005 (2005) Food safety management systems– Requirements for Any Organization in the Food Chain, ISO, Geneva.

ISO 9001:2008 (2008) *Quality Management Systems – Requirements*, ISO, Geneva.

ISO 9004:2009 (2009) *Managing for the Sustained Success of an Organization – A Quality Management Approach*, ISO, Geneva.

ISO/IEC 17021:2011 (2011) *Requirements for Bodies Providing Audit and Certification of Management Systems – Conformity Assessment*, ISO, Geneva.

ISO 19011:2012 (2012) *Guidelines for Auditing Management Systems*, ISO, Geneva.

Peri C. (2005) *Oltre i sistemi qualità*, Hoepli, Milano.

Peri, C., Lavelli, V. and Marjani, A. (2006) *Qualità nelle aziende e nelle filiere agroalimentari*, Hoepli, Milano.

21

The hygiene of the olive oil factory

Cristina Alamprese[1] **and Bruno Zanoni**[2]

[1]*Department of Food, Environmental and Nutritional Sciences, Milan, Italy*

[2]*Department of Agricultural, Food and Forestry System Management, University of Florence, Florence, Italy*

Abstract

The hygienic requirements of an olive oil factory are listed. These involve the external environment, buildings and internal logistics, plant and equipment, and personnel. The structure of a hygiene management system (HMS or HACCP) is thoroughly discussed. In Annex 21.1, the fundamentals and practical applications of hygienic design are presented.

21.1 Introduction

The hygiene of oil factories (either mills, oil storage facilities or bottling plants) is crucial to avoid oil contamination. Foreign bodies and contaminants can derive from: (i) contact between the oil and dirty surfaces or unsuitable material; (ii) particulate matter falling accidentally from buildings, equipment or people; (iii) transfer of volatile compounds or fine dust or smoke or aerosols from the atmosphere; (iv) faeces and body fragments of birds, rodents, pets, insects and so forth.

Unsatisfactory hygiene of an oil factory may lead to loss of oil quality and safety, which may then result in economic losses such as product recall cost, liability cost and loss of business.

At the same time, the hygiene of the olive oil factory is an indicator of management reliability that results not only from good cleaning practices, but also from educated and responsible personnel and orderly and timely carrying out of operations.

The Extra-Virgin Olive Oil Handbook, First Edition. Edited by Claudio Peri.

First glance evaluation of an olive oil factory

When visiting an olive oil factory, immediate attention should be given to the hygiene of the plant, the workers and the environment. A negative impression should be taken as an indication of poor organization and limited awareness of negative consequences. The following 12 points are suggested to evaluate at a glance the factory's ranking in management and reliability:

1. Dirty floors splattered with olive paste or soil and spilled oil.
2. Dirty walls and ceilings, corners, windows and doors. Defective closure of doors and windows.
3. Unprotected access to animals and the presence of their excrement (cats, dogs, birds, mice, rats, cockroaches and flying insects).
4. Medley of incompatible materials such as extra-virgin olive oil and packaging material or plant's spare parts, lubricating oil and grease, detergents, cleaning supplies and so forth. These materials must be stored in different rooms, in specific closets.
5. Improper process logistics, with mixing or lack of barriers between dirty areas (olive reception area and waste discharge area) and clean areas (the mill, the area of oil handling and bottling).
6. Confusion in receiving the olives without checking their integrity and cleanliness and without keeping a first-in-first-out order.
7. Improper clothing and inadequate personal hygiene of workers.
8. People eating or smoking in working areas.
9. Unplanned presence of visitors in the working areas.
10. Lack of daily cleaning of the plant or gaps in the cleaning programme of the buildings, floors and the factory environment.
11. Waste, fermented and rotted by-products on the outside of the premises, attracting animals, especially rodents and insects. Odour pollution.
12. Smoke from burning wood or paper or plastic or car exhaust. Smoke is readily absorbed by oil with negative effects on both its health-promoting and sensory quality.

21.2 Hygiene of the external environment and buildings

In order to facilitate a more detailed and systematic implementation of hygiene in an olive oil factory, requirements and recommendations concerning the external environment, the buildings and the internal environment, the plant and the personnel are listed below.

21.2.1 External environment

- The area of an olive oil processing facility should be at no risk of flooding, far from waste or garbage dumps as well as from sources of chemical or biological pollution, including stock farms. Air in the vicinity of the processing plant can be a source of biological, chemical, and particulate contaminants, including solvents, odours and smoke;
- The olive oil factory should be located away from areas prone to infestations by pests.
- The factory should be established in an area that is sufficiently large to facilitate access by suppliers and customers.
- An adequate supply of potable water must be available. If nonpotable water is used (e.g. for fire control, heating, refrigeration and other similar purposes where it would not contaminate food), it must be clearly identified and must not be connected to potable water systems.
- Protected and suitably equipped facilities should be available for waste discharge, accumulation and treatment. They should be isolated in order to avoid odour pollution and the attraction of rodents and other pests to the processing plant.
- The area outside the building must be in polished cement at least five meters wide around the building. Access should be through paved driveways.
- The area of olive reception and storage, with floor in polished cement, must be frequently cleaned and washed; it should be covered with a roof, and be large enough to store the olive containers in a first-in-first-out order, easily accessible for visual inspection.

21.2.2 Buildings and internal environment

There are four main areas in an olive oil factory:

1. The area for reception and storage of olives.
2. The processing area.
3. The oil storage, bottling and packaging area.
4. The shipping area.

Areas 1 and 4 are prone to having people from the outside moving around; these areas should therefore be physically separated from the processing, storage and bottling areas. Unauthorized people or customers or suppliers or visitors should not be allowed to enter the processing and bottling areas. Acceptable arrangements are shown in Figure 21.1 in which crossover of dirty and clean areas is avoided.

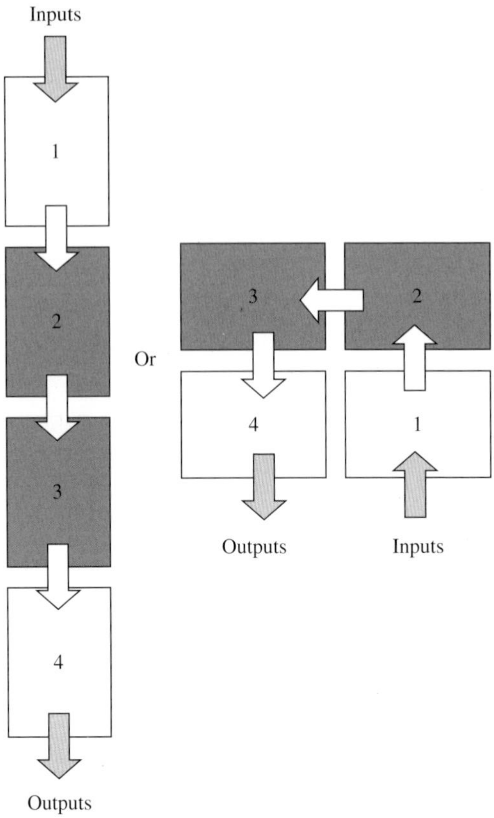

Figure 21.1 The logistics and separation of functional areas in an olive mill.

- Supporting services and functions must be separated from the four main areas, in particular:
  - administration offices
  - laboratory for chemical analyses and panel tasting room
  - toilets and locker rooms for workers
  - heating and conditioning plant and power station
  - repair and maintenance shop
  - storage of packaging materials
  - storage of chemicals and detergents
  - retail shop
- Buildings should be dry and waterproof because dampness facilitates contamination. Water leaks and stagnant water should be avoided.

- Floors should be self-draining, nonslippery, resistant to corrosion by cleaning and sanitizing agents, with rounded angles and edges to facilitate cleaning. Floors should be designed to minimize physical entrapment of soiling material.
- Walls should be smooth, coated with impermeable and washable material.
- Floor and wall openings for lines or pipes should be adequately protected to prevent any contamination.
- Doors and windows should be designed to prevent intrusion of contaminated air, rodents, birds and insects and they should be easy to clean and disinfect.
- The processing room should have a minimal number of windows because they are usually difficult to clean. Windows should be hermetically sealed during plant operation.
- Direct access from the outside or from dirty areas to the processing areas must be avoided.
- Natural or artificial lighting should provide appropriate light intensity (for example: 220 lux in the processing area, 540 lux in points needing inspection). Light bulbs should have safety protection to prevent contamination in the event of breakage;
- Condensed water dripping from the air conditioning system should be avoided. The conditioning system should be accessible for periodic inspection and cleaning as well as filter replacement.
- Air flow must be directed from the finished product and clean areas toward the raw material and dirty areas.

Rodents and insects are a constant challenge because they like olive oil and take advantage of any dirt and weakness in physical barriers. Direct and indirect measures to prevent their entrance to the oil factory must be implemented, for example:

- seal any cracks, scratches or fissures in the walls, doors and windows with appropriate material;
- carefully insulate ducts and piping of telephone and electric cables;
- use double automatic doors for access to the processing rooms from the outside;
- provide mosquito nets on all the windows;
- protect the outside opening of the sewage system from intruders with nets having 5 mm wide maximum openings;
- maintain an environment that does not attract rodents and other vermin – periodic cutting of wild grass and weeds on the outside of the factory;

- remove waste materials from external and internal areas. Regularly empty and clean waste containers;
- avoid conditions that create a warm, humid microenvironment inside the factory.

21.3 Hygiene of the plant

The hygienic design of a plant is such a critical requisite for product safety and quality that Annex 21.1 is devoted to a detailed presentation of this subject.

Equipment should be designed to ensure that it can be adequately cleaned, disinfected and maintained to avoid contamination of the oil. It should be durable and movable or capable of being disassembled to allow for maintenance, cleaning, disinfection and inspection for pests.

The following points should be considered:

- Before buying a new plant or part of it, the plant supplier should give all the information concerning its hygienic design and the best practice for maintaining a high hygienic standard during processing. The processing equipment must have built-in hygiene; it must be constructed with smooth, impervious, food-grade materials.
- Walls nearby and the floor around and under the equipment must be accessible for easy cleaning.
- Equipment must comply with the laws protecting workers' health and safety, in order to avoid mechanical or electrical accidents or exposure to excessive noise. Inappropriate ergonomics must be avoided, especially heavy loads in handling olives or oil containers.
- All parts of the plant must be cleaned daily. The cleaning-in-place system is the best solution, but cleaning out-of-place may be needed periodically for specific parts. The cleaning procedure should be described in a written document, readily available to the operators.
- All parts in contact with the oil should be automatically drainable, in order to avoid the presence of oil during periods of standstill.
- Systematic controls should be carried out in order to prevent parts of the plant (such as screws, gaskets or paint) from ending up in the product. Parts with surface roughness due to corrosion or rust must be replaced because they increase the rate of dirty deposition during processing and are more difficult to clean.
- All the oil containers must always be covered so that the oil cannot be accidentally contaminated by extraneous material from the plant or people or the environment.

- The daily washing procedure may be carried out with hot potable water (60 °C) with disassembling and manual washing of points inaccessible to the free flowing of warm water, while twice a week at least, washing should be carried out in four steps:
  - rinsing with hot potable water (60 °C)
  - washing correctly with appropriate detergents
  - rinsing with potable water to eliminate detergent residues
  - disassembling – manual washing – reassembling of points inaccessible to the free flowing of detergent and rinsing water.

The washing procedure of all the equipment must be carried out after treating batches of unhealthy, rotten olives or when switching from traditionally to organically grown olives.

21.4 Hygiene of the personnel

Workers and operators who come directly or indirectly into contact with the oil are not likely to contaminate it if they maintain an appropriate degree of personal cleanliness and behave and operate in an appropriate manner. In particular:

- All workers and operators must attend a short, practical course on personal hygiene and the hygiene management system. They should understand hygienic implications in terms of consumer safety and company's image and reputation.
- Good personal hygiene must be a habit for people working in an olive oil processing facility. Personnel should:
  - have a daily bath or shower, with washing of hair
  - keep nails trimmed and clean
  - cover possible cuts and wounds with suitable waterproof dressings.
- The clothing used during work must be clean and should not be used outside the working area and work time. Clothing should be regularly changed and washed. In order to avoid accidental dropping of items in the oil, clothing must not have breast pockets or buttons: clothing should be closed with adhesive tape.
- Plastic or rubber boots should be used in the working area.
- Hand washing and drying facilities should be accessible, easy to use, clean and functional.
- An adequate number of toilets of appropriate hygienic design should be available.

- Personnel should always wash their hands when personal cleanliness may affect oil safety, for example:
  - at the start of oil handling activities
  - immediately after using the toilet
  - after handling olives or any contaminated material.
- People engaged in oil handling activities should refrain from behaviour that could result in contamination of the oil, for example:
  - directly passing from a dirty into a clean area
  - smoking in the processing area
  - eating in the processing area
  - introducing pets.
- The repeated use of cloths should be avoided in cleaning surfaces, especially those that can be in contact with the oil; suitable paper should be used only once and thrown away after use.

21.5 Hygiene management system (HMS) and HACCP

21.5.1 General principles of HMS

Hazard analysis and critical control point (HACCP) is the name of a management system especially designed to ensure food safety through systematic, documented prevention of risks to food. First established in 1989 by the US National Advisory Committee on Microbiological Criteria for Food, in 1997 HACCP was proposed as the world reference system for food safety assurance by the UN/FAO Codex Alimentarius Commission, with the title 'Hazard Analysis and Critical Control Point (HACCP) System and Guidelines for Its Application'. Hazard analysis and critical control point is now being adopted worldwide and in some cases is mandatory.

In this handbook 'hygiene management system' (HMS) and HACCP are considered as synonyms.

Risk analysis has been presented in Chapter 18 as a general method for identifying points that are critical for achieving the company's objectives, not only hygiene and food safety objectives. On the other hand, management systems are presented in Chapter 18 as systems that organizations can apply to meet product and process requirements, including hygiene and consumer safety requirements.

Figure 21.2 is a version of Figure 18.1 (the structure of the management system) specifically adapted to the case of consumer safety and process hygiene. The company's policy about product safety and process hygiene (point 1 of Figure 21.2) can be defined in different ways. A minimalistic approach involves planning a system merely conforming to the compulsory hygienic standards. On the other hand, a

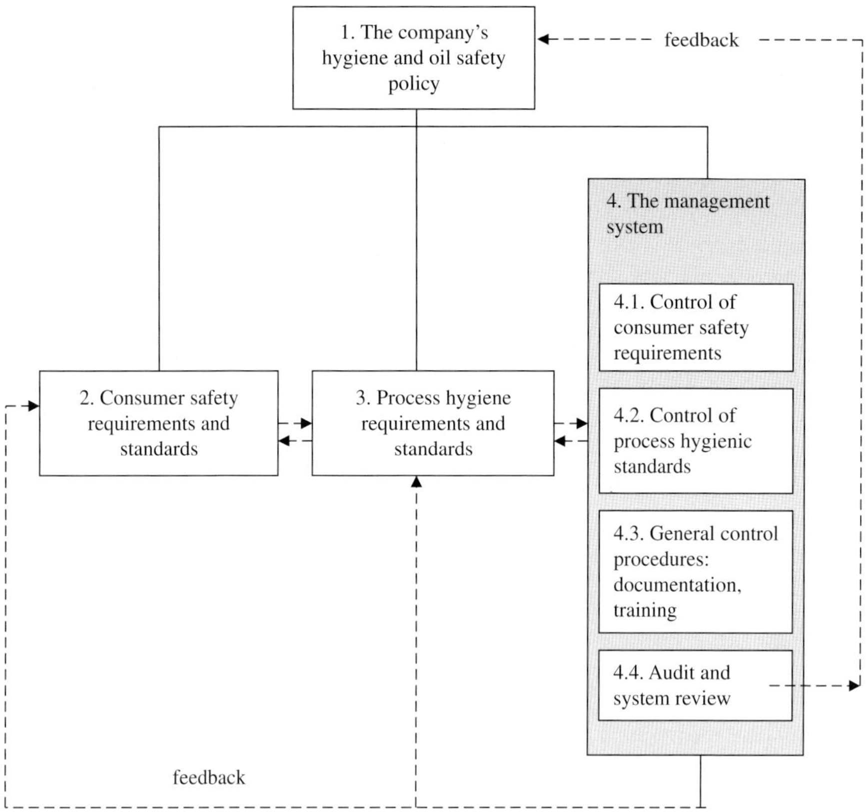

Figure 21.2 The structure of a Hygiene Management System (HMS) of an olive mill factory.

company can decide that oil safety and process hygiene also have an important value in terms of the company's reputation and consumer trust. This can be especially important in the case of direct selling of the oil at the factory and in the case of consumer and customer visits.

Consumer safety requirements (point 2 of Figure 21.2) are defined by laws. However, special attention can be given to product contamination, especially in the case of olive oil purchased from commercial partners. For instance, when buying a non-filtered extra-virgin olive oil, it may be useful to carry out filth tests in order to identify particulate contaminants. This analysis, which is fast and cheap, can offer valuable information about the hygiene of the supplying factory. Periodical analyses of environmental contaminants can be planned to test for environmental pollution.

The hygienic requirements of the process (point 3 of Figure 21.2) can be limited to legal prescriptions or stricter requirements can be defined, at the company's will, for the hygiene of buildings, plants and people. Special care should be given to the hygiene of the external surrounding space of the factory and to the prevention of bad odour pollution. The hygienic design of plants or the control of process water potability can be the object of systematic attention and control as these may be weak points of the process.

Consumer safety requirements – whether requested by the law or voluntarily defined by the company – should be the object of evaluation and control (point 4.1 of Figure 21.2) depending on risk. If the risk of nonconformity is low, analytical evaluation can be disregarded or carried out occasionally. If the risk of nonconformity is significant, evaluation and control should be planned on a more frequent and systematic basis.

The control of process hygienic standards (point 4.2 of Figure 21.2) should be defined through a careful risk analysis. This is the point where risk analysis contributes to the definition of the management system.

Table 18.9 in Annex 2 of Chapter 18 presents an example of evaluation of the hygienic risk gravity in an extra-virgin olive oil process. Table 21.1 presents an elaboration of the data in Table 18.9 with some further details and comments.

From the analysis of Table 21.1, decisions can be taken.

Finally, points 4.3 and 4.4 of Figure 21.2 must be applied as explained in Chapter 18.

Table 21.1 Critical points for hygiene in an olive oil mill (elaborated from Table 18.9).

Risk gravity	Critical point	Risk treatment
High	Overall hygiene condition of the milling factory	Prevent contamination from the environment, people, pets, and the atmosphere (particulate, volatile compounds, smoke, dust). Control of pest infestation. Standard Hygiene Procedures (SHP) may be required for the prevention of specific risks depending on the factory site and operation.
Moderate	The packaging operation	SHP may be required for preventing particulate contamination in containers, especially glass fragments in bottles. The risk can be eliminated by in-line equipment for air blowing of containers just before filling.
Moderate	Pesticide residues	Prevention of pesticide residues fully depends on pesticide treatments at the olive grove. Therefore, the preventive measure in this case is a contract agreement with olive producers, with suitable inspection and occasional analytical check. Pesticide and rodenticide treatments at the milling factory must be carried out by specialists according to well planned and carefully implemented SHP, with systematic recording of activities from trap capture to treatment.
Low	Environmental contaminants such as PCBs, PAHs, etc.	This was considered as a low gravity risk in Table 18.9. In this case only occasional analytical checks are planned. Depending on factory and olive-grove location, this point can require specific SHP.

Table 21.2 Decisions about consumer safety and process hygiene management.

Risk	Management decision
Hygiene of buildings	Risk prevention A written procedure for the cleaning and maintenance of buildings with well-defined responsibilities, frequency, materials and methods
Hygiene of plant and machinery	Risk prevention A written procedure for plant cleaning and maintenance with well-defined responsibilities, frequency, materials and methods
Rodent and insect	Outsourcing Contract agreement with specialized companies. An internal inspection should be systematically carried out.
Personal hygiene of workers and employees	Risk prevention A written procedure for personal hygiene with well-defined responsibilities, frequency, materials and methods
Waste handling and treatment	Risk prevention A written procedure for waste handling and treatment
Pesticide residues	Outsourcing. A contract agreement with olive producers; periodical inspection; occasional analyses of residues
Particulate contaminants in bottles and other containers	Risk elimination In-line air blowing of containers

21.5.2 Documentation of the hygiene management system (the HMS manual)

All information and decisions about hygiene management should be part of a more general document, the HMS manual. This manual may be organized to perfectly match the requirements of a HACCP manual.

Summary of the HMS manual

First section: presentation of the company, the factory, the process and the product

This part includes the company's name and address; the consumer safety and process hygiene policy; products, process and business; main available resources in terms of people with their respective roles and responsibilities, buildings, plants, and plant facilities. A process flow-chart and a plant layout can be presented to better represent the system and its technical features.

Second section: presentation of risk analysis results

This section is very important because it shows the company's approach and commitment to interested parties, including public officers, customers and consumers. A figure similar to Figure 21.2 and tables similar to Tables 18.9, 21.1 and 21.2, can be presented in this section in order to show how and why the final decisions about the critical control points were made.

Third section: standard hygiene procedures (SHP)

This part, representing the operating standards and procedures of the system, includes:

00. Occasional product or environmental analyses (to be defined)

01. SHP of cleaning and maintenance of buildings

02. SHP of cleaning and maintenance of the plant

03. Contract agreement for rodent and insect control (outsourcing)

04. SHP of personal hygiene of workers and employees

05. SHP of waste handling and treatment

06. Contract agreement for pesticide treatment in the olive groves and General Control Procedures:

07. Documentation

08. Training of employees and workers

09. Internal audits and system review

It is suggested that the standard hygiene procedures (SHP) listed in the manual should be presented in a standard format.

The possible scheme of a Standard Hygiene Procedure (SHP)

A SHP should include:

1. A code identifying the procedure.
2. The date of last updating.
3. The aim of the procedure and also an indication of what the consequences can be in case of a careless application of the procedure. Evaluation or analyses should be defined to evaluate the effectiveness of the procedure

in preventing the risk or in reducing its impact on the safety and hygiene objectives.

4. The person who is responsible for implementing the procedure and for verifying its effectiveness
5. The description of the activities to be carried out. Schemes and drawings can be added to explain in an easy-to-understand way what should be done. Notifications can also be placed at the work site to remind the operators what they are expected to do
6. The data to be recorded. A format may be suggested or preprinted for data registration. Formats for registration, observation, comments or suggestions by the operators can be included

Annex 21.1: Hygienic design

Hygienic design and correct installation and maintenance of a plant should:

- guarantee protection of food products from contamination and any equipment leaks;
- avoid dirt buildup, which may result in possible micro-organism and vermin proliferation in dead legs (components of a piping system that normally have no significant flow) or places that are inaccessible for cleaning and disinfection;
- facilitate cleaning and disinfecting operations to make them easier and more effective.

Faster cleaning allows processing times to be prolonged and plants to operate at lower costs. Cleaning of plants is an issue that cannot be tackled after plant building and installation, but must be specifically included in the planning stage. A detailed inquiry on this issue is recommended when purchasing.

The main sources of detail and thorough description about hygienic design of plant, fittings and equipment are: the 3-A Sanitary Standards, Inc. (3-A SSI, www.3-a.org) and the European Hygienic Engineering and Design Group (EHEDG, www.ehedg.org).

The leadership of 3-A SSI includes the Food and Drug Administration (FDA) and the US Department of Agriculture (USDA).

The EHEDG actively supports European legislation. Within the European Community, all fittings and equipment are required to comply with requirements specified in Chapter V of Annex II of Regulation (EC) No. 852/2004. They are to be: (i) effectively cleaned and, when necessary, disinfected; (ii) so constructed as to minimize any risk of contamination; (iii) installed in such a manner as to allow adequate cleaning of equipment and the surrounding area.

A number of basic application elements relevant to the oil sector have been drawn from several published standards and guidelines and are summarized in the following points.

General aspects

- Since the hygienic design of a plant depends on the constructor, it is advisable that the purchaser, when purchasing, requires both to view the relevant hygienic design manual and to receive a list of all building solutions to guarantee the hygienic conditions of the plant. This list can be a useful reminder for any changes and maintenance operations to be performed on plants. It is common for a plant that has been designed and built according to hygienic standards to become less hygienic and less safe over years, as a result of inappropriate service activities and changes.

- Equipment should be installed in such a way that it can be accessed from all sides and raised to allow for thorough cleaning of the underlying floor. All outer surfaces should be appropriate to minimize dirt residues, growth of micro-organisms, and avoid becoming shelters for insects and rodents. Plants must be, and appear to be clean, giving an impression of order and care even to outside observers and visitors.
- Platform-mounted plants are preferable, as they can be periodically (e.g., in case of annual extraordinary maintenance) removed to facilitate both inspection and cleaning of supports, drains, etc.
- Connections that require product transfer between machines must be as short as possible. The best solution may often be a vertical development of plants, using gravity conveyors.
- Machines must be connected in such a way as to allow the process to develop as linearly as possible, without any crossing over and flow recycling.
- Plants must be easy to disassemble and, if possible, allow for visual inspection of all surfaces, which come into contact with the product. It should be taken into account that, in specific cases, some plant parts must be disassembled and cleaned manually. Instructions for cleaning of plants must clearly indicate the plants involved, the service intervals, and so on.

Materials and surfaces

- All surfaces in contact with extra-virgin olive oil must be made of inert, consistent materials according to legal requirements.
- All surfaces in contact with food must be smooth and nonporous. Surface finishing specifications for stainless steel surfaces for food use are reported in ISO/R 2037 (1972) and ISO 468 (1982) standards. Equipment made of iron, copper, bronze, etc. must be absolutely avoided, as they may release oxidation catalysers to the oil.
- All gaskets must comply with legal requirements for food quality and be nonabsorbent, correctly installed, exposed to cleaning flow, and periodically inspected and replaced.
- All product contact surfaces must be resistant to all cleaning and antimicrobial agents at the full range of operating pressures and temperatures.

Hygienic design principles and devices

A large number of precautions are required to guarantee the hygienic conditions of each plant and plant detail. The following is a brief reference list of specifications that every extra-virgin olive oil business should take into consideration when both purchasing plants and implementing maintenance operations and changes.

- All product contact surfaces must be self-draining. All plant parts must be self-emptying. It must be possible to remove by gravity the oil contained in a plant, without building up residues in blind bottom portions and non-self-draining wells.
- The use of O-rings in contact with the product is not allowed in hygienic equipment and piping systems. Avoid metal-to-metal joints other than welding. Avoid misalignment of pipe connections.
- Avoid contact of product with screw threads and crevices. This can be achieved either by using appropriate gaskets or – even better – by creating outer joints (Figure 21.3). Checks must be performed regularly to avoid the risk of screws and joints loosening and dropping into the product.
- Sharp corners are not allowed. The minimum radius of corners is 3 mm.
- Clamps connecting flexible and rigid materials must be as close as possible to the product side to minimize entrapment of product between the rigid and flexible surfaces (Figure 21.4).

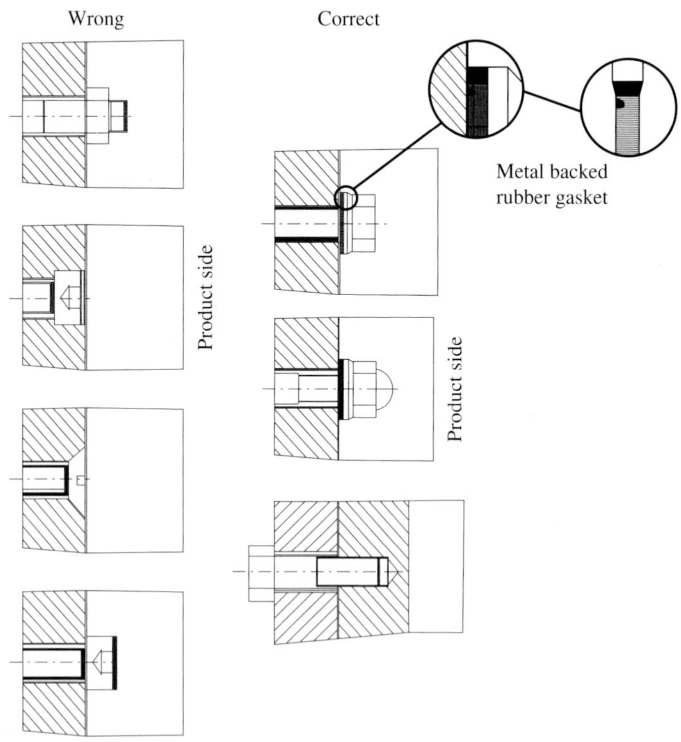

Figure 21.3 Avoid contact of product with screw thread and crevices because product and micro-organisms may accumulate in these non-cleanable places (Zanoni 1993).

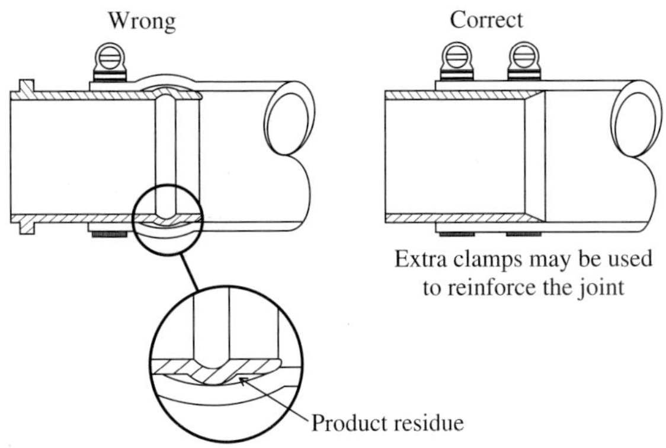

Figure 21.4 Clamps connections between flexible and rigid materials.

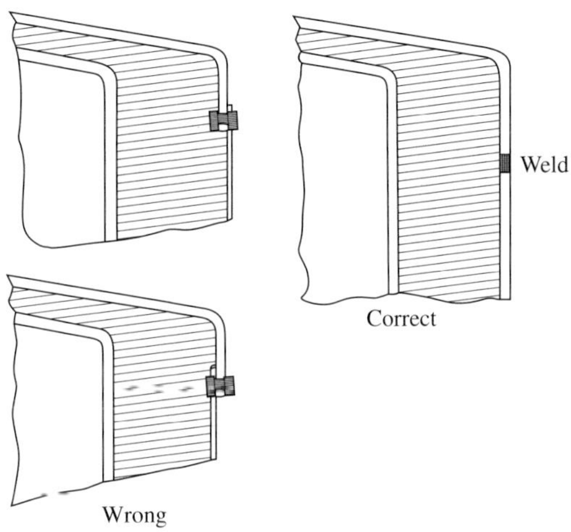

Figure 21.5 Insulation should be protected from entrance of water or product.

- Insulating layers, which are typically defined as nonporous, must be protected from entrance of water and product. The coating must be waterproof and welded, not jointed (Figure 21.5).
- The product path should be designed to avoid dead legs, where product residues may build up. Attention should be paid when installing measuring instruments, level indicators, agitators, etc. (Figure 21.6).
- All product containers, mixing and holding vessels, must have a lid.
- All pipes, on which condensation may form, must be kept distant from product handling points, however not located directly over processing equipment.

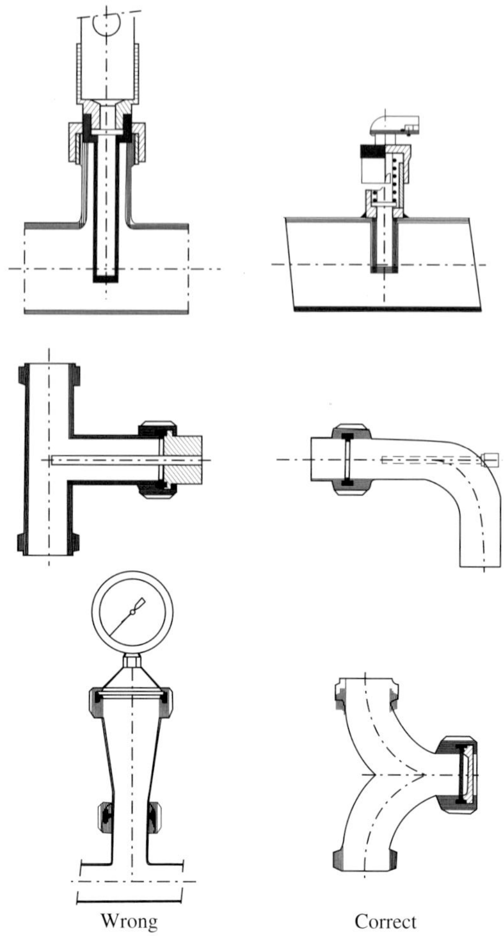

Figure 21.6 Examples of equipment with and without dead legs.

- Motors and gear cases must have an underlying sump to collect liquid lubricant to avoid it from dripping on the floor or plant parts.
- It should be mentioned that ball valves and horizontal pipes are more difficult to clean. Pipes must be appropriately sloped to allow for easy self-draining.
- Never make holes in walls where soil residues may be trapped and bacteria can accumulate and multiply.
- Promptly remove and renew cracked or peeled paint to avoid it from dropping into the product.
- Regularly check and remove any deposit, fouling and rust.

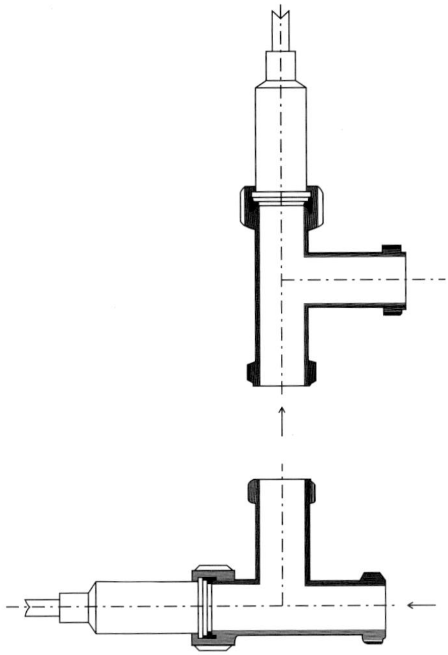

Figure 21.7 Recommended positions of unavoidable dead legs.

- Avoid misalignment of pipe connections and other devices in contact with equipment and the building structure to prevent soil residues from remaining trapped outside the plants.
- If dead legs are unavoidable, special care must be taken in the design and proper installation, which must be discussed and agreed upon in advance with the plant's supplier. The depth of the dead legs must be less than their diameter. Under no circumstances must downward dead legs be present, as they are not self-draining (Figure 21.7).
- Conveyor belts are often difficult to clean and can be sources of contamination. Rather, when possible, conveyor belts with PVC surface and polyester frames should be used, as they are easy to clean and involve minimal contamination. Drive chains must be easily loosened, and easy to clean and separated from the conveyor surface area.

Reference

Zanoni, C. (1993) "Igiene degli impianti". In: C. Peri. *SAGI – Sistema Aziendale di Garanzia dell'Igiene nelle industrie agro-alimentari*, pp. 41–56, Milano: Centro Studi sull'alimentazione Gino Alfonso Sada, Italy.

Further reading

CAC/RCP 1-1969, Rev. 4-2003 (2003) *General Principles of Food Hygiene. Hazard Analysis and Critical Control Point (HACCP) System and Guidelines for its Application (Annex)*. Codex Alimentarius Commission, Rome.

International Olive Oil Council (2006) *Quality Management Guide for the Olive Oil Industry: Olive Oil Mills*. International Olive Oil Council, Madrid.

ISO 22000:2005 (2005) *Food Safety Management Systems – Requirements for Any Organization in the Food Chain*, International Standards Organization, Geneva.

ISO/TS 22004:2005 (2005) *Food Safety Management Systems – Guidance on the Application of ISO 22000:2005*, International Standards Organization, Geneva.

Peri C., Lavelli V. and Marjani A. (2004) La sicurezza alimentare e l'HACCP, in *Qualità nelle Aziende e nelle Filiere Agroalimentari*, Hoepli, Milano, Italy, Ch. 10.

22

Olive mill waste and by-products

Claudio Peri[1] and Primo Proietti[2]

[1] *University of Milan, Milan, Italy*

[2] *Department of Agricultural, Food and Environmental Sciences, University of Perugia, Perugia, Italy*

Abstract

Material balances of the mill wastes and by-products resulting from two-phase and three-phase processing are presented. The main use of wastewater deriving from three-phase processing is for spreading on cultivated soil; the spreading load in hectolitres per hectare is suggested. Physical separation treatments allow the polluting charge of wastewater to be reduced by as much as 30–50%. The case of two- and three-phase pomace is presented. Oil and energy recovery, animal feed and composting are discussed. Annex 22.1 gives a detailed description of material balances and suggests how such calculations can be applied in practice for an effective control of mill performance.

22.1 Introduction

Waste is a product that is discarded or required to be discarded. In this handbook, wastewater produced in the three-phase processes is considered as waste.

By-products are secondary products derived from a process that can be useful and marketable or can be used in a secondary process to produce valuable and marketable products. In this handbook, pomace from three-phase and two-phase processes qualifies as a by-product.

Olive oil extraction requires a considerable amount of water and generates huge quantities of liquid and solid waste. Moreover, such huge quantities are concentrated in a limited period of time, corresponding to two-three months of olive harvesting. Finally, olive mill waste contains natural components with bacteriostatic and phytotoxic activity, which limit the possibility of microbial digestion or spreading on cultivated soil.

The Extra-Virgin Olive Oil Handbook, First Edition. Edited by Claudio Peri.

A clear understanding of the problems surrounding waste and by-products in extra-virgin olive oil mills requires that a material balance of process inputs and outputs be precisely known. An example of a material balance is given in Annex 22.1 with details about the calculation procedure and suggestions on how to put it into practice for process control.

Figure 22.1 presents a flow-chart of a traditional, first-generation three-phase process. Hypotheses of olive composition and calculated values of the main inputs and outputs in three-phase technology are reported in Tables 22.4, 22.5 and 22.6 of Annex 22.1.

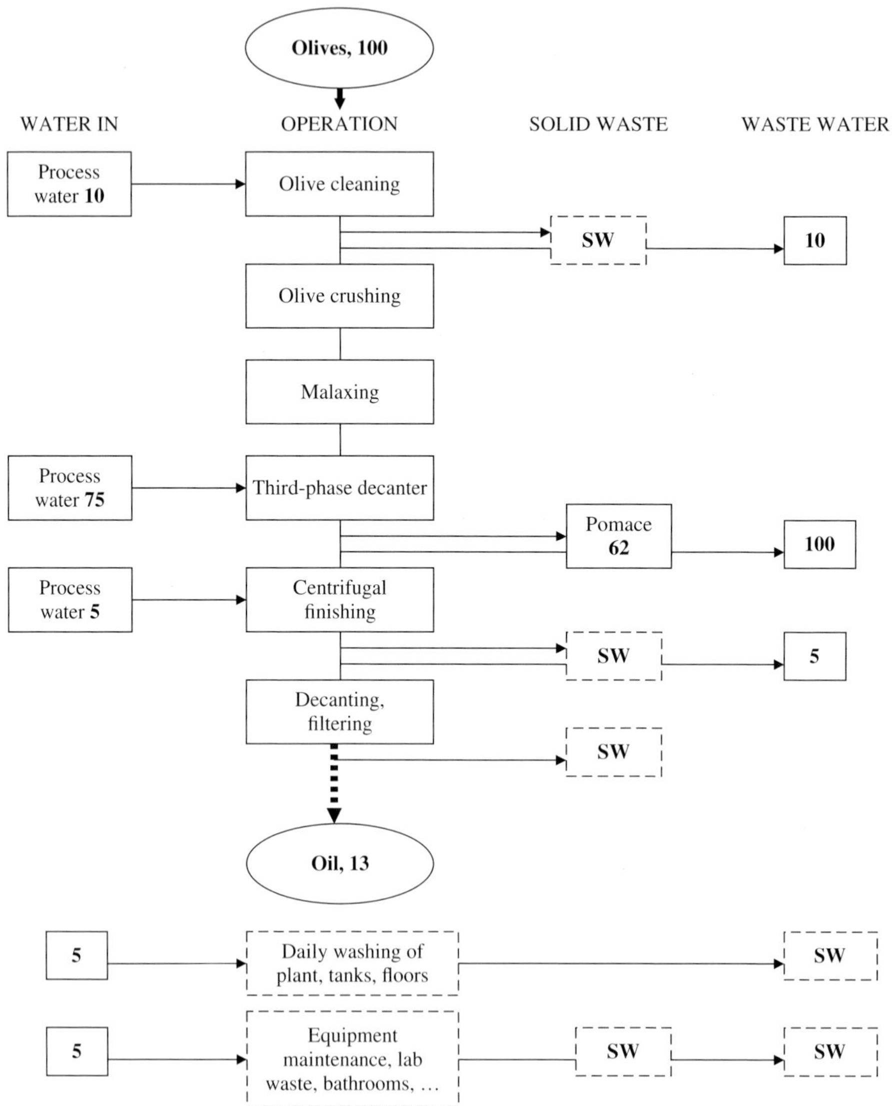

Figure 22.1 Water and waste balance in the three-phase process.

Quantities are in kg and material balances refer to the processing of 100 kg olives. The data reported in the figures and tables in this chapter are not the result of a specific process but average values based on literature data. With three-phase technology, the main waste is wastewater and the main by-product is pomace. The total consumption of 100 kg water per 100 kg olives as noted in Figure 22.1 can be considered as a conservative figure.

On the right side of the flow-chart, quantities of solid and liquid wastes are reported, along with an indication of special waste (SW), such as:

- the solid residues from olive cleaning (mainly olive leaves)
- the solid deposits from the clarifying centrifuges
- the solid residues from decanting or filtering operations (including exhausted filter pads)
- the liquid waste deriving from washing of the plant, tanks, floors, etc. (including detergents)
- the various liquid and solid wastes deriving from equipment maintenance (including lubricants), laboratory waste (including solvents, reagents, etc.), bottling and packaging waste, bathrooms, etc.

Precautions and documentation are required according to national and local regulations for the correct handling and disposal of these special wastes.

Figure 22.2 presents a flow-chart of a two-phase process. With two-phase technology, the main waste stream is pomace. Hypotheses of olive composition and calculated values of the main input and output streams in two-phase technology are reported in Tables 22.7 and 22.8 of Annex 22.1.

22.2 Composition, treatment and uses of olive mill wastewater

In Table 22.1 composition data are reported for wastewater from three phase technology. The literature presents extremely variable data (Roig *et al.* 2006); hence the values are the result of a comparison and rounding off of data from several sources.

The first point to be underlined is the very low value of total solids. This makes concentration by evaporation too slow when carried out in lagoons or too costly if carried out in evaporators. Recovery of chemicals or the application of costly separation techniques, such as membrane processes, are unsuitable because of the high dilution and the variable composition of this waste.

Some minor components reported in the lower part of Table 22.1 may have a relevant negative or positive impact. The high concentration of phenols could have a bacteriostatic effect on micro-organisms and some phytotoxic effect on cultivation. On the other hand, large quantities of phosphorous and potassium can give a significant contribution to the fertilization of cultivated soil.

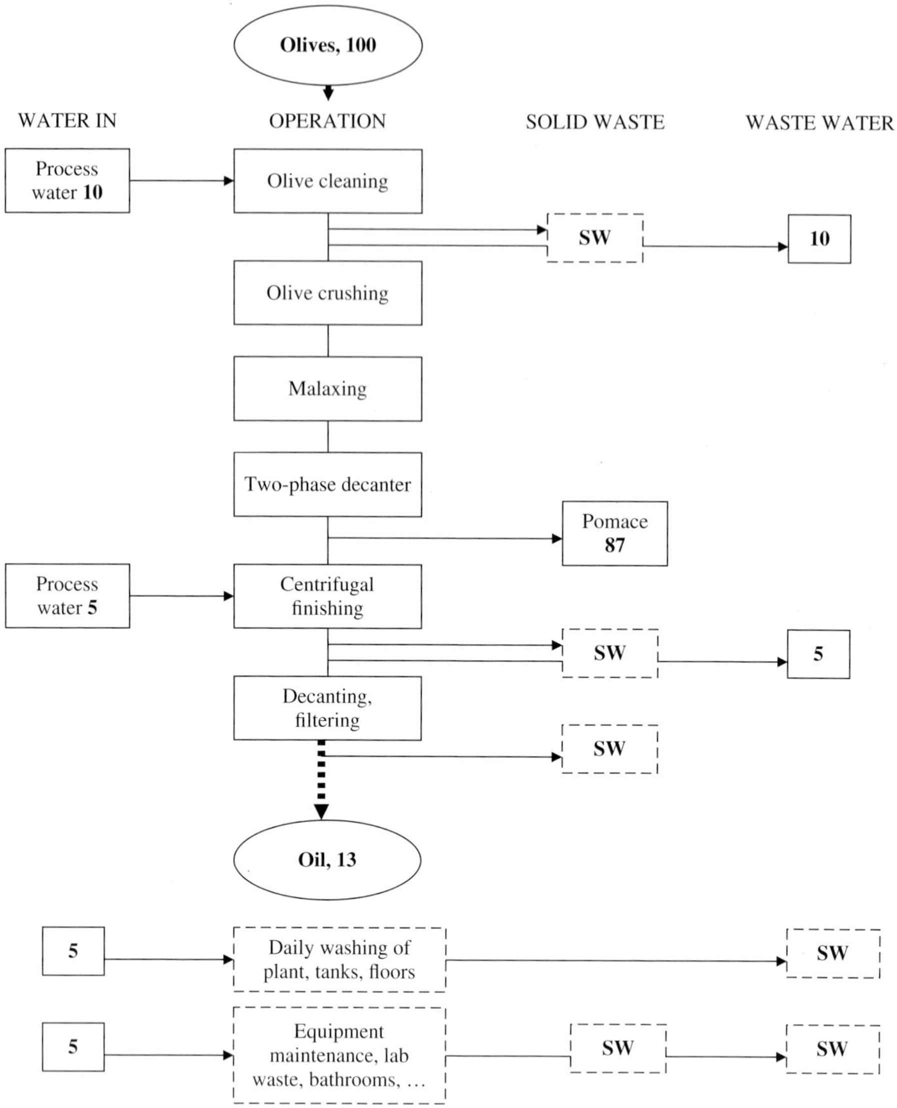

Figure 22.2 Water and waste balance in the two-phase process.

22.2.1 Wastewater treatment and use

Figure 22.3 summarizes the possible treatment and use of wastewater deriving from the three-phase process. It suggests that the most natural destination of wastewater is for it to be spread on soil. However, the presence of organic compounds, especially polymeric phenolic compounds with considerable antimicrobial and phytotoxic activity, the high content of salts, the slightly acidic reaction and the imbalanced C/N ratio, all suggest the need to manage this waste carefully.

Table 22.1 The composition of wastewater from the three-phase process.

Composition parameters	Range of variation	Comment
kg per 100 kg olives	90–140	The average value, according to Roig *et al.* (2006) is 115 kg /100 kg olives
pH	4.8–5.5	
Total solids (%)	2–5	Suspended mucilage and oil can cause a substantial increase (20–50%) of this value
Reducing sugars (%)	0.5–1.2	
COD[a], kg/m^3	20–60	Greatly influenced by organic compounds in suspension or oil in emulsion
BOD_5[b], kg/m^3	7–20	
C/N ratio	45–55	
	Minor components	
Total phenols, g/L	4–10	This figure is ten to fortyfold higher than the concentration of phenolic compounds in the oil
P_2O_5, g/L	0.1–0.2	
K_2O, g/L	0.5–1.0	

Notes: [a]Nearly all organic compounds can be fully oxidized to carbon dioxide and water with a strong oxidizing agent under acidic conditions. The conversion of ammonia to nitrate is referred to as nitrification. The Chemical Oxygen Demand (COD) is a measure of organic compounds in water and hence a measure of water polluting charge. COD is expressed in milligrams per Litre or kg per cubic metre, which indicates the mass of oxygen consumed per unit volume of water. [b]Biological oxygen demand or BOD is the amount of dissolved oxygen needed by aerobic biological organisms in a body of water to break down organic material present in a given water sample at a certain temperature over a specific time period. The BOD value is most commonly expressed in milligrams of oxygen consumed per Litre of sample during five days of incubation at 20 °C. It is also a measure of the degree of organic pollution of water.

The antimicrobial and phytotoxic activity is mainly due to the tendency of phenolic compounds to combine with other organic compounds, especially proteins, thus altering cell membrane permeability and intercellular transfer mechanisms. Discharge of wastewater into water streams is prohibited because it may result in a significant reduction of natural purification and damage to the aquatic flora and fauna.

The main precautions for spreading wastewater on the soil are:

- Spread only on agricultural land; avoid spreading in forests or in gardens or on uncultivated land. Agronomic practices and soil tillage accelerate oxidative degradation of polyphenols and facilitate interaction between organic material and soil.
- Avoid spreading in the germination period of a herbaceous plant's lifecycle. Phytotoxic activity is particularly dangerous to seed germination. It is preferable to spread wastewater on the soil 2–3 months before sowing, possibly followed by mixing with the surface layer of the soil. Mixing with soil accelerates organic matter-soil interaction with benefits to soil porosity.

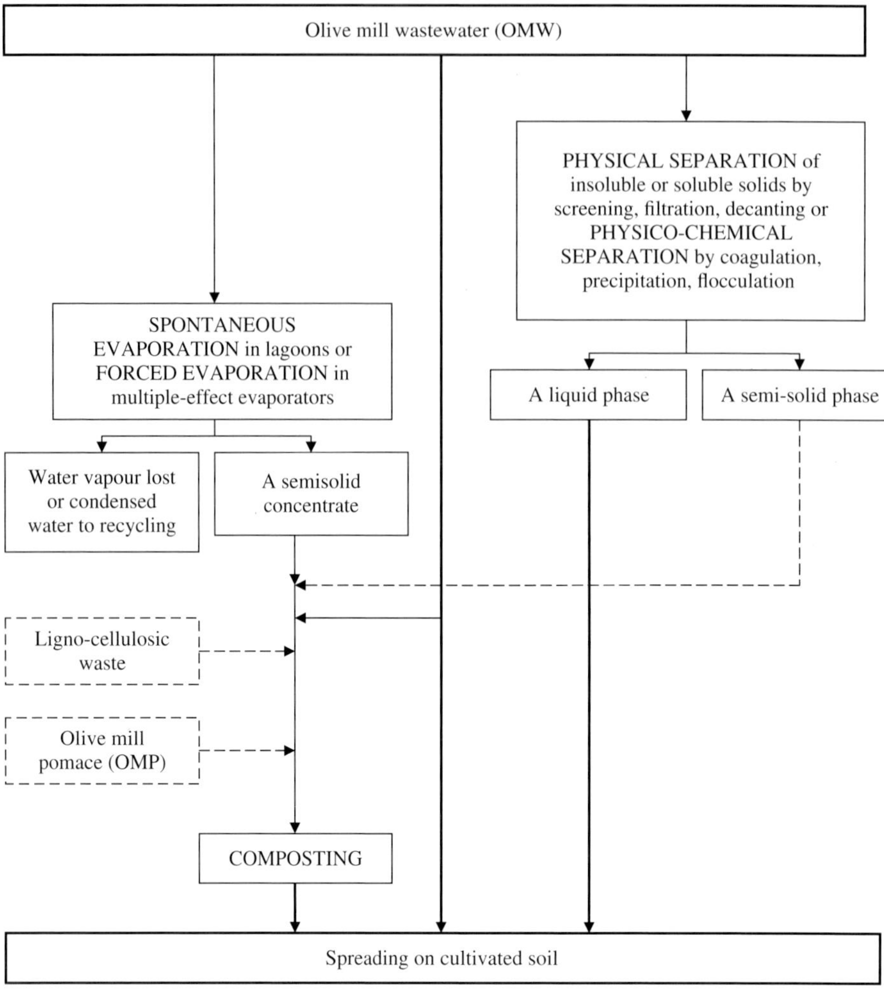

Figure 22.3 Possible uses of olive mill wastewater (OMW) from the three-phase process.

The are no time limits for spreading wastewater on tree crops, particularly olive groves.

- Avoid spreading on land with a slope > 15°. Surface flowing of wastewater is detrimental to soil stability and often results in runoff into creeks and water drains.
- To avoid contaminating water, keep a safe distance of at least 10 m from rivers and other water streams.
- To avoid odour pollution, keep a safe distance of at least 200 m from populated areas.

- Avoid mixing wastewater with other effluents and waste. The antimicrobial activity of wastewater hinders biological purification.
- Transfer wastewater in closed tank trucks to avoid spilling.
- Distribute uniformly on the soil by adapting manure spreaders.

22.2.2 Spreading load

The most critical decision in wastewater application to soil is determining 'spreading load' (SL), which is expressed as hectolitres per hectare (a hectare (ha) is a surface area of 10 000 square meters; to convert from ha to acres multiply by 2.471). In some countries the maximum SL allowed is established by the local or national laws. In the absence of such indication, a good suggestion is to establish a SL value depending on the polluting charge of the wastewater, which may be considered as proportional to the total solids content. The spreading load should be decreased with increasing total solids in the wastewater. The 'rule of 12' can be applied as a rough indication, when wastewater is spread on crop land (Table 22.2).

Table 22.2 The rule of 12 for OMW spreading loads (SL).

Total solids (%)	3	4	5	6	7	8	9	10
SL = hundreds of hectolitres (or tens of cubic meters) of wastewater per hectare	9	8	7	6	5	4	3	2

The sum of the two numbers in each column is 12. The first line indicates the total solids concentration in the wastewater, while the second line indicates the suggested quantity of wastewater that can be spread on the soil in hundreds of hectolitres per hectare. For example, if the total solids concentration is 5%, 700 hl or 70 m^3 can be spread per hectare.

The SL can be increased by as much as 50% in the case of spreading on tree crops, namely olive groves. The SL can also be substantially increased by treating the wastewater with physical or physical-chemical treatments that can partially remove the suspended solids and, to some extent, some of the dispersed macromolecules and emulsified oil. In general, the elimination of suspended solids also results in a significant reduction in polymeric phenolic compounds with attenuation of the antimicrobial and phytotoxic effect.

22.2.3 Physical and physical-chemical treatments of wastewater

Sedimentation is the separation of solid particles from the liquid in which they are suspended, under the action of gravity, due to difference in density. According to Stoke's law, sedimentation velocity is proportional to the square of the diameter of the particle and therefore larger particles settle more rapidly than smaller particles.

Particles of very small diameter (in the range of 10 μm or less) form stable dispersed systems and can be eliminated only through the application of flocculation or flotation or fine filtration.

Flocculation is a process of adhesion whereby the suspended solid particles form large-sized clusters. This phenomenon takes place spontaneously in olive mill wastewater, but at a very slow pace. It can be greatly accelerated by the addition of flocculating agents. These are synthetic polymers with charged molecular chains (polyelectrolytes) that are able to link both coarse and very fine particles or even single macromolecules into clusters. The effect of flocculation is a rapid increase in diameter of the suspended particles with consequent acceleration of sedimentation.

Flotation is a process in which air is finely bubbled through the wastewater. Tiny air bubbles adhere to the suspended materials due to surface adhesion so that the bubble-solid complexes have a low 'apparent' density and can readily surface as a floating air-solid layer. This layer is separated by a skimming and overflow mechanism.

Both flocculation and flotation tend to drag oil droplets along in their movement with further reduction of the polluting charge.

Coarse filtration through layers of sand or fine particulate media can also be used to remove the coarser suspended material. If charcoal or activated carbon are used in the filter media, polyphenols are absorbed and removed very effectively, with substantial reduction of antimicrobial and phytotoxic effects.

The solid residues from wastewater pre-treatments may be mixed with pomace in composting or spread on or tilled into the soil directly.

22.2.4 Other possible treatments

Spontaneous evaporation in ponds in the open ('lagoons') is one of the oldest methods of wastewater treatment and is still used in some countries with a warm climate, such as in southern Europe and northern Africa. Although lagooning is very simple and inexpensive, there are some serious drawbacks: (i) contamination of groundwater can take place if the lagoon is not properly insulated; (ii) decomposition of organic materials produces significant odour pollution; (iii) lagoons attract insects and contribute to insect (and rodent) infestation. Consequently, lagooning is rapidly disappearing (Roig *et al.* 2006).

Other treatments have been proposed and experimentally tested at a laboratory scale, such as:

- *processing with membrane* operations (by reverse osmosis or ultrafiltration)
- *multiple effect evaporation*
- *fine filtration* with filter sheets
- *selective extraction* of valuable components, especially phenolics

However, with the technologies used so far, these alternatives are too complex and costly to recover valuable components economically.

Table 22.3 Composition and calorific value of three-phase and two-phase pomace.

Pomace	Three-phase	Two-phase
kg per 100 kg olives	62	87
	Gross composition of 100 kg pomace	
Oil	3.2	2.3
Total solids + oil (%)	53.9	41.4
Water (%)	46.1	58.6
	Calorific value and balance of 100 kg pomace[a]	
Pit fragments (%)	34	24
Calorific[b] contribution of pit fragments, at 15 MJ/kg,	510 MJ (483480 BTU or 121686 Kcal)	360 MJ (341280 BTU or 85896 Kcal)
Calorific contribution of oil, at 30 MJ/kg	96 MJ (91000 BTU or 22900 Kcal)	69 MJ (65400 BTU or 16450 Kcal)
Energy needed for water evaporation, about	101 MJ	129 MJ

Notes: [a]MJ is the abbreviation of mega-joules (millions of joules). To convert from joule to BTU multiply by 0.000948. To convert from joule to Kcal multiply by 0.0002386. [b]The calorific value is the amount of heat produced by the complete combustion of a material.

22.3 Composition, treatment and uses of olive mill pomace

In Table 22.3, some composition data are reported for pomace deriving from three-phase and two-phase processes. Literature data on the composition of pomace is very scarce. The values reported in Table 22.3 were elaborated from the data in Tables 22.5 and 22.8 in Annex 22.1.

The first point to be considered is the different composition of the two types of pomace. The three-phase pomace has a higher solids and a lower water content. As a consequence, the oil concentration is higher. This is a crucial difference in favour of three-phase pomace for oil recovery.

The flow chart in Figure 22.4 presents the possible treatment and uses of pomace. An important detail of the flow-chart is the arrow reversibly connecting the three and two-phase pomace. It means that the treatments presented in the figure can be applied to both pomaces, however with a different degree of technical and economic feasibility.

At the bottom of the flow-chart, four main products are indicated as the possible result of treatment: oil, energy, animal feed and compost.

On the left side of the figure, the vertical flow concerning the recovery of oil is preferably applied to three-phase pomace. The second vertical flow concerns the recovery of thermal energy, which can be obtained by direct burning of pomace or, more efficiently, by burning dried pomace. The pomace itself can provide the heat needed in the drying operation. Pit fragments are a significant source of energy; they can be separated from pomace at low cost with mechanical-pneumatic equipment. This operation is becoming increasingly common in olive mills.

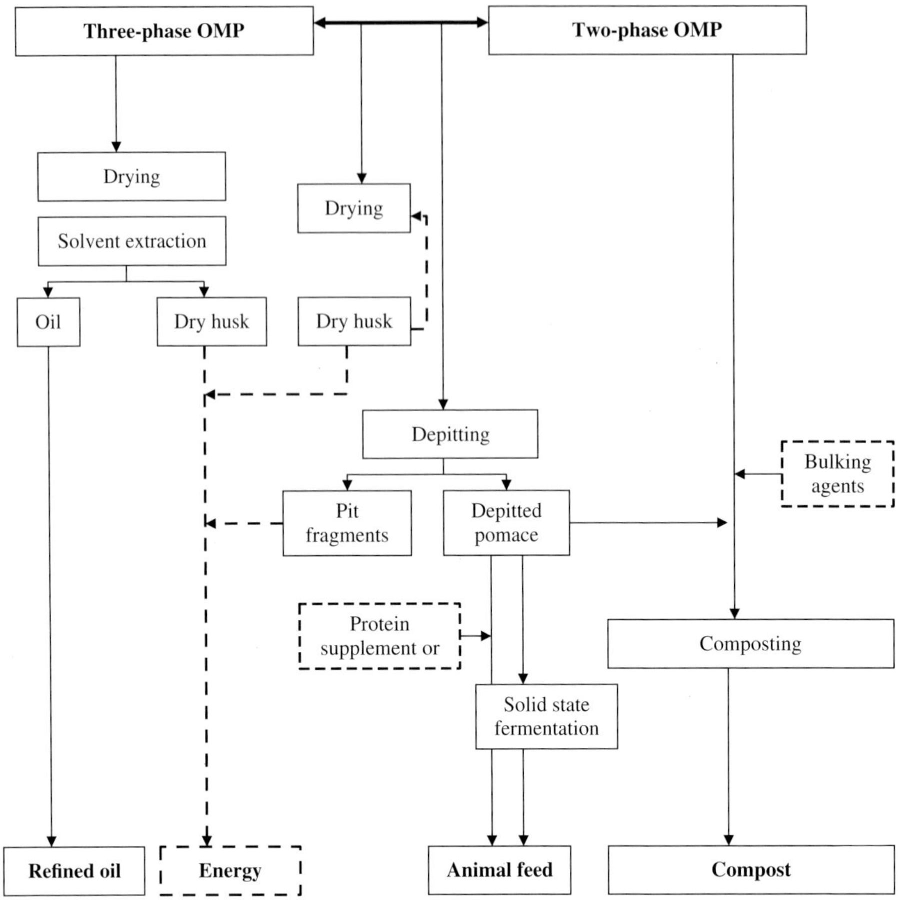

Figure 22.4 The possible use and treatment of olive mill pomace (OMP) from three- and two-phase processes.

In the third vertical flow, depitted pomace could become an animal feed component with addition of a protein supplement or an increase of cell protein through solid-state fermentation. Finally, the fourth vertical flow indicates the use of pomace for composting.

In technical terms, oil recovery can be best pursued by using three-phase pomace, while composting is the preferable use of the two-phase pomace. The diffusion of one or the other uses depends on economic opportunity. If the value of pomace oil is high, it may be interesting to direct not only the three-phase, but also the two-phase pomace, to oil recovery. If, on the contrary, the value of pomace oil is low, composting becomes a more favourable destination for two-phase as well as three-phase pomace.

The use of both types of pomace for energy production is becoming increasingly attractive.

The same considerations apply to pomace deriving from the most recent advances in three-phase technology, with intermediate water content and reduced water consumption.

22.3.1 Composting

The decreasing interest in oil recovery and the increasing diffusion of two-phase processes is fostering composting as a suitable low-cost strategy for recycling olive oil by-products and waste. Two-phase pomace has a suitable composition and water content for composting but three-phase pomace can be easily adjusted, if necessary, with some recycling of wastewater. Mixing pomace with other solid or semi-solid olive mill waste is also possible, including: leaves and small twigs from the olive cleaning step; the solid deposit discharged from centrifuges; filtration residues and exhausted cellulose filtration pads. Similarly, bulking agents such as straw, pruning residues, grape stalks, and so forth, can be used to achieve the optimal porosity of the compost pile.

The composting process is carried out through the following steps:

1. Pile preparation and building over time, which imply choosing, mixing and layering the composting material in order to achieve:
   - An optimal water content of 60–70% by weight. Excess water must be avoided because it may create local anaerobic conditions. Three-phase pomace may be diluted by using some wastewater.
   - An optimal porous structure in order to guarantee aeration and avoid anaerobic fermentation. Oxidation is, in fact, the microbial metabolic mechanism of material decomposition and mineralisation.
   - An optimal C/N ratio of 30 to 1. Carbon is the energy source for micro-organisms. They produce heat (+CO_2 and H_2O) by oxidation of organic carbon compounds. Nitrogen is needed by micro-organisms to grow and reproduce. Too high a value of the C/N ratio as in pomace and wastewater (C/N about 50) determines a loss of carbon with a proportional loss of fertilizing value. In this case, addition of a nitrogen source, usually urea, is needed.
2. Initial fermentation by mesophilic bacteria with decomposition of carbohydrates, protein and lipids. During this phase of intense biological activity, exothermal metabolic reactions tend to rapidly increase the temperature of the pile. The compost must be mechanically turned for cooling and aeration in order to keep the temperature below 60 °C (140 °F). At higher temperatures the microbial activity stops.
3. Due to the above temperature evolution, the initial fermentation by mesophilic bacteria is followed by a thermophilic phase with prevailing activity of thermophilic bacteria. During this phase, decomposition of cellulose takes

place along with mineralization, with production of CO_2 and H_2O. At high temperatures, water evaporates at a high rate and addition of water may be required. This is an opportunity for recycling olive mill wastewater.
During this thermophilic phase, many unwanted seeds and nearly all pathogens are destroyed. The sanitizing effect of composting is particularly desirable (a few days at temperature higher than 55 °C) when there is a high likelihood of pathogens, as in the presence of manure.

4. In the following 'stabilization' or 'curing' phase, fermentation slows down, the temperature decreases and recolonization by mesophilic bacteria takes place. 'Microbial pesticides' including thermophilic and mesophilic bacteria increase. The activity of some detritivores such as black-soldier larvae and redworms contribute to reducing many pathogens. Earthworms ingest partially composted material and continually create aeration and drainage channels as they move through the compost. Optimal results require fostering complex equilibria and synergies of living organisms such as bacteria, Actinomycetes, fungi and moulds, protozoa, rotifers, earthworms and insects.
5. Long-term stabilization produces further mineralization and increasing availability of nutrients for plant absorption.

At full ripeness, compost is a dark-brown product with an earthy smell, which is due to Actinomycetes. Compost provides a rich growing medium and an absorbent material that holds moisture and minerals. It is therefore a soil additive supplying humus and nutrients and a key ingredient in organic farming. Pomace compost showed no phytotoxic effects and is particularly rich in mineral nutrients.

22.3.2 Soil amendment

Pomace can be directly tilled into the soil as a soil conditioner. This is obviously a very simple and direct utilization and hence a very frequent one.

To a lesser degree than compost, pomace improves the physical quality of the soil and adds nutrients especially carbon and potassium but also nitrogen and phosphorous in a considerable proportion.

22.3.3 Pomace as an energy source

Energy and environmental considerations are making pomace an increasingly interesting energy source.

Three-phase pomace

If the data reported in Table 22.3 are taken as the basis for calculation, the calorific value of 100 kg of three-phase pomace is 606 MJ:

- This corresponds to 53.9 kg of dry matter (total solids + oil). Therefore 606/53.9 = 11.2 MJ is the calorific value per kg of dry matter.

- If in a dried pomace there is 8% residual moisture, the calorific value is 11.2 × 0.92 = 10.3 MJ per kg of dried pomace.
- If the overall efficiency of the process is 80%, the final net calorific value of 1 kg of dried pomace is 10.3 × 0.80 = about 8.2 MJ/kg (29 335 kcal/kg). This is a low but significant value (about 47% of the calorific value of compact wood and about 23% the calorific value of oil).
- The water to be evaporated in order to produce dried pomace is about 0.7 kg per kg of dried pomace, corresponding to a consumption of heat of about 1.5 MJ. This means that about 18% of dried pomace must be used to dry the pomace itself. The remaining 82% with a calorific value of 8.2 MJ/kg is an energy source of considerable interest.

Two-phase pomace

- The calorific value per kg of dry matter is 10.4 MJ (429/41.4).
- If the dried pomace has 8% residual moisture and the overall efficiency of the process is 80%, the true calorific value of 1 kg dried pomace is 7.65 MJ.
- The drying process requires the evaporation of 1.22 kg water for every kg of dried pomace, corresponding to a heat consumption of 2.61 MJ.

This means that about 34% of dried pomace must be used to dry the pomace itself. The remaining 66%, with a calorific value of 7.65 MJ/kg, can still be considered an interesting source of thermal energy.

The use of dried pomace as solid fuel in boilers may allow warm water to be produced for the needs of the milling process and for air conditioning of the milling factory. This is very frequent.

Excess heat can be used for other purposes, including the production of high-pressure steam and electric power generation.

More complex systems based on mixing pomace with other energy sources and integrated systems of thermal and electrical energy production are of some interest in areas of intense olive oil production and are the subject of pilot experiments. Their description is beyond the scope of this handbook.

22.3.4 Other uses of pomace

- *The production of biogas* (a mixture of CH_4 and CO_2) through the anaerobic digestion of pomace has been the subject of study and experiment. The low sugar and protein content and the inhibiting effect of polyphenols are significant limitations to this use.
- *The extraction and purification* of valuable compounds such as pectin or biophenols should be considered as difficult from a technological point of view (great variability in composition, low quantities available) and, at the moment, economically unsustainable.

Annex 22.1: Mass balance of the extra-virgin olive oil process

A hypothetical average mass balance

A mass balance is the balance of materials in the input and in the output of a process. In the absence of material accumulation, the sum of mass in input is equal to the sum of mass in output. This is the case in the extra-virgin olive oil process.

Masses are expressed in kg. Calculations are to be referred to some unitary quantity of an important input or output material. In the following calculations the reference mass is 100 kg olives in input. All other masses are referred to this quantity.

The data presented below are not the result of a specific process but average data based on values reported in the technical literature. Actual data from a specific process may differ greatly from those presented in this annex, depending on the plant, the olives and the operating conditions. However, if differences from the data presented here are too great, this should be considered as a warning signal of poor process control.

The assumptions and calculation procedure are explained step-by-step in detail, so that they can be used as a guide for carrying out mass balances in practical processes.

Water and waste balance in the three-phase process

A balance of input and output masses in a three-phase process are presented in Table 22.4.

The data in Table 22.4 are based on the following assumptions:

Input

1. The percentage composition of 100 kg of olives (or of olive paste after the milling operation) is assumed to be:

   - 15% oil;
   - 57% vegetation water containing 10.5% soluble solids (6 kg) and 89.5% water (51 kg)
   - 28% insoluble solids containing 25% cellular components (7 kg) and 75% pit fragments (21 kg)

2. The process water added to 100 kg olive paste is assumed to be 90 kg according to the flow-sheet of Figure 22.1, based on conservative literature values for traditional, first-generation, three-phase decanters (Roig *et al.* 2006). It is the sum of:

   - 10 kg for the olive washing operation
   - 75 kg for decanter separation
   - 5 kg for centrifugal finishing

Table 22.4 The input-output mass balance (kg) in a three-phase process (see also Figure 22.1).

	Input		Output		
	Olive and process water		*Pomace*	*Wastewater*	*Oil*
Vegetation water	Water + soluble solids	51 + 6 = **57**	30.6 + 1.4 = **32**	20.4 + 4.6 = **25**	–
Insoluble solids	Cellular components + pit fragments	7 + 21 = **28**	7 + 21 = **28**	Trace	–
Oil	Oil	**15**	**2**	Trace	**13**
Total of olive paste		**100**			
Process water		**90**		**90**	
Total		**190**	**62**	**115**	**13**

Process water is defined as the water that comes into contact with the product (the olives or the oil). It must conform to potable water standards. *Vegetation water* is the watery phase in the olive paste, which includes water and water-soluble components. *Wastewater* is the watery phase which is discharged from the process. It includes part of the vegetation water and part of the process water with their soluble solids content.

Input

1. It is assumed that:
    - insoluble solids are fully retained in the pomace fraction
    - they represent 45% of the pomace weight, which is based on good decanter efficiency.

    Consequently the total weight of pomace is 28/0.45 = 62 kg.
2. The recovery of oil is assumed to be 87% (15 × 0.87 = 13 kg), which means that 13% of the oil (2 kg) remains in the pomace. The amount of vegetation water in the pomace can be calculated by difference: 62 – (28 + 2) = 32 kg.
3. The aqueous phase that is fed into the decanter is the sum of the vegetation water (57 kg) plus the 75 kg of process water, totalling 132 kg. The soluble solids are therefore diluted to a final concentration of 6/132 × 100 = 4.5%. This is the concentration of soluble solids retained in the pomace; the composition of the aqueous phase in the pomace therefore consists of 32 × 0.045 = 1.4 kg of soluble solids and 32 – 1.4 = 30.6 kg water.
4. The wastewater resulting from the process is the sum of the water separated at the decanter (100 kg) plus the water coming from olive washing (10 kg) plus the water discarded from the finishing centrifuge (5 kg). It contains 4.6 kg of soluble solids plus a negligible amount of insoluble solids deriving from olive washing and the incomplete separation in the decanter and the finishing centrifuge. For the sake of calculation, the amount of total solids in the final stream of wastewater is rounded to 5 kg.

Table 22.5 The percentage composition of pomace from the three-phase process.

Component	Calculation	Percentage
Oil	2.0/62 × 100	3.2
Insoluble solids	28.0/62 × 100	45.2
Soluble solids	1.4/62 × 100	2.2
Water	30.6/62 × 100	49.4
Total		100.0

Table 22.6 The percentage composition of wastewater in the three-phase process.

Component	Calculation	Percentage
Oil	*Trace*	*Trace*
Soluble + insoluble solids	5.0/115 × 100	4
Water	110.0/115 × 100	96
Total		100

Notes: Differences in the oil content of olives have a great importance in economic terms but very little influence on the above balances.

In conclusion, Tables 22.5 and 22.6 summarize the resulting composition of the pomace and the wastewater deriving from a three-phase process.

Water and waste balance in the two-phase process

In the two-phase process, it is assumed that 100 kg of olives require 15 kg of process water (for the olive-washing operation and centrifugal finish). This amount is shown in the output as wastewater. It does not contain vegetation water as in the three-phase process; therefore it has a low polluting charge and can be discharged into the sewer.

The two-phase process results in a large output of pomace (Table 22.7) with the composition reported in Table 22.8.

Three-phase decanters with reduced water consumption

For decades, the wastewater deriving from press extraction or the three-phase decanter has been the most polluting and troublesome waste from olive mills. The introduction of two-phase technology is the most effective solution to this problem thus far.

On the other hand, substantial improvements have been introduced in the last generation three-phase decanters in order to reduce the quantity of water consumed and the wastewater produced. These systems are based on improvement of the design of the decanter and/or partial recycling of the wastewater. If the above calculation is applied to these new decanters and the hypothesis of water addition is 30 kg per

Table 22.7 The input-output mass balance (kg) in a two-phase process (see also Figure 22.2).

	Input		Output		
	Olive paste + process water		*Pomace*	*Wastewater*	*Oil*
Vegetation water	Water + soluble solids	51 + 6 = **57**	51 + 6 = **57**	trace	trace
Insoluble solids	Cellular components + pit fragments	7 + 21 = **28**	7 + 21 = **28**	trace	trace
Oil		**15**	**2**	trace	**13**
Total of olive paste		**100**	trace		
Process water		**15**	trace	**15**	trace
Total		**115**	**87**	**15**	**13**

Table 22.8 The percentage composition of pomace from the two-phase process.

Component	Calculation	Percentage
Oil	2.0/87 × 100	2.3
Insoluble solids	28.0/87 × 100	32.2
Soluble solids	6.0/87 × 100	6.9
Water	51.0/87 × 100	58.6
Total		100.0

100 kg olives, it can be seen that the wastewater produced is 70 compared to 115 kg (Table 22.4) having a total solids content of about 6% instead of 4% (Table 22.6).

For process control, it is recommended that mass balances be adapted and verified in practical situations according to the procedure described above.

Evaluation of mass balances in practical processes

The evaluation of mass balances is not an academic exercise, but a most valuable tool of process control. If mass balances vary from one milling batch to another or depend on casual decisions of the mill operator or depend on inaccurate plant regulation, the process is not under control and can lead to negative consequences on oil quality and yield and on processing costs.

Oil millers should plan and carry out periodical mass balance controls similar to those presented here. This requires some direct evaluation of quantities as well as some simple analysis of material composition. The evaluation procedure is suggested in the following tables. Masses are indicated with the symbol M and a progressive numerical index.

Table 22.9 presents the simplest overall mass balance in an olive mill. Generally, this is the only mass balance precisely known to mill operators.

The second step of the procedure consists in the analysis of the gross composition of the olives. This can be carried out on a homogeneous and representative sample

Table 22.9 The overall mass balance of the olive-milling process.

Identify one batch of olives representative of the normal working conditions (the 'reference batch'). The kg of olives in the reference batch is the first mass to be recorded:	M_1
The kg of oil actually obtained from the reference batch of olives is the second mass to be recorded:	M_2
The actual yield obtained in the process can be calculated from the previous data and recorded as olive mill yield (OMY):	OMY is obtained dividing M_2 by M_1 and multiplying by 100

Table 22.10 Mass balance of the olives in input.

A few grams of a homogeneous sample of olive paste taken at the olive mill are carefully weighed and the weight is recorded as:	M_3
The sample is dried to constant weight in a laboratory oven under standard analytical conditions. The weight of the dried sample is recorded as:	M_4
The water content of the olive paste (which corresponds to the water content of the olives) is calculated as:	$M_5 = (M_3 - M_4)$ divided by M_3 and multiplied by 100.
The dried sample is submitted to the analytical procedure for oil determination (the Soxhlet method). The weight of the extracted oil is recorded as:	M_6
The oil content of the olive paste (and of olives) is calculated as:	$M_7 = M_6$ divided by M_3 and multiplied by 100
The total solids content of the olive paste can be calculated by difference as:	$M_8 = 100 - (M_5 + M_7)$
Important information may be derived from these analyses by comparing the OMY value of table 9 with the oil content of the olives. The ratio of OMY (the kg of oil extracted from 100 kg olives) to M_7 (the oil content of 100 kg olives) is a measure of olive mill efficiency (OME):	OME is obtained dividing OMY by M_7 and multiplying by 100

of olive paste resulting from the milling operation. The analytical determinations should be carried out in reliable laboratories. The sample of olive paste taken at the olive mill should be kept at low temperature – preferably frozen – in order to avoid spoilage and fermentation. Methods are available in specialized laboratories for a fast and cheap evaluation of data presented in this annex.

Analytical procedures and calculations are summarized in Table 22.10.

An olive mill efficiency (OME) higher than 85% is an indicator of very good efficiency of the milling plant and operation. An OME lower than 75% is an indication of poor performance.

The third step of the procedure consists in analysing a pomace sample as summarized in Table 22.11. From the total mass of pomace and its percentage composition, the mass of water, oil and solids corresponding to 100 kg olives can be calculated.

It may be considered that the results of these calculations are within a range of 5–10% precision, which is a more than an acceptable level of precision for the use

Table 22.11 Mass balance of the pomace in input.

Weight of a homogeneous sample of pomace:	M_9
Weight of the dried pomace sample:	M_{10}
Water content of the pomace is calculated as:	$M_{11} = (M_9 - M_{10})$ divided by M_9 and multiplied by 100.
The oil content, determined according to the Soxhlet method is:	M_{12}
The oil content of the pomace is:	$M_{13} = M_{12}$ divided by M_9 and multiplied by 100
The total solids content of pomace is:	$M_{14} = 100 - (M_{11} + M_{13})$
Considering that all the solids in 100 kg of olives pass into the pomace, knowledge of the olive and pomace composition enables the mass of pomace corresponding to 100 kg olives to be defined:	M_{15} is obtained dividing M_8 by M_{14} and multiplying by 100

Table 22.12 Mass balance in the wastewater of the three-phase process.

A sample of vegetation water is obtained by centrifugal separation of the water from the olive paste, followed by fine filtration so that insoluble solids are completely removed:	M_{16}
The weight of the clear vegetation water sample is:	
The weight of soluble solids in the vegetation water determined after drying is:	M_{17}
The percentage concentration of soluble solids in the vegetation water is:	M_{18} is obtained dividing M_{17} by M_{16} and multiplying by 100
A sample of wastewater is obtained by filtering wastewater through a fine filtering medium. The weight of the clear wastewater sample is:	M_{19}
The weight of soluble solids in the wastewater determined after drying is:	M_{20}
The percentage concentration of soluble solids in wastewater is the basic value for calculating the Spreading Loads according to the 'rule of 12':	M_{21} is obtained dividing M_{20} by M_{19} and multiplying by 100
M_{18} and M_{21} can be used to calculate the mass of wastewater considering that the M_{21}:M_{18} ratio corresponds to the dilution ratio due to the addition of process water to vegetation water. If, for instance, M_{21} is found to be 3%, while M_{18} is 9%, the dilution ratio is 1 to 3. M_5 being the water content of the olive paste, the dilution ratio suggests that a quantity of 2 M_5 kg of water has been added to the paste fed into the decanter. Given that the mass of water in the pomace is M_{11}, the total mass of wastewater is:	$M_{22} = 3\,M_5 - M_{11}$

of these data. Unfortunately, in actual practice in olive mills, these data are seldom available, which is a clear indication of poor control over the process.

Finally, an evaluation can be carried out to calculate the mass of wastewater in the three-phase processes. The suggested method is based on the analysis of soluble solids in the vegetation water of the olive paste and in the wastewater. Data and calculations are reported in Table 22.12.

In conclusion, evaluation of mass balances and comparison of results in different operating conditions can suggest critical control points and optimal operating conditions of the milling process.

Reference

Roig, A., Cayuela, M.L. and Sánchez-Monedero, M.A. (2006) An overview on olive mill wastes and their valorization methods. *Waste Management* **26**, 960–969.

Further reading

Niaounakis M. and Halvadakis C.P. (2006) *Olive Processing Waste Management*, Elsevier, Amsterdam, p. 498.

23

The production cost of extra-virgin olive oil

Enrico Bertolotti
BTS Business & Technic Systems srl, Milan, Italy

Abstract

A methodology and an example are presented for the calculation of the production cost of a unit package of extra-virgin olive oil. Three cost centres are evaluated concerning: (i) olive cultivation and harvesting, (ii) olive milling, and (iii) oil storage and packaging. The example refers to small olive grove companies and small-to-medium sized mills. Application of this methodology should allow readers to calculate costs in specific practical conditions. Final costs of 0.5 litre bottles, 0.75 litre bottles and 5 litre metallic containers are compared. The influence of the cost of the olives varies from 64 to 86% of the total final cost of package units and the cost of harvesting accounts for 50% or more of the olive cost.

23.1 Introduction

In a small business producing and selling extra-virgin olive oil, the control of product cost should be considered not only as an indicator of the health of the company, but also as an essential tool of process management and control.

A thorough, analytical cost report is a central document in the yearly review system. The relationships between fixed and variable costs, the production and sale volumes and the projected and actual profit should be continuously updated and monitored.

This chapter presents an example of cost calculation "from the field to the bottle" of a selling unit of extra-virgin olive oil. The distribution and selling costs are not considered.

The example refers to small-sized enterprises for two reasons: in the first place they are the most frequent case in the extra-virgin olive oil world and therefore will probably be the highest number of readers of this handbook. In the second place,

The Extra-Virgin Olive Oil Handbook, First Edition. Edited by Claudio Peri.

medium and large-sized companies have advanced accounting systems and do not need to learn from this handbook how to set up a suitable cost analysis.

Directions for the reader

The accounting system is presented in three steps which represent three well identifiable cost centres:

1. Olive cultivation and harvesting
2. Olive milling
3. Oil storage and (primary) packaging

The calculations presented in this example are based on the hypothesis that the olive-cultivation and olive-milling companies are independent of one another in terms of management and business. These two parts of the olive oil chain are, in fact, totally different in the technology used, economies of scale and optimization criteria.

The production cost, which will be referred to as a packaging unit of extra-virgin olive oil, is calculated by adopting the point of view of a company that has direct and operative responsibility over processes 2 and 3, while buying the olives from external growers or suppliers. However, the method allows cost calculation of any possible combination of responsibilities and ownership over the three steps.

Readers should consider this chapter as an exercise. Hypotheses and values adopted in calculations are realistic but do not correspond to a specific situation. Readers are encouraged to substitute the figures presented in this example with data and hypotheses corresponding to real situations that interest them. After achieving some familiarity with the exercise, readers can ask the publisher of this handbook for a free copy of the excel program that has been set up by the author of this chapter. It is available for improvement and application to different case studies.

In order to facilitate reading this chapter, tables are divided into three series, corresponding to the three steps or processes presented above:

Tables identified with letter A (A0, A1, A2, etc.) refer to the calculation of the cost in the first step of olive cultivation and harvesting;

Tables identified with letter B (B0, B1, B2, etc.) refer to the calculation of the cost in the second step of olive milling;

Tables identified with letter C (C0, C1, C2, etc.) refer to the calculation of the cost in the third step of oil storage and packaging;

Tables with number zero – A0, B0 and C0 - present the structural and operating hypotheses concerning steps A, B and C, respectively. Tables labelled with Roman numerals I and II at the end of the chapter summarize the final total costs per unit package.

23.2 Concepts, terms and definitions

Costs can be classified as follows.

Fixed and variable costs

Fixed costs are the costs associated with the product that have to be paid regardless of the quantities that are produced and sold. Fixed costs are, for example, rent for building or space, depreciation of equipment, insurance, payroll of employees and permanently hired workers and so forth.

Variable costs are directly related to production. They include raw materials, energy consumption for processing operations, personnel in temporary hiring and so forth.

Semivariable costs. Some costs, like electric power or logistics have some components that are fixed and others that are variable. In calculations, the fixed proportion should be allocated to fixed costs and the variable portion to variable costs.

Direct and indirect costs

In the evaluation of production costs, an important distinction is between direct and indirect costs. Direct costs are raw material or labour or the cost of the plant or equipment, associated with the work for producing the product.

Indirect costs are those that affect the entire company, not just the product. These are, for example, advertising, depreciation, general supplies for the firm and accounting services. Indirect costs are usually called 'overheads'. Indirect costs may be tax deductible items.

The breakeven point

A company's breakeven point is the point at which its sales exactly cover its costs. If it sells more, then it makes a profit. If it sells less, it takes a loss. A graphic presentation of the breakeven point is given in Figure 23.1. The line of total costs is the sum of fixed costs plus variable costs. The line of sales revenue is the product of sales volume times the unit selling price.

The price of the product is set by the company by looking at the production costs of the product, and marking it up. The markup is the percentage increase that is added to the product's production cost to obtain the selling price.

If fixed costs correspond to the total fixed costs for the firm, while price and variable costs are stated per unit cost, the breakeven point in units of product (number of packages to be sold) can be calculated according to the following formula:

$$\textit{breakeven point in units} = \textit{total fixed costs}/(\textit{unit price} - \textit{unit variable cost})$$

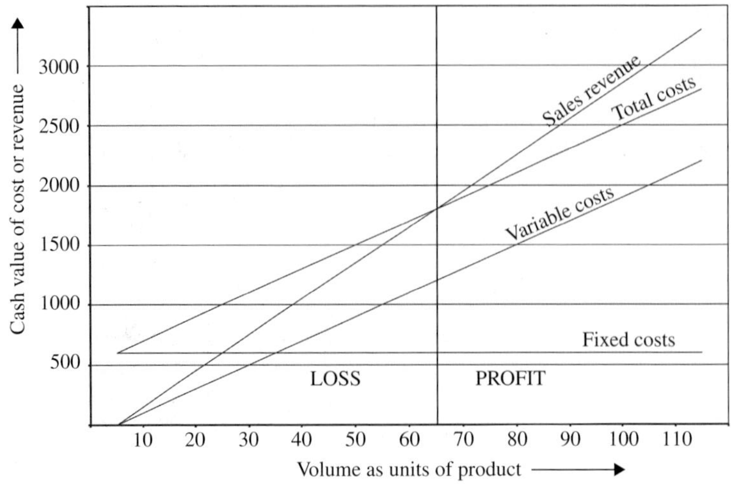

Figure 23.1 A graphical presentation of break-even point (BEP).

The denominator (unit price *minus* unit variable cost) is called the 'contribution margin' and it represents the amount per unit of product sold that the company can contribute to paying its fixed costs. It can also be written that:

$$\textit{breakeven point in units} = \textit{total fixed costs}/\textit{contribution margin}$$

This underlines the importance of the contribution margin. In fact, it defines the difference in slope between the sales revenue and the total costs. The greater this difference, the lower the amount of product needed to achieve the breakeven point.

For the owner of a small business, any decision about pricing of the product determines the relationship between the cost of the product and the selling target.

The difference (contribution margin *minus* fixed costs per unit product) is defined as the 'net operating profit or loss'.

Figure 23.2 illustrates the structure and relationships of the various cost components. The figure represents the piling up of the various costs. Variable costs have been divided into two components: raw materials and other variable costs. The difference between costs and price, which is defined as profit in the report of an actual situation (price refers to actually sold product), is called markup in business planning (price refers to product to be sold).

23.3 Hypotheses for the cost analysis

In the following tables, the hypotheses for cost calculation adopted are presented for each of the three cost centres identified in Section 23.1.1.

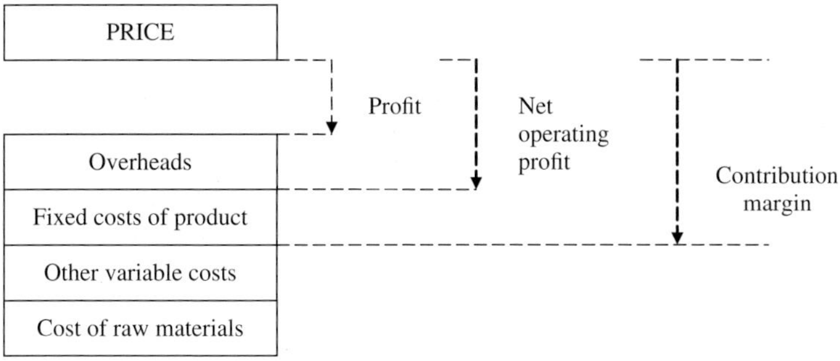

Figure 23.2 The structure and relationships of cost components.

23.3.1 Step A: the olive grove (cultivation and harvesting)

This example concerns small olive groves with a few thousand olive trees, which are most interested in niche production of excellent extra-virgin olive oil. According to our definition, a medium-size olive grove has tens of thousands olive trees and a large olive grove has hundreds of thousand olive trees.

Table A0 shows the basic hypotheses and data for cost calculation. It is worth noting that two harvesting systems have been considered, with a different level of mechanization: case 1: hand-held harvesting machines and nets, and case 2: Trunk shakers and inverted, wrap around umbrella.

Table A0 Step A: structural and operating conditions of olive cultivation and harvesting.

Structural and operating conditions	Hypotheses for cost calculation
The structural characteristics of the olive grove	
Olive grove surface area	6.0 hectares
Value of the land	18 000 euros per hectare
Olive tree density	333 olive trees per hectare
Total number of olive trees (rounded off)	2000
Cost of planting	5.0 euros per tree
Useful olive tree lifetime	40 years
Productive cycle	8 years' starting period, 32 years at full productivity
Production at full productivity (average of last 5 years)	18 kg olives per olive tree
Total yearly production of the grove	18 × 2000 = 36 000 kg per year
Agronomic operations	
Frequency of pruning	Once every two years
Number of pruners	4
Pruners' working time	67 hours per pruner
Pruners' cost	15 euros per hour
Various agronomic treatments (including fertilization, irrigation, etc.)	750 euros per hectare per year

(continued overleaf)

Table A0 (*continued*)

Structural and operating conditions	Hypotheses for cost calculation
	Harvesting
Harvesting period (at optimal olive maturity)	(3–4 weeks)
	Harvesting systems
First harvesting system	Hand-held harvesting machines and nets. Olives are put into plastic crates 25 kg net weight
The harvesting team	6 people: 4 for harvesting with hand-held harvesting machines and 2 for managing nets and crates
Cost and time in the first system	Harvesting is carried out 7 hours a day at a cost of 10 euros per person per hour
Harvesting rate	40 kg olives per hour per person
Harvesting productivity	$4 \times 7 \times 40 = 1120$ kg olives per day, corresponding to 62 trees per day
Second harvesting system	Trunk shakers and inverted, wrap around umbrella. Olives are put into plastic bins with 300 kg capacity
The harvesting team	3 people: 1 is the operator of the trunk shaker, 2 for complementary work and bin operation
Leasing cost of the trunk shaker including the trunk operator	60 euros per hour
Harvesting productivity	300 kg olives per hour, 2100 kg per day, corresponding to 117 trees per day
	Transportation of olives to the olive mill
Transportation of olives to the oil mill	One person with tractor 2 h per day

23.3.2 Step B: The olive mill

The example concerns an olive mill with a capacity of 2000 kg olives per hour. Hypotheses of one and two working shifts are considered. Table B0 shows the basic hypotheses and data for cost calculation.

23.3.3 Step C: oil storage and packaging

Table C0 shows the basic hypotheses and data for calculating oil storage and packaging costs.

23.4 Cost calculation

Data presented in tables A0, B0 and C0 are the inputs for the calculation of the cost of a unit package of extra-virgin olive oil.

Table BO Step B: Structural and operating conditions of oil extraction.

Structural and operating conditions	Hypotheses for cost calculation
The structural characteristics of the olive mill	
Buildings: include the olive reception area, the olive mill, the waste disposal area; the warehouse for oil storage; the storage room for packaging materials and other materials, the administration, locker room for workers; water, electricity and room conditioning services; customers reception, etc.	600 m^2 with a total cost of 500 000 euros, two-thirds of which are for the production needs. The rest is to be ascribed to marketing, promotion, direct selling and administration
Depreciation time of the buildings	20 years
Plant and equipment directly related to the production process	Total cost of 180 000 euros
Depreciation time of plant and equipment	10 years
The operating conditions of the olive mill	
Processing capacity	2000 kg olives per hour
Daily working time	Two shifts of 16 hours with 14 hours effective milling time and 2 hours for plant cleaning and washing
Manpower	3 people per shift
Cost of manpower	15 euros per hour per person
Overall cost for electricity, water and other supplies	13 500 euros per year
Average extraction yield	15 kg oil per 100 kg olives

Table CO Third cost centre: structural and operating conditions of oil storage and packaging.

Structural and operating conditions	Hypotheses for cost calculation
Oil storage	
Storage volumes	It is assumed that the available storage volume should be 80% of the total yearly production, 20% of the oil being sold or taken back by olive producers during the harvesting period.
Storage conditions	Hermetically sealed, stainless steel tanks equipped with an inert gas blanketing system, CIP (cleaning-in-place) system and direct connection with the bottling line
Storage time	Nine months a year with three months available for maintenance and repair
Personnel	20% of the time of an operator, otherwise committed to other duties as, for example, oil packaging, olive mill maintenance, direct selling, and so on.
Cost of tanks	50–60 euros per hectolitre of storage capacity

(continued overleaf)

Table C0 (*continued*)

Structural and operating conditions	Hypotheses for cost calculation
Depreciation time of tanks and other storage-related equipment	15 years
Average cost for analyses and certification	600 euros per certified lot
Average cost for additional material, workers, energy, oil handling, maintenance of equipment and building	10% of the total cost of the plant and equipment
Packaging	
Containers	The example calculation is based on the following hypotheses: – 1/3 of production is packaged in 0.5 litre bottles – 1/3 of production is packaged in 0.75 litre bottles – 1/3 of production is packaged in 5 litre tin-free steel containers
Cost of containers including closure, pouring and tamper-evident devices plus label	– bottles 0.5 litre: 1.15 euros – bottles 0.75 litre: 1.19 euros – 5 litre containers : 1.65 euros
The bottling plant	Semi-automatic with in-line closing and labelling, with a capacity of 350 bottles per hour for both 0.5 and 0.75 litre bottles
Cost of bottling plant	9000 euros
Depreciation time of bottling plant	8 years
The filling plant for metal containers	Semi-manual with closing and labelling out-of-line, with a capacity of 150 (5 litre) containers per hour
Cost of the filling plant	4500 euros
Depreciation time of the filling plant	8 years
Personnel	One person full time for 9 months a year plus others with part-time commitment based on the operating hours of the packaging lines

23.4.1 Calculation of cost items of step A

Table A1 Cost of land.

Land	Data	Value	Years of use	Depreciation
Surface area	hectares	6.0		
cost	euros/ha	18 000.00	40.00	2700.00

Table A2 Cost of planting.

Olive planting	No of plants	Planting cost	Years of use	Depreciation
N/ha	333			
Total	2000			
euros		10 000	40.00	1500

Table A3 Cost of cultivation (various operations).

Surface area	Cost	Cost/year
6.0 ha	750 euros/ha	4500

Table A4 Cost of pruning.

Pruning	Data	Notes
Capability	120 trees/day	Four people
Trees/year	1000	Pruning once every two years
Days required	8.33	
Hours required	67	Days × 8 hours/day
Cost/h	15	
Total cost	4000	Hours × (cost/hour) × 4

Table A5 Cost of harvesting (case 1).

Hand-held machine	Data	Notes
Yield	160 kg/hour	Standard team of four people with hand-held machine plus two people with nets
Production per day	1120 kg	7 hours' work
Total production	36 000 kg	18 kg per plant
Manpower days	32	Total production/production per day
Manpower needed to comply with harvesting time: 21.5 days	8	1.5 team: six people with hand held-machines plus two people with nets
Cost per hour	10	
Manpower cost	13 760	8 hours × 8 persons × days
Machine costs	3000	N. 6 hand-held machine

Table A6 Cost of harvesting (case 2).

Trunk shaker and wrap around	Data	Notes
Lease with operator	60 euros/h	
Yield	300 kg/h	two people and operator
Production per day	2100 kg	7 h work
Total production	36 000 kg	18 kg per tree
Manpower days	18	Total production/production per day
Cost per hour	10	
Manpower cost	2880	8 h × 2 people × days
Lease cost	8640	8 h × lease cost × days
Total cost	11 520	2880 + 8640

Table A7 Cost of transportation to the olive mill: cost of personnel.

Personnel	Data	Note	Data	Notes
Hours	40.00	20 days × 2 h	43.00	21.5 day × 2 h
Cost/hour	10.00	One person	10.00	
Total cost	400.00		430.00	

Table A8 Cost of transportation to the olive mill: cost of vehicle.

Vehicle	Cost	Years of use	Depreciation	Days/year	Cost/day
Tractor	30 000	10.00	3000	200	15

Table A9 Cost of consumables.

Item	Cost/day	Cost/day × 20 days	Cost/day × 21.5 days
Tractor	15.00	300	322.50
Consumables	5.00	100	107.50
Total		400	430.00

From the data in Tables A1–A9, the total costs in Table A10 are calculated for step A in euros per kg of olives.

Table A10 Cost of olives (step A).

Typology	Cost item	Case 1	Case 2	Notes
FC	Land	2700	2700	
	Olive grove planting	1500	1500	
	Cultivation	8496	8496	Sum of cultivation and pruning
	Total fixed costs	**12 696**	**12 696**	
SVC/VC	Machines and vehicles	3430	400	
	Personnel	14 190	11 950	
	Total variable and semi-variable costs	**17 620**	**12 350**	Costs of harvesting
TC	TOTAL COSTS	**30 316**	**25 046**	
	Production, kg olives	36 000	36 000	
CI	Cost/kg	**0.84**	**0.70**	
	Mark up + 30%	1.09	0.90	

Notes: FC: fixed costs; SVC: specific variable costs; VC: variable costs; TC: total costs; CI: cost index.

In comparing case 1 and case 2, the advantage of using the trunk shaker is evident. In fact, the unit cost of olives harvested with the hand-held harvesting machines is about 20% higher. The cost of harvesting accounts for 58% of the total cost of olives in case 1 (hand-held harvesting machines and net) and 49% in case 2 (trunk shakers and inverted umbrella).

We are considering the calculations from the point of view of the milling company, so a 30% markup is the supposed margin required by the olive grower. The markup value obviously depends on the grower's marketing ability and the selling price of local competitors.

23.4.2 Calculations of cost items of step B

Table B1 Working period in days/year.

Working days/year	80
Maintenance days/year	10
Total working days/year	90

Table B2 Manpower hours/day.

Manpower	2 shifts
Hours in production	14
Hours in maintenance	2

Table B3 Production capacity and manpower working hours.

Shift	Total production time (in 80 days)	Working hours in production – 3 people per shift	Working hours in maintenance (in 80 days) and 3 people per shift	Total production, kg – full capability (total hours × 2000 kg olives/hour)
2	80 days × 14 hours net production time = 1120 h	1120 × 3 people = 3360 h	2 shifts × 2 hours × 3 people × 80 days = 480 h	1120 × 2000 = 2 240 000 kg

A full production capability is unreal: 85% of full capability is assumed in the following calculations. In these conditions, the olive milling costs are:

Table B4 Olive milling costs.

Olive mill	Reference value	Unit cost (euros)	Time of use	Cost (euros)	Notes
Olive mill surface area	600 m^2	900	20 years	18 000	+ 9000 structural cost of other company's functions
Plant and machinery	Complete olive mill	180 000	10 years	18 000	
Personnel	three people per shift, two shifts	15 euros/h	3360 hours	50 400	
Water, conditioning and various				8500	+ 4500 structural cost of other company's functions
Electricity	250 kW/h	0.125	952 hours	29 750	85% of capability
Maintenance		15 euros/h	720 h	10 800	480 h + 10 days for the opening and closing of the mill, with 3 workers (10 × 3 × 8)
Production (olives)					
Production capacity per hour, kg/h	2000				
Total production capacity, kg/year	2 240 000		85%	1 904 000	
Machinery working hours	1120		85%	952	
Production (oil)					
Oil production (kg)	285 600				15% yield
Oil production, litre	312 473				

Finally, the milling cost per kilogram of olives and per kilogram of oil are given in Table B5.

Table B5 Table of milling costs.

Typology	Cost item	Costs	Notes
FC	Building	18 000	
	Machinery	18 000	
	Total specific fixed costs	**36 000**	
SVC/VC	Personnel	61 200	Sum of 50 400 working in production and 10 800 working in maintenance
	Consumables	38 250	Electricity and water, room conditioning, etc.
	Total variable and semi-variable costs	**99 450**	
TC	TOTAL COSTS	**135 450**	
	Production, olives (kg)	1 904 000	85% of full capability
CI	Cost/kg olives	**0.071**	
	Production, oil (kg)	285 600	
CI	Cost/kg oil	**0.474**	
	Production, oil (L)	312 473	
CI	Cost/litre oil	**0.433**	

Notes: FC: fixed costs; SVC: specific variable costs; VC: variable costs; TC: total costs; CI: cost index.

23.4.3 Calculation of cost items of step C

Detail of calculation of the cost of oil storage

Table C1 The amount of oil to be stored.

Amount of oil to be stored	80% of total theoretical production. (It allows a 15% reserve of storage volume)	294 092 L

Table C2 Storage tanks.

Tank capacity	10 000 litres	Number of tanks: 29
Cost (euros)	5500 per tank	159 500
Nitrogen blanketing system	350 euros every four tanks	2800
Valves and piping	50	1450
Total cost		163 750

Table C3 Costs per 180 working days/year.

Warehousing	Number	Unit cost	Time of use	Cost	Note
Machinery	29	5647	15 years	10 917	
Personnel	1	15 euros/h	0.20	4320	20% of a person × 180 days × 8 hours
Maintenance				1092	10% of machinery cost
Consumable				1500	
Certification and analyses	2	600 euros/lot		1200	
Total cost				19 028	

Table C4 Table of storage costs.

Typology	Cost item	Costs	Notes
FC	Machinery	10 917	
	Maintenance	1092	
	Total specific fixed costs	**12 008**	
SVC/VC	Personnel	4320	
	consumables	2700	Includes quality and certification
	Total variable and semi-variable costs	**7020**	
TC	TOTAL COSTS	**19 028**	
	Production oil (L)	312 473	the whole oil production must be processed
CI	Cost/litre oil	**0.061**	

Notes: FC: fixed costs; SVC: specific variable costs; VC: variable costs; TC: total costs; CI: cost index.

Detail of calculation of the cost of oil storage

Table C5 Hypothesis on containers.

Hypothesis on containers	1/3 oil in 0.5 L bottles, 1/3 in 0.75 L bottles and 1/3 in 5 L metal containers

Table C6 Hours required for packaging and maintenance.

	0.5 L bottles	0.75 L bottles	5 L containers	Total
Litres	104 158	104 158	104 158	
Number of containers	20 315	138 877	20 832	
Machinery capacity	350	350	150	
Hours required	595	397	139	1131
Maintenance (10%)	60	40	14	113
Total hours	655	436	153	1244
Days equivalent	82	44	15	125

Table C7 Packaging costs.

Packaging	Number	Unit cost	Time of use	Cost
Bottling machinery	1	9000	8 years	1125
Metal container filling machinery	1	4500	8 years	563
Personnel	1131 h	15 euros/h		16 963
Maintenance	113	15 euros/h		1696
Consumables				1500

Table C8 Table of packaging costs.

Typology	Cost item	Case 1 0.5 L bottles	Case 2 0.75 L bottles	Case 3 5 L container	Notes
FC	Machinery	675	450	563	Costs are a function of hours of use
	Total specific fixed costs	675	450	563	
SVC/VC	Personnel	8928	5952	2083	Costs are a function of hours of work
	Maintenance	893	595	208	Costs are a function of hours of work
	Consumables	650	650	300	Costs are a function of hours of work
	Total variable and semivariable costs	10 471	7197	2591	
TC	TOTAL COSTS	11 146	7647	3154	
	Oil production (l)	104 158	104 158	104 158	
CI	Cost/litre oil	0.101	0.069	0.025	
CI	Cost per package	0.050	0.052	0.124	

Notes: FC: fixed costs, SVC: specific variable costs; VC: variable costs; TC: total costs; CI: cost index.

23.5 Total cost

The example calculation was based on the hypothesis of an olive milling company directly operating the olive mill as well as the storage and packaging of the oil, while buying the olives from outside suppliers. The olive milling company is supposed to buy the olives at a price given by the cost of olive production plus the markup chosen by the olive growers (30% in the example). The cost of the three types of containers, including closure, pouring and tamper-evident devices, can be considered as follows:

- 0.5 litre bottle: 1.25 euros
- 0.75 litre bottle: 1.29 euros
- 5 litre metal container: 1.75 euros.

Table I Total cost per litre and per packaging unit.

Cost items	0.5 L bottle	0.75 L bottle	5 L metal container
Cost of olives per litre of oil	5.52	5.52	5.52
Cost of milling per litre of oil	0.43	0.43	0.43
Cost of storage per litre of oil	0.061	0.061	0.061
Cost of packaging per litre of oil	0.101	0.069	0.025
Total cost of production per litre of oil	6.11	6.08	6.04
Cost of containers	1.25	1.29	1.75
Cost per unit package	4.31	5.85	31.93

Table II Cost composition of packaging units.

	0.5 L bottle		**0.75 L bottle**		**5 L container**	
Cost of olives	2.76	*64.1%*	4.14	*70.7%*	27.59	*86.4%*
Cost of the olive mill	022	*5.0%*	0.33	*5.6%*	2.17	*6.8%*
Cost of storage	0.03	*0.7%*	0.05	*0.8%*	0.30	*1.0%*
Cost of packaging	0.050	*1.2%*	0.052	*0.9%*	0.12	*0.4%*
Cost of container	1.25	*29.0%*	1.29	*22.0%*	1.75	*5.5%*
Total	4.31	*100*	5.85	*100*	31.93	*100*

Table I shows the results of the final calculations.

From the absolute values reported in Table I, the percentage cost composition has been calculated in Table II for an easier comparison of differences among the three packaging solutions.

Conclusions are very clear and simple to draw. The final cost of the (oil and package) unit is determined by the cost of the olives in a proportion of 64, 71 and 86% in the 0.5 L bottle, 0.75 L bottle and 5 L metal container, respectively. More than 50% of the olive cost is determined by the harvesting cost (Table A10). The second item of cost is the package cost, which accounts for 29%, 22% and 5.5%, in the three containers respectively. Despite these differences, but considering the discussion of Table 16.1 and point 16.1.1. in Chapter 16 about the use of containers for oil in families and restaurants, 0.5 L or 0.75 L bottles may still be considered as more interesting for high-quality extra-virgin olive oil.

Further reading

Biondi, A., Borghetti, N., Castellanza, V. *et al.* (2008) *L'analisi del settore dell'olio di oliva in Italia*, www.iulm.it/wps/wcm/connect/fda509004a9b69bcb-133b7627c63c64a/Economia%20dei%20settori%201.pdf?MOD=AJPERES (accessed 11 October 2013).

Commissione Europea, Direzione Generale dell'Agricoltura e dello Sviluppo Rurale (2012) *Analisi economica dell settore oleicolo*, Brussels, www.docstoc.com/docs/150818533/economic-analysis_it (accessed 11 October 2013).

Gabrielli, F., Gucci, R., Polidori, R. *et al.* (2008) Riduzione dei costi in olivicoltura – soluzioni tecniche economiche. *L'Informatore Agrario* **37**, 27–45.

Vieri, M., Rimediotti, M. and Daou, M. (2007) A pettine o pneumatico, test su tre modelli di agevolatori. *Olivo e Olio* **2**, 18–20.

Zerilli, V. (2013) *Frantoi aziendali a confronto – pregi e difetti*, www.teatronaturale.it/strettamente-tecnico/l-arca-olearia/3685-frantoi-aziendali-a-confronto-pregi-e-difetti-caratteristiche-e-qualche-riflessione-per-compiere-una-scelta-consapevole-che-sia-davvero-a-misura-da.htm (accessed 11 October 2013).

24

The culinary uses of extra-virgin olive oil

Alan Tardi

University of Gastronomic Sciences, Pollenzo, Italy

Abstract

The cultivation of the olive tree (*Olea europaea*) and the extraction of olive oil dates back to the beginning of Western civilization. Initially, its primary use as a source of light and body ointment developed an important mystical/ritual significance. During the time of the ancient Romans, olive oil, while retaining its ritual significance, became increasingly important as a foodstuff, forming one of the fundamental elements of the Mediterranean diet. In modern times, however, with more sophisticated technology and deeper understanding of the perishability of olive oil, an entirely new level of production quality may be attained resulting in a vastly superior product with increased sensory characteristics and cultivar-based diversity, which also opens up an entirely new world of culinary applications and considerations.

24.1 A brief history of the olive

The olive is one of the earliest plants to be cultivated by man and one of the most significant. Though the actual place of origin of the wild olive tree (*Olea oleaster*) is not certain, recent research pinpoints the northern Levant area around the border of present-day Syria and Turkey as the site of the olive's earliest domestication and cultivation, which took place prior to 6000 BCE (Besnard *et al.* 2013). The domesticated olive tree (*Olea europaea*) has larger, less bitter olives with a greater amount of oil than the wild one. Olive pits found in archaeological sites date back approximately 8000 years and the earliest evidence of olive oil production, in the form of mortars and stone presses, was found in present-day Israel, dating to around 4500 BCE. Three genetic 'hotspots' of the cultivated olive tree have been identified: the Near East (including Cyprus), islands in the Aegean Sea and the Strait of Gibraltar.

The Extra-Virgin Olive Oil Handbook, First Edition. Edited by Claudio Peri.

From there, the domesticated olive began to move westward with the expansion of human civilization, mutating and diversifying in the process.

The domesticated olive was important to the societies that cultivated it from the very beginning but it took on even greater significance once it arrived in Greece, both on the mainland and on the islands. Olive trees were planted widely throughout the island of Crete and commerce in oil likely helped fuel the development of the Minoan civilization which dominated the Mediterranean from around 2700–1500 BCE. The Minoans developed techniques of brining to preserve olives, which made the fruit easier to export as well as more palatable to eat.

According to Greek mythology, in a competition held by Zeus, Athena (his daughter) was made patroness of the city of Athens for providing the people the gift of the cultivated olive. The olive became an increasingly important element of the developing Greek society and was subsequently used as a colonizing tool as Greek civilization expanded throughout the Mediterranean area. It is probable that the olive was already present on the Italian peninsula when the Greeks arrived there, as they began establishing outposts of *Magna Grecia* in southern Italy and Sicily (as well as southern France) in the early eighth century BCE, they brought with them more sophisticated methods of olive cultivation and oil extraction, as well as an enhanced appreciation of both the practical and symbolic significance of the fruit and its various uses.

The cultivated olive sank its roots deeply in the propitious environment of the central Mediterranean area and spread throughout the Lower Peninsula. The Greek predilection for the olive was adopted by the ancient Romans (along with most of the other key aspects of Greek culture), becoming in turn a colonizing tool of their expanding empire and an increasingly significant economic commodity, as well as an important part of Ancient Roman cuisine.

As cultivation of the olive tree spread to numerous different climates and terrains, the plant continued to mutate and diversify into a multitude of distinct cultivars.

24.1.1 Significance of the olive in ancient times

Unlike other fruits and vegetables that attained a place of importance in both the cuisine and the culture/mythology of a given human society, the olive was initially prized for qualities other than its taste and nutritional value. The olive (especially the wild *Olea oleaster*) contains an extremely high amount of oil. One of the earliest and most important uses of olive oil was as fuel for lamps. The property of preserving the flame, thus harnessing and humanizing the energy of fire to produce light, undoubtedly conferred an elite status to this plant. In ancient Judaism, oil obtained by using only the first drop from a squeezed olive consecrated by the priests, was used to light the lanterns in the Temple.

Thanks to its natural beneficial properties, another early use of olive oil was as an ointment for the body and hair. Olive oil was used as a basis for perfume worn by the upper classes in ancient Egypt and was also used in the preparation of mummies (due to environmental conditions, Egypt was not able to grow olive trees and imported much of its oil from Palestine and Crete). In ancient Greece, athletes (who enjoyed an almost religious status) were ritually rubbed with olive oil before practice and competition, and winners of Olympic contests were crowned with a wreath made of wild olive branches.

Practical applications with mystical implications were combined in the widespread use of olive oil for anointing, which represents the introduction or existence of a divine influence in a person.

Olive oil has long played an important role in the three principal religions of the Western world. In Judaism, the festival of Hanukah celebrates the miracle of a small quantity of oil sufficing to keep the Temple lanterns lit for eight days until more could be obtained. Anointing with oil stems from the Old Testament and indicates that someone or something is being set apart for a sacred task or duty. In Christian rituals, priests are anointed with chrism oil during ordination, children (and anyone else entering the faith) are anointed at baptism, and the dying are anointed with blessed oil. In fact, the name 'Christ' literally means the 'the anointed one' and both the Old and New Testaments are filled with references to the olive and olive oil. They are significant for Muslims as well: 'God is the light of the Heavens and the Earth. His light is like a lantern inside which there is a torch; the torch is in a glass bulb which is like a bright planet lit by a blessed olive tree, neither Eastern nor Western, its oil almost glows, even without fire touching it, light upon light' (Al-Hilali and Khan 2013).

In ancient Middle Eastern societies, culinary uses of the olive were trumped by practical (fuel, ointment) and religious (anointing, mythology) ones but this began to change in ancient Greece, where the more favourable environment combined with more sophisticated methods of olive cultivation and transformation led to products of a higher quality and more pleasing taste. At the same time, an appreciation of cuisine and gastronomy took a big leap forward. Rather than being reserved exclusively for a small group of nobles, in ancient Greek society, festive meals were enjoyed by a greater number of people than ever before and a distinctly Mediterranean culinary culture began to take shape.

24.1.2 Culinary applications of olive oil in the Mediterranean Basin, from ancient times to present

In ancient Greece, olive oil was used as a cooking medium. Mixed with honey, salt and vinegar, it was used as a dressing for vegetables. Salt-cured olives were also regularly consumed.

As the cultivation of the olive spread throughout the expanding Roman Empire – first Italy, then France, Spain and the entire Mediterranean Basin – the culinary use of olive oil began to expand as well. Both *De Agri Cultura* by Marcus Porcius Cato (234–149 BC) and *Natural History* by Pliny the Elder (circa AD 77–79) contain detailed information about the cultivation of olive trees and the making and uses of olive oil. *De Re Rustica*, the 12-volume compendium of farming written by Lucius Junius Moderatus Columella (AD 4–70 circa), a Roman born in Spain who owned several farms in Italy, includes extensive information about olive tree cultivation as well as a number of recipes, many of which include olive oil (Faas 2003). *De Re Coquinaria*, compiled sometime around the late fourth to the early fifth centuries AD and attributed to Apicius, is considered to be the earliest cookbook in existence. While here, as in Columella, olive oil is a frequent ingredient that pops up in many of the recipes, it is used principally in one of two ways: in the process of cooking, to add fat to and help emulsify ingredients, or as a dressing, usually mixed with other ingredients like honey, vinegar and pulverized herbs or spices, and poured over cold (either cooked or raw) vegetables. It is not indicated for use as a condiment at table. The chief condiment of ancient Roman cookery was the ubiquitous Garum, a pungent liquid made from anchovies fermented in salt. A modern version of the ancient Roman Garum called 'colatura di alici' is still made and used to this day in and around the town of Cetara in the region of Campania, Italy.

24.2 Old versus new: expanded culinary possibilities offered by excellent extra-virgin olive oil

The olive oil has played a significant role in health, nutrition and religion/mythology in the Middle East and Western Europe from the dawn of human civilization. As the cultivated olive tree spread throughout the Western Mediterranean, the importance of its oil as a fuel and religious/mythological symbol declined and its use as a foodstuff increased. In the ancient Roman era, the olive – both fruit and oil – became increasingly important as an edible food. Olive oil occupied a key role in ancient Roman cooking and its culinary use expanded along with the Roman Empire, laying the foundation for what we now refer to as the Mediterranean diet.

As we can see from Pliny's *Naturalis Historia*, by the first century the Romans were already aware of numerous different varieties of olive and their particular characteristics, and understood that some of them lent themselves to different uses. They could appreciate the difference between a good olive oil and a bad one and, moreover, were quite cognizant of many of the basic requirements for producing a good one. Pliny clearly states that harvesting the olive at just the right moment (that is, when the olives are just beginning to turn black) is essential to produce high-quality oil. He also indicates that olives should be hand-picked from the tree rather than waiting for them to drop or beating them off the tree and that the olives should be milled immediately after harvest rather than allowing them to sit around and 'sweat'. He also says that the first oil to come off the press is the best and that the oil should be consumed as fresh as possible, and in any case within one year. Pliny does not, however, discuss how to store olive oil or how to use it.

With the decline and fall of the Roman Empire, much of the knowledge about olive oil (as well as knowledge about many other things) was lost. But, unlike most other scientific and intellectual pursuits which began to make a comeback during the Renaissance and Enlightenment periods, insight into the production and utilization of olive oil remained largely stuck in the Dark Ages. For the most part, despite tremendous advancements in technology, commerce, communication and gastronomy, it remains there to this day. People throughout the world are using more olive oil than ever before but the vast majority of them have little idea what it really is or how to tell a good one from a bad one, much less how to use it. Some of this has to do with culture. A certain amount of ignorance about the product can be expected on the part of American or British consumers who have never seen an olive tree and for whom olive oil is a fairly recently addition to their diet. But this explanation has its limitations. A scientist who specializes in olive oil and comes from an olive-producing part of Italy confessed that, as a child, a large barrel of oil was kept in the basement of his family home with a loosely fitting lid on top. 'We would go down every day to fill up our small bottles. As the quantity diminished, the oil became increasingly rancid and thoroughly disgusting. I know now it was completely spoiled but back then we all thought that was just how olive oil was.' Surely, even in olive-producing regions, this was not – and is not – an unusual scenario.

In fact, there is just about as much confusion about olive oil in the Old World as there is in the New, in the latter due mostly to a lack of familiarity and in the former perhaps to too much. Some of this confusion may also be attributed to unscrupulous producers or bottlers attempting to make an inferior product from inexpensive ingredients (some of which are not even derived from the olive), which they can sell at a high price, and even merchants who don't know how to properly store olive oil and so sell a spoiled one or, whether intentionally or not, sell olive oil way past its prime. Existing official regulations for the designation of extra-virgin olive oil (and other grades) are minimally sufficient, but the application of these regulations and enforcement of controls is often difficult or altogether impossible to impose. Even when applied, the existing regulations cannot control the provenance of the oil much less all the numerous other factors that critically impact how it was made. Labels on most bottles of extra-virgin olive oil are practically meaningless and in most cases consumers know little or nothing about what is actually inside the bottle.

It is safe to say that, for one reason or another, a tremendous amount of the extra-virgin olive oil being marketed and consumed today is inferior if not downright bad.

In the midst of this dismal situation, a growing number of producers are stepping up to the plate and making a voluntary commitment to produce a truly excellent extra-virgin olive oil; distributors and retailers are following their lead by paying attention to delivering this high-quality (yet extremely perishable) product to consumers, and educators, journalists and chefs are spreading the word. An olive oil producer's commitment to quality combined with the resources of modern technology (minimal time between harvesting and milling, sanitation in the mill, bottling under nitrogen, rapid temperature-controlled transportation, etc.) permits the production and commercialization of olive oil at a consistently higher level of quality than has ever been possible before.

And this is where things get really exciting.

If the practices of cultivation and production outlined in this handbook are followed, an entirely new and very colourful world of extra-virgin olive oil opens up. One finds a tremendous diversity in these excellent products and a notable difference in olive oils made from different cultivars and from different regions. These oils offer a whole new level of complexity and flavour; what was once an anonymous undifferentiated commodity becomes a multitude of different characters, each with its own distinct personality and sensory profile. Spicy, fruity, robust, delicate, floral, and herbaceous are just some of the many adjectives used to describe how they taste and, moreover, each oil has its own unique sensory impact on the palate: some feel rich and heavy, almost creamy, while others feel light and fresh; some are immediately aggressive on the palate while others seem unassuming at first with a slow burn that gradually emerges at the back of the throat (Chapter 4).

The language used to describe this intense sensory experience and wider diversity of flavour profiles offered by excellent extra-virgin olive oils, begins to resemble the kind of language that is generally associated with wine, though the descriptors are usually different and olive oil is rarely tasted on its own.

This vast new world of excellent extra-virgin olive oils brings with it many new possibilities of how to most appropriately use them. But before exploring these possibilities, it would be useful to look at how olive oil has traditionally been used in cooking.

The culinary use of olive oil takes four basic forms: as a preservative, as a cooking medium, as an ingredient and as a condiment. (Two or more of these forms may be used in the preparation of a single dish.)

24.2.1 In the pantry: olive oil as a preservation agent

The use of olive oil as a medium for preserving food probably began as an extension of olive oil as a cosmetic balm for the skin. Just as covering the body with olive oil protects and enriches the skin, submerging food items in oil creates a hermetic environment, which helps prevent microbial and oxidative spoilage while at the same time contributing to its flavour. The natural antioxidants in extra-virgin olive oil make it especially suited to this use. Of course, in order to be effective, the food products to be preserved must be free of bacteria to begin with and must be completely submerged in the oil. Moreover, the food / oil must be placed in a sterile, close-fitting container with an airtight lid and stored in a cool dark place.

Once a container is opened, it is generally best to use the contents as quickly as possible. Besides acting as a preservation agent, the oil affects the flavour of the food and the flavour of the oil is, in turn, affected by the flavour of the food. In the process of preserving food in oil there is a natural co-mingling of the flavour components. Some of the oil could be used in the subsequent preparation of the dish, however the inherent flavour properties of the olive oil have been somewhat compromised (assuming it has not gone rancid by over-exposure to air, light or drastic changes of temperature).

It should be noted that the primary function of the oil here is not so much flavour enhancement as preservation. For use as a preservation agent, a refined olive oil or an olive oil composed of refined and virgin olive oils (Chapter 1) may be the best choice. Alternatively, for some food products (sun-dried tomatoes, for example), an excellent extra-virgin olive oil with a mild sensory profile, which complements the sensory profile of the product to be preserved, can offer a special and good alternative.

Example 1: In preparing a typical Salade Niçoise with anchovies preserved in olive oil, a very light, practically neutral olive oil should be used to preserve the anchovies. However, in assembling the finished dish, which involves mixing the anchovies with other ingredients such as string beans, tomatoes, olives, hard-boiled eggs, cucumber, onions, lettuce, the very best and freshest extra-virgin olive oil should be used.

Example 2: Vitello Tonnato, a traditional dish of the Piedmont region of northwestern Italy, consists of cooked (either boiled or roasted) thinly sliced veal topped with a mayonnaise flavoured with tuna preserved in oil and capers. In this dish the delicately flavoured meat acts as a foil for the full-flavored sauce. In this case, the excess oil of the preserved tuna can be used to prepare the mayonnaise (along with additional light-flavoured oil, if necessary), in which the oil functions both as an emulsifier and an aid to really get the flavour of the tuna into the sauce. Using a top-quality extra-virgin olive oil with pronounced flavour profile could overwhelm the subtler flavours of this dish and may conflict with the flavour of the preserved tuna.

24.2.2 In the kitchen: olive oil as cooking medium

According to palaeontologist David Wrangham, cooking – that is, the transformation of raw ingredients into cooked food through heat – played a key role in the development of the human species (Wrangham 2009). Using olive oil as a cooking medium is understandably one of the earliest culinary uses of olive oil. As olive oil was initially used as a fuel, the distance from flame to frying pan is quite short.

As a cooking medium, olive oil can act as a lubricant to prevent foods from sticking to the cooking surface when grilling or pan frying. In marinades, besides helping to lubricate foods for cooking, olive oil adds a flavour element to the food to be cooked and transmits other flavouring agents such as seasonings, spices or herbs.

In pan-frying or sautéing, olive oil acts as a mean for of transferring heat from the heat source to the food. In sautéing, besides preventing the food from sticking and enhancing its flavour, searing the food in hot olive oil helps create a golden-brown crust around it. This enhances the visual appeal of the cooked food and makes it tastier; the caramelized surface has an appealing flavour / texture in and of itself and also helps seal moisture and flavour inside the food. A similar thing happens with high-temperature roasting, either in an oven or on a spit.

Besides sautéing and roasting, meat is typically seared in olive oil before braising. Here too, in addition to creating a more appealing appearance, the surface crust helps the meat absorb the braising liquid more slowly, resulting in the ideal texture of the finished product while also keeping more flavour in the meat itself.

In addition to pan-frying (or Wok-frying, in Asian cooking), many cultures have developed a cooking technique called deep frying in which the food items are completely submerged in bubbling hot oil. Food to be deep fried is typically dredged in flour or dipped in a wet batter (*pastella*, in Italian or *pâte a frire* in French). This creates a thick, uniform crispy crust around the food; when properly done, the food inside is moist and flavourful both inside and out. For good results, the oil must be extremely hot (190 °C or 375 °F) so that the food cooks and a thick crust forms as quickly as possible. If the oil is not hot enough, the finished product will not have the desired crispiness and will be greasy.

Sautéing is best done at a very high temperature. For such types of cooking, a refined olive oil or a vegetable oil are often the best options. (In Asian Wok frying, refined soya or sesame oil is generally used.) Another excellent possibility for the high-temperature sautéing of certain types of ingredients is a high-quality extra-virgin olive oil that has been well-filtered to avoid the presence of suspended material and excess water.

In braising and roasting, the oil affects the flavour of the final dish and the formation of the crust, however the particular characteristics of a fine extra-virgin olive oil are, for the most part, lost in the cooking process. For this reason, either a light flavoured extra-virgin olive oil or an olive oil composed of refined and virgin olive oils are the best choices.

Deep-frying requires a large amount of oil that can be heated to a very high temperature and maintained at a constant temperature for an extended period of time. As the food is completely immersed in the oil, the flavour should be somewhat neutral so as not to completely overwhelm the flavour of the food. Many restaurants use a mechanical deep fryer with large baskets to hold the food under the oil and a controlled temperature gauge, while in homes and small restaurants deep-frying may be done in a pot or pan on the stovetop. In the first instance, a refined olive oil or refined vegetable oil is best; the flavour is neutral and it can take the high temperature without burning. Note: the oil in deep-frying does not contribute flavour to the food but rather acts as a cooking medium (much like water in boiling) to transmit heat and form the distinctive crispy crust. It goes without saying that oil used for deep-frying can only be used for a limited period (the actual amount depends on how heavy the use is) before it must be discarded, the fryer tub cleaned and refilled with fresh oil.

In this quick survey of olive oil as a cooking medium, we have seen that the use of an excellent extra-virgin olive oil for some of the most common cooking methods – roasting, braising, grilling – is something of an unnecessary excess, while for others – sautéing, deep-frying – may even have negative results. A good and suitable alternative to the use of vegetable and seed oils, however, is refined olive oil, either used on its own or in combination with mild, extra-virgin olive oil.

There is, however, a cooking method that would make perfect sense, adapted for these new excellent extra-virgin olive oils. Poaching is a technique of cooking food by submerging it is a gently simmering shallow pan of water or stock. Poaching is much gentler than boiling which is usually done in a large amount of rapidly boiling water. Eggs are typically poached in simmering water with a little vinegar added to help them coagulate. In French cuisine, fish (usually filleted) is poached in a delicate fish stock with white wine, lemon and bay leaves (*nage*), and there is a technique in coastal areas of Italy of cooking fish in *acqua pazza* ('crazy water') of fish stock, white wine, tomatoes, garlic and herbs. In this case the cooking broth becomes a central part of the finished dish.

Adapting the technique of poaching to these new excellent olive oils, a fish filet could be slow-poached by being covered with extra-virgin olive oil with a few sprigs of fresh thyme, coarse salt, cracked pepper, lemon zest and sliced garlic, and heated over a low flame to a temperature of 71–85 °C (160–185 °F) until cooked through. The olive oil does not reach an exceedingly high temperature, so its special qualities are not compromised, but rather fully retained, its flavour melding with the aromatic ingredients and the fish, which is slowly cooked to perfect doneness.

The poaching oil is poured over the cooked fish (accompanied perhaps by boiled sliced potatoes) in a shallow bowl, thus becoming a central flavour element in the finished dish as well the medium for cooking it.

24.2.3 In the kitchen: use of extra-virgin olive oil as an integral ingredient in the preparation of a dish

As we have seen in Apicius, olive oil has long been used as a component in cooking, baking and the preparation of sauces. Besides flavour, olive oil has chemical characteristics that may make it a useful ingredient in a recipe. Let's take a look at some of these considerations.

In dressings, such as those typically used for salad or cold vegetables, the oil is not cooked, allowing it to retain its flavour. Many dressings call for oil to be mixed with other ingredients such as vinegar, lemon, salt, pepper, fresh/dried herbs or even salted anchovies. In such cases it makes sense to use an excellent extra-virgin oil, the flavour of which will remain an important and noticeable factor in the finished dish. In other cases, however, strong-flavoured ingredients in a dressing like Dijon or coarse mustard, olive paste or spicy peppers may compete with or completely overwhelm the flavour of the oil. Sometimes it is the other way around.

Olive oil is used in many sauces. The classic Ligurian pesto sauce is a room-temperature mixture of olive oil blended with fresh basil leaves, garlic and pine nuts mixed with warm pasta, potatoes and string beans. In this sauce, the olive oil acts as both binder and foil for the other ingredients, all of which must be in perfect balance in order for it to be successful. The most important of these ingredients is the fresh basil with its intense flavour and aroma. An olive oil with too much flavour could easily overwhelm the basil, throwing the final dish out of balance. In this instance, a fresh, light extra-virgin olive oil would be preferable.

Another sauce employing olive oil is the *bagna cauda* ('warm bath') in which sliced garlic and anchovies are cooked slowly in olive oil over a low flame until the garlic completely breaks down into something resembling a puree into which raw and boiled vegetables are dipped. The garlic, due to the long cooking time, does not have the strength or pungency of raw garlic but, instead, is sweet and creamy. Once again, the oil here is a foil for this sweet garlic. It must not stand out on its own or compete for attention but rather remain in the background letting the sweetness of the garlic and saltiness of the anchovies come through.

Oil is often used as an emulsifier in cooking, that is, it acts as a binder in which other ingredients (usually liquid or protein) are suspended, producing an un-cooked sauce with a thicker, creamier consistency. One such example of this is mayonnaise, in which oil is slowly whisked into egg yolks with a touch of vinegar or lemon juice. In mayonnaise, the primary function of the oil is as an emulsifier, not as a flavour component. While a discreet addition of a little bit of full-flavoured extra-virgin oil can add an extra touch to a mayonnaise, the exclusive use of a full-flavoured extra-virgin olive oil can completely take over the flavour of the sauce and whatever it is going to accompany. Therefore, a light-flavoured extra-virgin olive oil or even a neutral refined olive oil (with an optional teaspoon of a full-flavored extra-virgin) are best in making mayonnaise. Another excellent option would be to use a neutral oil to prepare the mayonnaise, but drizzle a full-flavored extra-virgin olive oil over the finished dish.

Vinaigrette is a classic dressing for salads composed of three parts olive oil slowly added to one part lemon juice or vinegar with a pinch of salt while vigorously mixing. If done properly (some cooks now use a food processor or blender) the oil will be suspended in the liquid to create a homogenous consistency. In cases like this, where the nature of the olive oil is not significantly changed or compromised and its inherent flavour properties are retained, it makes sense to use the very best and most flavourful excellent extra-virgin olive oil available.

There are many other recipes where olive oil plays a central role in the flavour profile of the dish. In French cuisine, for example there is a very simple rustic sauce called *jus tranché* typically used for roasted or spit-roasted meat, usually game or fowl, in which the drippings or roasting juices are collected at the end of the cooking and cut with a good amount of olive oil to create a 'broken' (i.e. not emulsified) but extremely tasty sauce. In such a case, an excellent extra-virgin olive oil would be the best choice.

In baking, too, dishes are turning up which highlight excellent extra-virgin olive oil. These include olive oil cake (a kind of dense, moist pound cake), olive oil ice cream, and olive oil biscotti. In such recipes, both the richness and the enhanced flavour of an excellent extra-virgin olive oil contribute to the success of the final dish. In order to really highlight the flavour of the oil, a small amount could be drizzled over ice cream or cake. In baked goods such as breadsticks, many bakers have begun to substitute olive oil for lard, which certainly makes the breadsticks lighter and healthier. The special flavours of the oil, however, usually seem to get lost in the baking process therefore in these cases it would be best to use a good refined olive oil.

24.3 Excellent extra-virgin olive oil as a condiment at the table and in the kitchen

Though initially, non-culinary applications of olive oil were the most significant, with the expansion of the Roman Empire and the formation of the Mediterranean diet, olive oil became much more important as a food product and was used in a number of different ways, primarily as a cooking medium and as a component in recipes.

Over time olive oil also began to be used as a condiment, that is, as an addition to a finished dish of food to add flavour and shine. The traditional cuisine of Tuscany offers some good examples of this: *Ribollita* is a thick twice-cooked vegetable and bread soup, which calls for a drizzle of extra-virgin olive oil on top, as does *pasta e fagioli* (pasta and bean soup) and *fagioli al fiasco*, white beans cooked in a flask. *Fettunta* (literally 'greasy slice') is toasted bread rubbed with a garlic clove, drizzled or brushed with extra-virgin olive oil and sprinkled with sea salt. There's the classic *bistecca alla Fiorentina*, a thick steak that is cooked rare, sliced, and liberally drizzled with extra-virgin olive oil, and *pinzimonio*, in which assorted raw vegetables are cut into bite-sized pieces and dipped into a bowl of extra-virgin olive oil with salt and pepper.

These are just a few examples from one particular olive oil-producing region, but they have a number of important things in common: In these dishes the oil is not being used as a cooking medium or as a functional ingredient in the preparation of the dish, but rather as a flavour enhancement. The oil is not heated or blended with other ingredients, but rather added in its unadulterated state just before the dish is to be consumed.

Typically, the very best extra-virgin olive oil is used in these dishes, especially the fresh, green, just-pressed oil after the harvest. In such instances, the consumption of these dishes with the new oil becomes a sort of seasonal ritual. Assuming the olives were picked at the right time and milled shortly after the harvest, the fresh olive oil has not yet run the risk of spoiling and is at its best.

When olive oil is used as a condiment you can see it, you can smell it and you can taste it; the inherent flavours of the oil combine with but are not obscured by the other flavours of the dish, and the high quality of the oil and its unique sensory profile come to the forefront, making a huge contribution to the finished product. It is as condiments that these excellent extra-virgin olive oils really shine.

As stated previously, when procedures to insure the highest standard of quality are followed, the result is not only an extra-high quality olive oil but one that has a much more individually distinctive character expressing the particular cultivar (or cultivars) used as well as the skill and know-how of its producer. While there is tremendous complexity and diversity in this level of olive oil, certain basic general categories can be established: Robust or Full-Bodied; Delicate, Soft or Mellow; Spicy; Fruity; Herbaceous; Astringent or Sharp; Green.

Drizzling an excellent extra-virgin olive oil over food creates a dynamic relationship between condiment and food, and this relationship changes with the nature of the finished dish. In more complicated dishes the food has its own balance and complexity, and layers of flavour and textural differentiation; the dish is clearly the centre stage and the olive oil yet one final supporting character, a last bit of ornamentation that can nonetheless add an entirely new dimension to a dish. In more simple dishes (e.g. *pinzimonio* or the chicken breast, pasta or fish in salt crust), the olive oil is a leading character sharing – perhaps even dominating – centre stage with the central element actually acting as a canvas for the olive oil.

This activity of pairing an appropriate extra-virgin olive oil to a dish of food resembles to a certain extent that of pairing wine, with the important difference that whereas wine is served side-by-side with food and the matching takes place in alternating tastes of food and sips of wine, olive oil is poured directly on the food and the two are ingested together, creating a more intimate blending of flavours. Moreover, the considerations when pairing olive oil and food are considerably different from those that guide food and wine pairing. Both, however, have at least one fundamental thing in common: in a successful pairing, the food and the wine / olive oil should complement one another without losing their individual integrity or, put another way, both food and wine / oil should taste better together than they would on their own.

As with wine, there are different strategies that can be taken to arrive at a favourable pairing. For example, one could pair a big bold olive oil with a full-flavoured heavy dish in order to stand up to the food or a light, delicate, aromatic oil to undercut the richness of the food. But, unlike wine, with olive oil it is possible to have a number of oils (two or three would be ideal) available to 'play' with. Exploring and tasting the way different extra-virgin olive oils interact with the same dish is one of the exciting new activities offered by this new world of excellent olive oil. Given the rich diversity of extra-virgin olive oil, the flavour complexity of many foods and the subjective nature of taste, the possibilities are truly limitless and exploring them is fun, interesting and delicious. However, some basic observations about pairing extra-virgin olive oils with food may be made. Some examples are given in Table 24.1.

24.4 Putting excellent extra-virgin olive oils to work

The activity of selecting an appropriate extra-virgin olive oil is almost as significant a part of the culinary experience as eating it. Chefs may select a particular oil to complement the flavour profile of a particular dish much in the way they would select a certain herb or spice, and the oil may be added to the dish either by cooks before it leaves the kitchen or by a waiter or maitre d'hôtel at the table. If it is added at the table, the server can explain where the oil comes from and why the chef chose it for this dish, adding an enjoyable educational dimension to the dining experience.

Another possibility might be for a member of the service staff to function as a sort of olive oil sommelier, providing guests with a menu of available excellent extra-virgin olive oils along with descriptions (either written and / or verbal) and suggestions of what might go particularly well with the food ordered. In this scenario, the diners at a table might select a few olive oils with different flavour profiles from different places and cultivars to accompany their meal. Serving portions would be small – perhaps 2 fluid ounces as a unit – and priced accordingly. In this way, guests could sample a number of different extraordinary olive oils with a variety of dishes to see which combination gives them the most pleasure. This would offer diners an entirely new sensory experience as well as a heightened awareness of what olive oil is and how it can add to the enjoyment of eating.

In the restaurant, offering this type of service would distinguish the establishment and increase customer loyalty. It would also offer an additional revenue centre.

Table 24.1 Some examples of extra-virgin olive oil–food pairing.

Dish[a]	Type of olive oil[b]	Pairing rationale
Sole meuniére	A full-bodied, fruit-driven excellent extra-virgin olive oil with minimum astringency; tart green apple, green pepper, fresh grass, slight nuttiness.	A classic dish of French cooking in which a skinless sole is floured, sautéed until golden-brown and finished with a sauce of brown butter, capers and parsley. Fresh lemon is typically squeezed on top. Here the fish could be sautéed in olive oil (a refined olive oil or a light-flavoured extra-virgin olive oil would be fine). Instead of the butter sauce, the fish would be drizzled with the selected extra-virgin oil and a squeeze of fresh lemon (capers and chopped parsley are optional). The full-body of the oil would stand up to the richness and firmness of the caramelized filet. The low astringency of the oil would not compete with the sweetness and delicacy of the fish while the green apple/vegetal/herbaceous flavours would articulate it. Tart acidity would complement that of the lemon and the nuttiness would take the place of the brown butter.
Red snapper* baked in salt crust (*or any other white-fleshed fish)	A light, delicate, elegant, excellent extra-virgin olive oil with floral perfume and subtle flavours of exotic fruits and wild herbs. A robust, full-bodied but mellow excellent extra-virgin olive oil with minimal bitterness and pronounced spiciness.	In this dish a whole fish is baked under a crust of coarse sea salt, producing an effect similar to other techniques such as baking in clay or burying food over a fire in a pit in the earth. Encasing and baking the fish in a tight-fitting enclosure is a form of steaming/baking which keeps all the flavour, moisture and aromatics in the fish. Using coarse sea-salt for the baking element is like returning the salt-water fish to its natural environment. Usually this cavity is filled with fresh herbs such as rosemary and thyme. The finished dish is served with fresh lemon for acidity. The delicate sweet flesh of the fish here acts as a foil for the intense, complex flavours of the oil. The markedly different characters of the two oils function equally well to bring out the subtle yet rich flavours of the fish while showcasing the particular qualities of the oil.
Seared sea scallops	Sweet, fruity, excellent extra-virgin olive oil with fresh citrus flavours; full-bodied and buttery.	The sweet fruitiness and buttery texture match the sweet flavour and dense texture of the scallops while the citrus flavours and acidity help counterbalance it. The full-bodied oil matches the caramelized crust on the scallops. A sprig of fresh chervil would provide the perfect embellishment to the olive oil–scallop combination.
Grilled sardines	A medium-bodied, herbaceous, dry (i.e. nonfruity) excellent extra-virgin olive oil.	The dryness of the oil balances the natural oiliness of the fish while the green herbaceous elements complement the rich, pungent, saline flavours.

(continued overleaf)

Table 24.1 *(continued)*

Dish[a]	Type of olive oil[b]	Pairing rationale
Chicken curry	Spicy, high-toned, excellent extra-virgin olive oil with pronounced notes of dried orange rind and preserved lemon.	The intensely aromatic and complex flavours of the curry are the driving force of this dish; in fact the chicken, while providing substance, acts mostly as a prop to showcase them. The oil, too, drizzled over the dish, is a final ornamentation of this exotic spiciness and an addition of yet one more component to accentuate the colourful bazaar of flavours.
Bistecca alla Fiorentina	Full-bodied, intensely spicy, recently processed green, excellent extra-virgin olive oil.	A thick steak of Chianina beef grilled rare over charcoal embers. While the steak is cooking, it is typically brushed with extra-virgin olive oil using a branch of rosemary. After cooking, the steak is sliced, sprinkled with coarse salt and cracked pepper and liberally drizzled with olive oil. Here the full body of the oil stands up to the meaty intensity of the steak; the astringency of the new oil complements the charred, smoky flavour of the grill, while the spiciness echoes the black pepper and articulates the rich flavour of the beef.
Roast leg of lamb	Robust, full-bodied, excellent extra-virgin olive oil with spicy, green apple flavours and firm astringency.	The full body stands up to the pronounced flavour of lamb. Spiciness adds interest to the meaty texture and astringency cuts through the fat.
Boiled potatoes or tagliatelle pasta	Most any excellent superior extra-virgin olive oil.	The bland richness and starchiness of both the boiled potatoes (sprinkled with coarse sea salt and black pepper) and egg noodles (tossed with grated Parmigiano cheese) offer perfectly unassuming partners to showcase the special characteristics of just about any excellent extra-virgin olive oil, while the complexity of an excellent extra-virgin olive oil would bring these simple earthy products to life. The tagliatelle might benefit from a richer, fuller bodied oil while the potatoes could support a bit more spicy burn.

Notes: [a]These are basic types of dishes that do not take into account all the various secondary ingredients that might play a part in the actual preparation. Key ingredients are listed where appropriate. [b]Basic organoleptic profiles of olive oils are given here, not specific cultivars and / or specific regions.

Chefs are inherently creative: the more chefs discover the true nature of olive oil and, especially, the exciting new range of culinary possibilities offered by this new breed of excellent extra-virgin olive oils, the more they will develop new dishes and new culinary techniques to showcase them. While trends are transitory by nature, they do often leave behind them a change of consumer consciousness with a lasting impact (Drescher 2010). As restaurateurs are always looking for ways to distinguish themselves and to capitalize on the next dining trend, surely new ways of marketing extra-virgin olive oil will emerge, whether in the form of olive oil service in a restaurant setting or retail shops specializing in olive oil or perhaps even olive oil bars, which would be similar to the recent proliferation of wine bars that offer numerous wines by the glass and small plates of food to accompany them.

Often, restaurants and professional chefs set trends that then expand outwards to the general populace and there is also no reason why such trends should not extend into the home. Once consumers develop an awareness of the richness and diversity of these excellent extra-virgin oils, it will be difficult for them to go back to a bland (not to say rancid) commercial product. There may come a time when it will not be uncommon for people who care about food to have a selection of oils at home: a light-flavoured extra-virgin olive oil for sautéing, a refined oil for deep-frying (for those who wish to deep fry), two or three different excellent extra-virgin olive oils of markedly different flavour profiles and perhaps even a good-quality unrefined walnut or sesame oil to drizzle over certain types of dishes.

24.5 Education and communication: revolutionizing the perception of olive oil one drop at a time

What we are talking about here is nothing less than a revolution in the way extra-virgin olive oil is perceived and used. In order for this revolution to spread a few basic things must take place:

- Producers must continue their commitment to making extra-virgin olive oils at this high level of quality and more producers must join them so that such oils are more widely available. Moreover, the high level of quality must be traceable and verifiable with each year's harvest/production.
- Once the excellent extra-virgin oil has been produced, bottlers, shippers, importers, distributors and retailers must make sure that it is not adulterated or compromised in any way. Bottlers must assure that nothing but olive oil is in the bottle and the label accurately reflects what is inside; shippers and distributors must make sure that during transport the olive oil does not spend too much time in hot warehouses or containers where it may spoil; and retailers must ensure that the bottles of oil are not displayed in store windows or for extended periods on shelves in rooms with extreme temperature fluctuations. Furthermore, they must ensure that stocks are rotated and that the olive oil is sold while still in optimum condition.

In the absence of any single organization that represents quality producers of extra-virgin olive oil and without the application of any consistent set of regulations and procedures to guarantee quality and transparency, these controls must be left up to the individuals concerned.

- Once these excellent olive oils are in the market, they should reach chefs who are on the lookout for such high-quality products and know what to do with them. Information about these oils would most likely come from the wholesale distributor, so it is important for a quality olive oil producer to have a good importer who can properly represent these products in the marketplace. Chefs have a direct relationship, through food, with their customers, who often discover new ingredients, new dishes and new food trends in restaurants. Increasingly, chefs are also becoming entertainment celebrities on television, which greatly expands their influence.
- Consumers, either at home or in restaurants, are the end users of extra-virgin olive oil. While they do not necessarily need to become experts, they must become aware of the special nature of this product and appreciate the difference between an excellent extra-virgin olive oil and a mediocre, spoiled or false one. They must be aware of the perishable nature of extra-virgin oil, know how to detect the special characteristics of excellent extra-virgin olive oils and have some idea how best to use them. (It would also be helpful to have dates of harvest, milling and expiration on the bottle and other tools to permit traceability of the product.)

Without the ability to differentiate a good olive oil from a bad one and faced with the numerous products on the market, most consumers will naturally opt for the one that costs less.

- Consumers may develop this awareness from chefs/restaurants and from salespeople in food shops specializing in high-quality products. It may also come from journalists in print media and on the Internet and from educators, both in the classroom (cooking classes) and on television. In this sense, journalists (Internet bloggers included) and teachers are like chefs: they pass along information to consumers, help set food trends and can impact change in eating and cooking habits. While importers can pass information on to retailers and chefs who in turn pass it on to their customers, it is up to individual olive oil producers, or one of the various associations of quality olive oil producers, to channel information to journalists and educators in order to solicit coverage in the media that will reach consumers.

Once this information has reached the consumer and an interest in and awareness of the nature of extra-virgin olive oil has been kindled, an excellent next step would be to bring the consumer into the olive orchard during harvest time and into the mill during processing. Obviously, this would only be possible in areas where olives are grown and oil produced and would not be feasible on a large scale. However, seeing the care involved in the proper harvesting of olives and the rush to get them to the mill for processing and, finally, tasting the deliciously fresh green oil that emerges, would validate their interest and make them true ambassadors of excellent extra-virgin olive oil.

Waging the olive oil revolution is a long and complicated process that must take place on all the different fronts mentioned above, it must begin with the solidification and expansion in the production of excellent extra-virgin olive oil. But it is a revolution that can – and I believe will – be accomplished. Indeed, it is already well on its way.

References

Al-Hilali, M. and Khan, M. (trans.) (2013), *The Noble Quran, Surat Al-Nur 24-35*, Dar-us-Salam Publications, Houston, TX.

Besnard, G., Khadari, B., Navasqués, M. *et al.* (2013), The complex history of olive tree: from Late Quaternary diversification of Mediterranean lineages to primary domestication in the northern Levant. *Proceedings of the Royal Society B* **280** (1756), 1471–2954.

Drescher, D. (2010), The Super-Premium Olive Oil Experience: Developing a Culture of Flavour Discovery, *Beyond Extra-Virgin, The Fourth International Conference on Olive Oil Excellence*, organized by Association 3E (Milan, Italy), the Academy of Georgofili (Florence, Italy), The Culinary Institute of America (St Helena, California) and the Olive Center of the University of California Davis, Verona, 22 September 2010.

Faas, P. (2003), *Around the Roman Table*, St Martin's Press, New York.

Pliny the Elder (circa. AD 77–79) *Natural History* (English translation in the Loeb Classical Library), Harvard University Press, Harvard MA, Book XV, Chapters 2–6.

Wrangham, R. (2009) *Catching Fire – How Cooking Made us Human*, Basic Books, New York.

25

An introduction to life-cycle assessment (LCA)

Stefano Rossi
Life Cycle Engineering, S.r.l., Torino, Italy

Abstract

Life-cycle assessment (LCA) has become a recognized instrument to assess the ecological impact due to industrial processes or, more generally, to human activities. In the context of a global awareness of climate change, the carbon footprint (CFP) is used as a simple way to sensitize the purchasing behaviour of consumers, and public opinion in general. Limitations of CFP arise if one decides to expand the outlook to include other environmental impacts, which are commonly evaluated in LCA. This chapter is a simple guide to understanding LCA. It describes the history and methodological approach of LCA and how to develop credible environmental communication. Particular focus is placed on the Environmental Product Declaration (EPD) for food, and virgin olive oil in particular.

25.1 Introduction

The application of life-cycle thinking began in the 1960s with a focus on the physical behaviour of industrial systems. Despite some alarmism about the possible results of these studies, it is now generally accepted that with increasing population, industrialization and consumption of limited reserves, the earth's resources will one day be exhausted. It is therefore important to introduce practices that use the available resources in the most effective and efficient manner.

Two unavoidable consequences result from processes in the manufacturing industry: energy consumption and waste production.

Minimizing the use of energy and reducing waste has been the task of engineers in manufacturing industries for over a century, because energy use and waste generation directly affect the profitability of an enterprise. However, during recent decades,

The Extra-Virgin Olive Oil Handbook, First Edition. Edited by Claudio Peri.

further pressure has been exerted by governments and public opinion, so that the environmental issue has become a crucial leverage for market positioning.

Initially, a focus on the use of energy, especially fossil fuels, was often referred to as energy analysis. Like the calculations required to build flow-charts with material and energy balances, the consumption of raw materials and the generation of waste were also calculated. At this point, some analysts referred to this approach as 'resource and environmental profile analysis'. The term life-cycle assessment (LCA) was proposed for the first time in 1990 at the first conference in this field, in the state of Vermont, USA.

During the 1990s, the International Standards Organization (ISO) set up a technical committee to agree on formal standards for LCA. They were then reviewed in 2006, finally leading to ISO standards 4040 and 14044.

25.2 Methodological approach

A life cycle considers all operations, starting with the production of raw materials and following the sequence of operations involved in manufacture and use until the materials are finally disposed of. Such a sequence is often referred to as 'cradle-to-grave'. According to this procedure, the consumer has to be treated as an integral part of the sequence.

Producers of raw materials such as steel or plastics can trace their operations from the extraction phase until their products are ready for dispatch to the next processing operation. The same applies to agricultural companies producing raw vegetable or animal materials for transformation into foods. Such analyses are important as building blocks of LCA. There is yet a third type of analysis known as gate-to-gate. This assessment is usually carried out by manufacturers who take in raw materials and transform them into products that are the raw material of further transformation in a following process. A typical example is beer-can manufacturers. They take in sheets of steel or aluminium and convert them into cans, which are then sent to beer fillers. Such manufacturers have no control over the producers of their inputs or the users of their outputs and usually have no detailed quantitative knowledge of their operations. Many of the unit operations for which data have to be collected when carrying out an LCA are of this gate-to-gate type. Different types of possible analyses are shown in Figure 25.1 with reference to the case of extra-virgin olive oil.

The chain 'from field to bottle' is the most frequently applied concept in product certification.

The chain segment 'from gate to gate' is the most easily controlable segment in the industrial production and marketing of olive oil.

The chain segment 'from field to table' corresponds to the definition of product chain traceability. It can only be applied in short chain systems.

The chain 'from cradle to grave' or 'from soil to soil' recommended in the LCA approach is very difficult to apply for the enormous complexity of food chains that

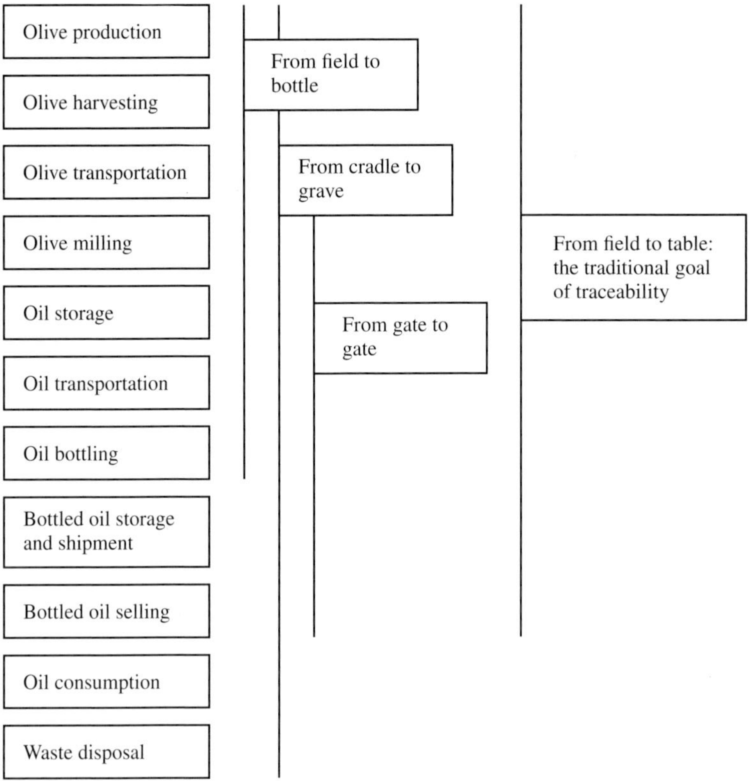

Figure 25.1 Virgin olive oil chain segments to be considered for LCA.

start from numerous operators in the first step and end at much larger numbers of users in the last step of consumption.

From an organizational point of view, LCA phases are classified according to the ISO standards scheme in Figure 25.2.

Goal definition and scoping is the preliminary phase in which the study group defines the goal and scope of the study, functional unit, system boundaries, requirement of data, hypothesis and limit.

Life-cycle inventory (LCI) is the phase of the study related to the life cycle of the product. The purpose is to make an energy and material flow model throughout the entire life cycle of the product, including transport and process.

Life-cycle impact assessment (LCIA) aims at assessing the environmental impact (emissions and resources consumption) related to the company's processes and activities. Data refer to a production functional unit.

Life-cycle interpretation aims at proposing changes that could reduce the environmental impact due to the company's process and activities. This phase should continue as an iterative process that allows the company to constantly detect changes and spur improvement.

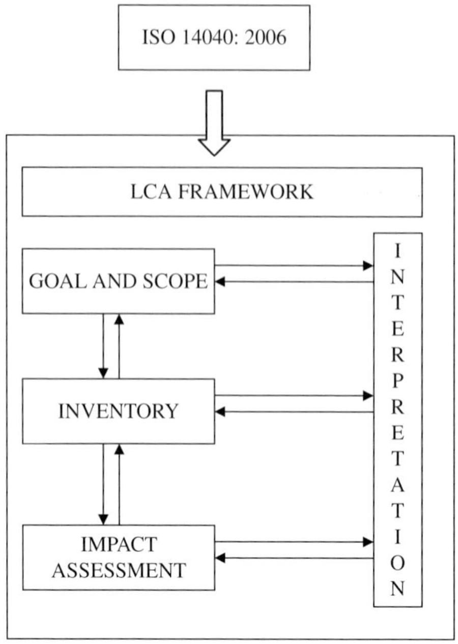

Figure 25.2 Life-cycle assessment structure proposed by ISO 14040:2006 (2006).

Table 25.1 List of indicators for LCA impacts.

Scale	Impact
Global	Global warming
	Ozone depletion
	Nonrenewable resource consumption
Regional	Acidification
	Eutrophication photochemical ozone creation
	Chronic toxicity
Local	Local area depletion
	Human toxicity

The final goal of LCA is the quantification of a set of indicators describing the whole environmental performance of a product on different scales according to the list in Table 25.1.

25.3 Limits and advantages of the carbon footprint

Certification or eco-labelling programmes should encompass a wide range of effects of production on the environment and society. This could help avoid any unintended change in the behaviour of producers and consumers that results from focusing on one single aspect of the environmental impact. There are many examples in which the limit of measuring only the carbon footprint is evident: nuclear power electricity generation, battery cages for laying hens, corn-based ethanol, PET bottles versus glass bottles. All the processes mentioned above have a low carbon footprint compared to their equivalent, but this does not prove that they are more sustainable in a wider context.

Certification and eco-labelling are based on existing, well-established and widely used methods, standards and guidelines such as the International Life Cycle Reference Database Handbook, ISO 14040-44, ISO 14064, PAS 2050, BP X30, WRI/WBCSD GHG protocol, the Sustainability Consortium approach, ISO 14025, Ecological Footprint, Global Reporting Initiative, WRI GHG Protocol, CDP Water Footprint, DEFRA guidance on GHG reporting, and ADEME Bilan Carbone. Others will be available in 2015 relating to:

- the product environmental footprint (PEF), a method based on LCA for calculating the environmental performance of a product.
- organization environmental footprint (OEF), a method based on LCA for calculating the environmental performance of an organization.

25.4 Environmental communication strategies

The International Organization for Standardization (ISO) has developed standards for three types of environmental claims on goods and services:

- type I based on third-party certification for specific goods and services (ISO 14024)
- type II based on self-declarations (ISO 14021)
- type III based on life-cycle impacts (ISO 14025).

Consumers are increasingly eager to buy goods and services with a reduced environmental impact. Producers, service providers and the advertising industry are aware of this trend and thus seek to attract clients through the environmental benefit of their goods and services. Misleading, false, meaningless or unclear environmental claims must be avoided. They may result in consumers losing trust in environmental claims and in generating unfair business competition.

On the other hand, a verified and credible environmental claim can lead to several benefits:

- Companies can benchmark their performance within their sector or product category. They can understand how their environmental performance is in comparison to their peers and can better target their improvement efforts.

- Benchmarking is a strong reputational incentive. For many companies, being a good environmental performer is part of their business values and strategy. Product category and sector benchmarks create a drive for strong improvements and have the potential for shifting the performance of the whole sector or product category upwards.
- Environmental claims enable consumers to take better informed purchasing decisions by comparing the performance of products in the same category.
- Investors can better target their decisions knowing how companies perform in comparison to peers in their sector. They can better assess the level to which a company deals with relevant environmental impact.
- Governments can better target their incentives by knowing the performance of beneficiaries within their sector. They can provide incentives for sustainable consumption, focusing on reliable green alternatives and avoid environmentally harmful subsidies.

An EPD® is a certified Environmental Product Declaration developed in accordance with the ISO standard 14025. It is based on the use of internationally accepted and valid methods for LCA identifying the most significant environmental aspects in a holistic perspective and leading to robust and continuous improvement. Another requirement of the International EPD® system concerns the aspect of critical review, approval and follow-up by an independent verifier.

25.5 The food sector

In the last decade, life-cycle assessment has become an important methodology for the analysis of the environmental performances of agri-food products. About 80 EPD have been published in this sector during recent years. Life-cycle assessment in the food sector is challenging because there are many processes involved. The product category rules (PCRs) play an important role in assuring that the EPD of products belonging to the same category are based on the same methodology. To address this issue a 'network' of PCRs is being developed in order to link PCRs, as shown in Figure 25.3.

With this approach the whole food sector, made up of 92 groups and 266 classes of products, is covered with only 25 PCRs, and inconsistencies among PCRs are avoided.

The extra-virgin olive oil sector

Eight EPDs of extra-virgin olive oil, made by five different producers, were published according to the international EPD system (www.environdec.com). Olive oil thus appears to be one of the products most frequently using this type of declaration. The system boundaries of the studies made for the EPDs are shown in Figure 25.4.

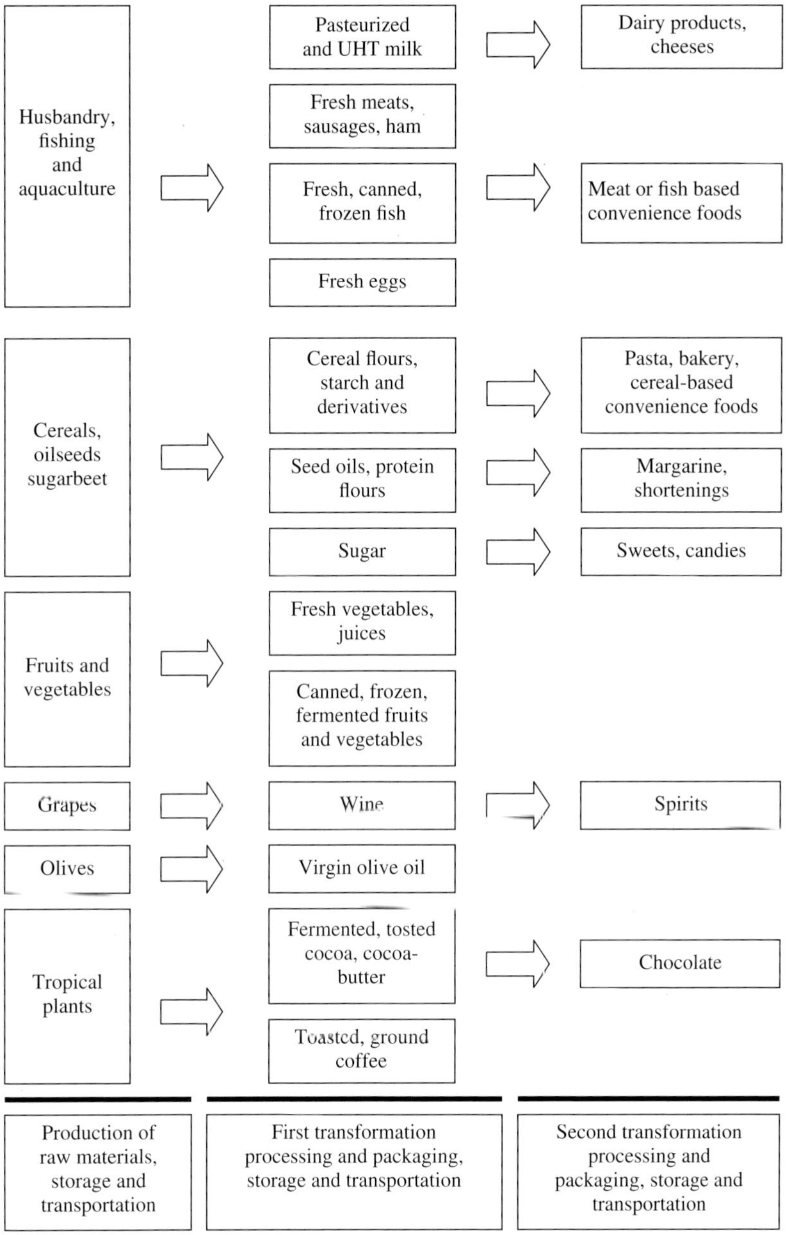

Figure 25.3 Product category rules for foods (*Source*: Peri *et al.*, 2006. Reproduced by permission of The International EPD Consortium (IEC)).

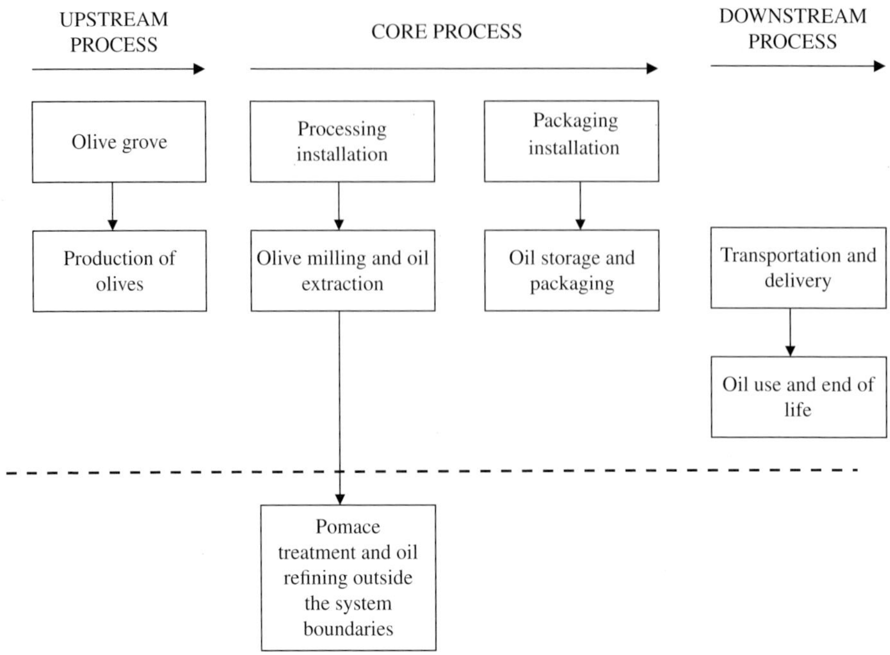

Figure 25.4 Product category rules for virgin olive oil (*Source*: Peri *et al.*, 2006. Reproduced by permission of The International EPD Consortium (IEC)).

The main environmental impact of the whole life-cycle of virgin oil is, in almost all the cases, related to olive cultivation, with:

- operations for the changing of land use, if the olive-grove lifetime is expected to be less than 25 years;
- operations for the establishment of the olive grove, including the irrigation system, if the olive grove life time is expected to be less than 25 years;
- the production of olives (CPC 0145), in which the following inflows are considered:
  - inputs of fertilizers and plant protection products;
  - waste management;
  - wood use as a by-product of pruning or end of life of olive trees;
  - external transportation of inputs;
  - extraction and use of water;
  - auxiliary materials for harvesting (nets, crates, etc.);
  - generation of energy (fuel and electricity) used at the farm;
  - external transportation of olives to the milling facility.

Another important aspect is packaging in glass bottles or aluminium cans, which have a high environmental impact especially in terms of global warming potential.

References

Baldo, G.L., Marino, M. and Rossi, S. (2008) *Analisi del Ciclo di Vita LCA*, Edizioni Ambiente, Milan.

Boustead, I. (2003) *An Introduction to Life Cycle Assessment*, Boustead Consulting Ltd, Horsham.

EPD System (2013) *International EPD System – General Programme Instructions*, version 2.0, www.environdec.com/Documents/GPI/General_programme_instructions_2_01_20130918.pdf (accessed 11 October 2013).

ISO 14040:2006 (2006) *Environmental Management – Life Cycle Assessment – Principles and Framework*, ISO, Geneva.

Appendix

A.1 Conversion table of physical parameters

The units most frequently used are given in Table A.1. Column A gives the reference units, which should be used in making calculations. Column B reports practical units of the metric and British systems. In the middle column the conversion factors are reported for converting units from column A to column B and vice-versa.

The conversion table should be used as follows:

- To convert a measure from a reference unit to a practical unit, data expressed in the reference unit must be multiplied by the conversion factor that is reported in the middle column of the table. For instance, to convert a length in metres into feet, it must be *multiplied* by 3.28084.
- To convert a measure from a practical to a reference unit, the value expressed in the practical unit must be divided by the conversion factor. For instance, to convert a length in feet into metres, it must be *divided* by 3.28084.
- To convert a practical unit to another practical unit, the operation requires a double transformation. For instance, to convert a measure in inches into centimetres, the first operation consists in dividing by 39.37 (from inches to metres) and the second in multiplying the result by 100 (from metres to centimetres); in practice inches are to be multiplied by 2.54 (100/39.37).

A.2 Weight-to-volume conversion of oil quantities

It is often necessary to convert extra-virgin olive oil weight into volume or vice versa. In general, the amount of harvested olives and of the oil obtained from the milling process are expressed in weight units, while the oil stored in tanks is often expressed in volume units. When the oil is sold in bottles, quantities are generally expressed in volume. When it is sold in bulk, quantities are generally expressed in weight. In conclusion, technical as well as business operators must be able to convert weights into volumes and vice versa. For this purpose it is necessary to know the value of oil density.

The Extra-Virgin Olive Oil Handbook, First Edition. Edited by Claudio Peri.

Table A.1 The most common units of physical quantities used in the extra-virgin olive-oil process.

Column A (reference units)	Conversion Factors (from column A to column B *multiply by* - from column B to column A *divide by*)	Column B (practical units)
Length		
Meters (m)	100	centimetres (cm)
	3.28084	feet (ft)
	39.37	inches (in)
	0.001	kilometres (km)
	1000	millimetres (mm)
	1×10^{-6}	microns (μ)
	1×10^{-9}	millimicrons (mμ)
Surface area		
Square meters (m^2)	0.0002471	acres
	0.0001	hectares (ha)
	10.7639	square feet (ft^2)
Volume		
Litre (cubic decimeter)	0.264179	gallons (US liq.) (gal)
	0.01	hectolitres (hl)
	0.001	cubic metres (m^3)
	33.81497	ounces (US, fluid) (oz)
	100	centilitres (cl)
	1000	millilitres (ml)
Mass		
Kilograms (kg)	2.2046	pounds (lb)
	0.01	quintals
	0.001	metric tons (t)
	1000	grams (g)
	1×10^{6}	milligrams (mg)
Pressure		
Bars	0.986923	atmospheres (atm)
	14.5038	pounds/square inch (psi)
	10.207	metres of water at 15 °C
	760	mm Hg
Energy		
Joules (J)	0.00023866	kilocalories (kcal)
	0.00094709	(B.t.u.)
	2.7777×10^{-7}	Kilowatt hour (kWh)

A.3 Density

Density (symbol ρ) is a measure of how much material is contained in a given unit of volume: density = mass/volume.

In international units, density is expressed as kg/m^3; in practice – in extra-virgin olive oil material balances – it is suggested to use kg/l. The density of olive oil is 0.917 kg/l at 20 °C. Therefore:

- to convert a quantity of oil at 20 °C from kg to litre, divide by 0.917
- to convert a quantity of oil at 20 °C from litre to kg, multiply by 0.917.

Density ρ decreases linearly with increase in temperature. The following equation can be used:

$$\rho = (925.59 - 0.41757 \times \mathrm{T})/1000$$

with ρ being expressed in kg/l and T being the oil temperature in °C.

For ease of consultation, density values are reported in Table A.2 for the range of temperatures that are most common in extra-virgin olive-oil processes.

(Water has a density of 1 kg/l at 20 °C (exact value: 0.998203).)

In making material balance calculations in the extra-virgin olive oil process, it is suggested:

- To use weight not volume units because weight, unlike volume, does not change with temperature
- To use kg as the weight unit and make conversions when data are expressed in different units
- Not to mix units when calculating yields. For instance the oil yield of a given quantity of olives should not be expressed as a volume-to-weight ratio (litres or gallons per ton), but preferably as a weight-to-weight ratio (kilogram per kilogram or per 100 kg or per ton).

Table A.2 Olive oil temperature-density relationship.

Temperature		Density (kg/l)
°C	°F	
5	41	0.9235
10	50	0.9214
15	59	0.9193
20	68	0.9172
25	77	0.9151
30	86	0.9131
35	95	0.9109
40	104	0.9089

A.4 Concentration

Concentration is the amount of a substance contained in the unit weight of a product. Concentration is a dimensionless number and therefore its value is the same independent of the units used. It may be expressed as:

- *Percentage concentration*. Saying that olives contain 20% oil means that 100 kg of olives contain 20 kg of oil or that 100 lb of olives contain 20 lb of oil or that 100 g of olives contain 20 g of oil, and so on. The remaining 80% is made up of water and solid components of the olive fruit.
- *Mass fraction*. The concentration of 20% can be expressed as mass fraction as 20/100 = 0.20. In this case the quantity of oil (0.20) refers to the unit mass of olives.

The above expressions are unsuitable for compounds that are present at very low concentrations, such as phenolic compounds, sterols, or the volatile compounds of flavour. In this case, concentration is often expressed as parts-per-million (ppm, mg/kg) or parts-per-billion (ppb, mg/t).

A.5 Yield

Yield is an index of the efficiency or productivity of a process and is obtained by dividing the quantity of the product as output by the quantity of a resource as input. Therefore yield is a number that gives how much product is obtained from a unit quantity of resource. Often the ratio is multiplied by 100 to obtain percentage yield. Table A.3 summarizes some of the yield values that can be calculated in the extra-virgin olive oil process.

A.6 Viscosity

Viscosity μ is a measure of the resistance of a fluid to flow under shear stress. In sensory evaluation, it is commonly described as 'thickness'. Often consumers erroneously use the adjective 'dense' to describe a sensation that they should identify as 'viscous'. Water viscosity at 20 °C is 1 centipoise (1 cP), while the viscosity of olive oil at 20 °C is about 84 cP. Olive oil is therefore much more viscous than water and this is why olive oil flows so slowly, silently and smoothly.

The centipoise (cP) is a practical unit, much more frequently used than the Standard International (SI) unit, which is the decapoise (= 10^3 centipoises). The use of centipoises is preferred because of the coincidence of water having a viscosity of 1 cP.

Table A.3 Various yield calculations in the extra-virgin olive-oil process.

Yield	Calculation	Comment
Average yield of olives per tree	Divide the mass (kg) of harvested olives by the number of trees.	It may be useful to evaluate the yield by cultivar or by olive grove location.
Average yield of olives per hectare	Divide the mass (kg) of harvested olives by the hectares of the olive grove.	It may be useful to evaluate the yield by cultivar or by olive grove location.
Average raw oil extraction yield	Divide the mass (kg) of raw oil at the output of the centrifugal finishing centrifuges by the mass (kg) of olives put in. In practice this yield is expressed as kg of oil per 100 kg of olives by multiplying the above ratio by 100.	This calculation should be applied to each milling batch as an index of the mill's efficiency and reliability. Then it should be applied to the total production of the company as an index of the overall production performance.
Milling efficiency index (%)	Divide the quantity of extracted raw oil by the true content of oil in the olives determined with a reliable analytical method (for example the Soxhlet extraction method). Multiply by 100.	An acceptable milling efficiency index is greater than 80% and a good value is greater than 85%. Values less than 75% are an indication of poor control of the milling plant and of the operating conditions.
Oil handling efficiency (%)	Divide the mass (kg) of oil sold (in bottles or in bulk) by the mass (kg) of raw oil obtained at the outlet of the finishing centrifugal separator. Multiply by 100. This procedure can be applied to single operations (for example filtering or bottling) for suitable control and optimization.	In the process of oil milling, storage and bottling, a series of operations may cause oil losses due, for example, to filtering, decanting or because of accidental or careless mistakes. An overall oil-handling efficiency lower than 97% should be considered as an indication of poor control of the oil handling operations.

Viscosity of extra-virgin olive oil decreases with increasing the temperature. The following equation can be applied:

$$\mu = 1.55 \times 10^{-7} \times \exp\,(32167/8.314 \times \mathrm{T})$$

with μ in centipoises and T, the absolute oil temperature, in K (Abramovic and Klofutar 1998; Bonnet *et al.* 2011).

For ease of consultation, viscosity values are reported in Table A.4 for the range of temperatures usually applied in the extra-virgin olive oil process.

Handling viscous products such as extra-virgin oil must be done at very low shear-rates (low flow velocity), avoiding turbulence. Turbulence, in fact, may cause air

Table A.4 Olive-oil temperature-viscosity relationship.

Temperature		Viscosity
°C	*°F*	*Centipoises*
5	41	155
10	50	130
15	59	105
20	68	84
25	77	69
30	86	56
35	95	44
40	104	38

bubble formation and dispersion in the oil, negatively affecting oil stability and sensory quality. Positive, rotating, low-speed, low-shear pumps should be used to transfer oil, and very low pressure should be applied in filtration. Valves and sharp changes in diameter and direction of piping should be minimized (see Annex 15.1).

A.7 Water activity

Water activity a_w is the ratio of p, the vapour pressure of water in a product and p_0, the vapour pressure of pure water at the same temperature:

$$a_w = p/p_0$$

The water activity value is a critical condition for food storage because microbial or enzymatic or chemical degradation closely depend on water activity. The activity of pure distilled water is 1. It may be roughly considered that the relationships of degrading phenomena and water activity are as given in Table A.5.

Table A.5 Water activity levels inhibiting some degradation phenomena in extra virgin olive oil.

The following degradation phenomena …	are inhibited at a water activity
Development of bacteria	lower than 0.90
Development of yeasts	lower than 0.80
Development of moulds	lower than 0.70
Enzymatic activities	lower than 0.40
Oxidative reactions	are not inhibited but, on the contrary, are favoured by the absence of water

In food products a_w is always less than 1, due to the presence of compounds that establish some links with water and therefore reduce its vapour pressure p. Hydrophilic compounds like sugars, starches, and proteins tend to lower p and to reduce the a_w value. This property is also called 'hygroscopicity'. On the other hand, oils and fats and, in general, foods containing lipophilic compounds do not link with water and therefore have a higher water activity. Extra-virgin olive oil is not only lipophilic but is obtained by separation from an aqueous medium and is therefore an almost water-saturated product. The water saturation threshold of extra-virgin olive oil is 300–400 mg per kg and a water content higher than this value corresponds to a water activity very close to 1. In terms of water activity, extra-virgin olive oil would be susceptible to all kinds of microbial and enzymatic degradations. The only way to preserve extra-virgin olive oil from biological and enzymatic degradation is to eliminate any fermentable substance and enzyme. To do this, a very through filtration is needed.

If this condition is met, a water activity close to 1 can have a positive effect on oil quality and stability because it allows for the presence of interesting polar, water-soluble antioxidants (for example phenolic compounds).

A.8 Temperature

The unit of temperature is the Centigrade degree (or Celsius degree, symbol °C), which is 1/100 the difference between the temperature of melting ice and that of boiling water under standard atmospheric pressure. The Fahrenheit degree symbol °F is 1/180 the difference between the temperature of melting ice and that of boiling water under standard atmospheric pressure.

The temperature of melting ice is 0 °C and 32 °F. The temperature of boiling water is 100 °C and 212 °F.

Degrees Fahrenheit can be converted to degrees Celsius by applying the following equation:

$$°C = 5/9(°F - 32)$$

Degrees Celsius can be converted to degrees Fahrenheit by applying the following equation:

$$°F = 9/5(°C + 32)$$

Absolute temperature is sometimes used in scientific formulas. It is measured using the Kelvin scale, the units of which are abbreviated as K. These units are the same size as the Celsius degree but the scale starts at absolute zero, which is −273.15 °C. In the Kelvin scale the temperature of melting ice is 273.15 K and the temperature of boiling water is 373.15 K.

Table A.6 can be used for rapid conversion of temperatures in the range most commonly applied in extra-virgin olive oil processing and storage.

Table A.6 Temperature conversion table.

°C	°F	°C	°F
0	32	20	68
1	33.8	21	69.8
2	35.6	22	71.6
3	37.4	23	73.4
4	39.2	24	75.2
5	41	25	77
6	42.8	26	78.8
7	44.6	27	80.6
8	46.4	28	82.4
9	48.2	29	84.2
10	50	30	86
11	51.8	31	87.8
12	53.6	32	89.6
13	55.4	33	91.4
14	57.2	34	93.2
15	59	35	95
16	60.8	36	96.8
17	62.6	37	98.6
18	64.4	38	100.4
19	66.2	39	102.2
		40	104

A.9 Specific heat

Specific heat is the quantity of heat (expressed in joules) required to increase the temperature of a 1 g mass of a substance 1 °C. The specific heat of extra-virgin olive oil is 2.0 J/(g) (°C), while the specific heat of water is 4.18 J/(g) (°C). Compared to water, olive oil has a low specific heat. This means that a lower (less than half) amount of heat is needed to increase the temperature of a given mass of oil compared to that needed for an equal increase in temperature of an equal mass of water.

A.10 Boiling and smoke point

The boiling point of virgin or refined olive oil at atmospheric pressure is 299 °C (570.2 °F). Smoke point is defined as the temperature at which a cooking oil begins to break down. The oil smokes and gives food an unpleasant taste. A high smoke point is a critical condition for the best use of oil in frying. The smoke point of a good extra-virgin olive oil is 210 °C (410 °F). The smoke point is higher in good extra-virgin olive oil and lower in low-quality virgin olive oil.

Frying in extra-virgin olive oil

The ideal temperature for frying products with a high water content such as vegetables, potatoes and fruit, is 130–145 °C (266–293 °F), while for small, quickly fried products, the frying temperature should be 175–190 °C (347–374 °F).

As can be seen from the above values, the ideal frying temperature is much lower than the smoke point of olive oil, therefore it can be concluded that good-quality extra-virgin olive oil is perfectly suited for frying. Furthermore, extra-virgin olive oil contributes to flavouring the fried products and preventing thermo-oxidation due to its high content of antioxidants.

A physical-chemical description of frying

From a physical-chemical point of view, frying is a drying-and-cooking operation. When a piece of food is plunged into an oil bath at a frying temperature, water is immediately evaporated from the surface of the product. Intense bubbling of water vapour in the oil and a particular sizzling noise occur. Three effects take place in sequence:

1. Very rapid drying at the surface with formation of a dried skin. The role of this skin is essential because it is gives crispiness to the fried food.

2. The dried skin partially isolates the inside of the product from the oil, thus reducing the heat flow toward the inside and the loss of water vapour from the inside. In conclusion, a fried product is a product that has been cooked inside a very thin container made of its own skin, formed when it was plunged into the oil bath. Only the skin reaches a high temperature, close to the oil temperature, while the inside of the product rarely exceeds the water boiling temperature.

3. In the conditions described above, two very different types of transformation take place.

    1. At the surface, due to the high temperature and the low water content, nonenzymatic reactions take place with formation of a brown colour and a typical flavour. Under these conditions, hydrophobic groups are formed at the surface and therefore more intense interactions take place with the oil components;

    2. In the inside, cooking takes place at a relatively low temperature (in the range of 85–95 °C) with changes in texture, starch gelification, protein denaturation and so forth. The decrease in water content on the inside is very limited and therefore the consistency of the inside remains soft. The original taste of the food on the inside is maintained, but perhaps even accentuated by the loss of water. Hence, it may be

concluded that heat damage (loss of vitamins and essential amino acids, loss or lower digestibility of nutrients, and so forth) is much lower in fried than in boiled or roasted foods. The softness of the inside and the crispness of the outside are an essential sensory feature of fried foods.

In conclusion, it may be said that extra-virgin olive oils are perfectly suited for frying and that fried foods are as healthy as they are tasty, provided that:

- the extra virgin olive oil is of a high quality, with a low free acidity and a good content of phenolic antioxidants;
- the oil is used or reused for a limited length of time in order to avoid chemical changes due to triglyceride transformation and to interaction of triglycerides with other food components under conditions of high temperature in the frying bath.
- the oil is well drained and dried from the surface of the product after frying in order to avoid an excessive amount of triglycerides in the diet.

A.11 Fatty acids of olive oil

Triglycerides represent from 97 to 99% of the weight of extra-virgin olive oil.

The fatty acid composition varies greatly depending on cultivar and environmental conditions. Table A.7 reports the range of concentration of fatty acids in olive oil (both virgin and refined) (Codex Stan 33-1981 2001).

Table A.7 Distribution of fatty acids in olive oils.

Names	Conventional description	Percentage of total fatty acids
Saturated		
Myristic acid	(C 14:0)	< 0.1
Palmitic acid	(C 16:0)	7.5 – 20.0
Heptadecanoic acid	(C 17:0)	< 0.5
Stearic acid	(C 18:0)	0.5 – 5.0
Arachidic acid	(C 20:0)	< 0.8
Behenic acid	(C 22:0)	< 0.3
Lignoceric acid	(C 24:0)	< 1.0
Monounsaturated fatty acids (MUFA)		
Palmitoleic acid	(C 16:1)	0.3 – 3.5
Heptadecenoic acid	(C 17:1)	< 0.6
Oleic acid	(C 18:1)	55.0 – 83.0
Polyunsaturated fatty acids (PUFA)		
Linoleic acid	(C 18:2)	3.5 – 21.0
α-linolenic acid	(C 18:3)	< 1.5

A.12 Minor components of extra-virgin olive oil

Minor components represent from 1 to 3% of the weight of extra-virgin olive oil.

The minor components fraction varies greatly depending on cultivar and environmental conditions. Table A.8 reports the ranges of concentration (Owen *et al.* 2000; Codex Stan 33-1981 2001; Servili and Montedoro 2002).

Table A.8 The range of concentration of minor components in extra-virgin olive oil.

Names	Range of concentrations	Comment
Squalene	From 2 to 9 g/kg	The content of squalene of refined olive oils is 20–30% lower than in extra-virgin olive oils. The squalene content in olive oil is higher than in the other vegetable oils.
Alpha-tocopherol	From 150 to 250 mg/kg	The addition of alpha-tocopherol is allowed in refined olive oils in order to restore the natural tocopherol lost in the refining process. The concentration of alpha-tocopherol in the final refined product, however, must not exceed 200 mg/kg.
Phytosterols	From 1 to 2.5 g/kg	A distinctive feature of olive oils, both virgin and refined, is that their sterols are composed of practically pure beta-sitosterol (>93% of total sterols). Cholesterol is absent.
Phenolic compounds	From 120 to 600 mg/kg	The most abundant phenolic compounds in extra-virgin olive oil are secoiridoids, which are exclusively present in plants belonging to Olearaceae. Secoiridoids account for more than 90% of the phenolic compounds in olive oil (Servili and Montedoro 2002) , the remaining 10% being composed of simple phenolic acids, phenolic alcohols and lignans. Phenolic compounds are not present in refined olive oil. However, as virgin olive oil must be added to 'olive oil composed of refined and virgin olive oil' and to 'olive pomace oil', variable concentrations (usually less than 100 mg/kg) of phenolic compounds can be found in these oils.

References

Abramovic, H. and Klofutar, C. (1998) The temperature dependence of dynamic viscosity for some vegetable oils. *Acta Chimica Slovenica* **45**(1), 69–77.

Bonnet, J.P., Devesvre, L., Artaud, J. and Moulkin, P. (2011) Dynamic viscosity of olive oil as a function of composition and temperature : a first approach. *European Journal of Lipid Science and Technology* **113**, 1019–1025.

Codex Stan 33-1981 (2001) *Codex Standard for Olive Oil, Virgin and Refined, and for Refined Olive-Pomace Oil*, Codex Alimentarius Commission, Rome.

Owen, R.W., Mier, W., Giacosa, A. *et al.* (2000), Phenolic compounds and squalene in olive oils: the concentration and antioxidant potential of total phenols, simple phenols, secoiridoids, lignans and squalene, *Food and Chemical Toxicology* **38**(8), 647–659.

Servili, M. and Montedoro, G. (2002) Contribution of phenolic compounds to virgin olive oil quality. *European Journal of Lipid Science and Technology* **104**, 602–613.

Further reading

Coupland, J.N. and McClements, D.J. (1997) Physical properties of liquid edible oils. *Journal of the American Oil Chemists' Society* **74**(12), 1559–1564.

Haynes, W.M. (ed.) (2013) *CRC Handbook of Chemistry and Physics*, 94th edn, CRC Press, Boca Raton, FL.

Peri, C. and Zanoni, B., (1994) *Manuale di Tecnologie Alimentari*, CUSL, Milan, Part 2.

Index

Note: Page numbers in *italics* refer to Figures and those in **bold** to Tables.

The Extra-Virgin Olive Oil Handbook, First Edition. Edited by Claudio Peri.